Modeling, Design and Optimization of Multiphase Systems in Minerals Processing

Modeling, Design and Optimization of Multiphase Systems in Minerals Processing

Special Issue Editor

Luis A. Cisternas

MDPI • Basel • Beijing • Wuhan • Barcelona • Belgrade • Manchester • Tokyo • Cluj • Tianjin

Special Issue Editor
Luis A. Cisternas
Universidad de Antofagasta
Chile

Editorial Office
MDPI
St. Alban-Anlage 66
4052 Basel, Switzerland

This is a reprint of articles from the Special Issue published online in the open access journal *Minerals* (ISSN 2075-163X) (available at: https://www.mdpi.com/journal/minerals/special_issues/minerals_processing_systems).

For citation purposes, cite each article independently as indicated on the article page online and as indicated below:

LastName, A.A.; LastName, B.B.; LastName, C.C. Article Title. *Journal Name* **Year**, *Article Number*, Page Range.

ISBN 978-3-03928-400-9 (Pbk)
ISBN 978-3-03928-401-6 (PDF)

Contents

About the Special Issue Editor

 **Luis A. Cisternas**, Professor, joined the Department of Mineral Processing and Chemical Engineering, Universidad de Antofagasta, in 1988 as an assistant professor, wherein 2001, he was appointed a professor. Professor Cisternas obtained a Ph.D. (Chem. Eng., 1994) from the University of Wisconsin Madison (USA). He has been visiting professor at CAPEC, DTU in Denmark and Aalto University in Finland. Professor Cisternas's principal research interest is the use of a systems approach to solving problems in mineral processing. In particular, his research combines the development of systematic (computer aided) methods and tools and experimental works for solving problems in the mining industries, which can be classified in terms of the following topics: modeling and optimization; design and analysis of minerals processes; water resources; critical material and circular economy. Professor Cisternas has published widely in journals, conference papers, book chapters, and books. By January 2020, eighteen MSc students and twelve PhD students had submitted and successfully defended their theses under the supervision of Professor Cisternas.

Preface to "Modeling, Design and Optimization of Multiphase Systems in Minerals Processing"

Mineral processing deals with complex particle systems with two-, three- and more phases. The modeling and understanding of these systems are a challenge for research groups and a need for the industrial sector. This Special Issue aims to present new advances, methodologies, applications, and case studies of computer-aided analysis applied to multiphase systems in mineral processing. This includes aspects such as modeling, design, operation, optimization, uncertainty analysis, among other topics. We have developed this Special Issue dedicated to the modeling, design, and optimization of multiphase systems in mineral processing to promote discussion, analysis, and cooperation between research groups. The Special Issue contains a review article and eleven articles that cover different methodologies of modeling, design, optimization, and analysis in problems of adsorption, leaching, flotation, and magnetic separation, among others.

This Special Issue considered different problems in several areas of multiphase systems in mineral processing. Thus, various strategies and tools were presented to solve or face those problems. On the whole, I hope that this Special Issue will contribute to a superior understanding of multiphase phenomena and will promote future research in Modeling, Design, and Optimization of Multiphase Systems in Minerals Processing.

Authors' contributions from China, Chile, Canada, Germany, Iran, Mexico, and Spain were received. I thank all of them for their contributions that helped the achievement of this Special Issue. Finally, I would like to thank the referees and editorial staff of Minerals for their valuable effort that contributes to the success of this initiative.

Luis A. Cisternas
Special Issue Editor

Editorial

Editorial for Special Issue "Modeling, Design and Optimization of Multiphase Systems in Minerals Processing"

Luis A. Cisternas

Departamento de Ingeniería Química y Procesos de Minerales, Universidad de Antofagasta, Antofagasta 1240000, Chile; luis.cisternas@uantof.cl; Tel.: +56-552-637-323

Received: 31 January 2020; Accepted: 1 February 2020; Published: 3 February 2020

The exploitation of mining resources has been fundamental for the development of humanity since before industrialization. After hundreds of years of exploitation of mining resources, the demand for these resources has continued to increase, and without a doubt, will be maintained and increased in the future to face the great challenges of engineering [1] and society [2]. Not only will traditional materials be needed but new mining resources, such as those classified as critical materials, will be required as well [2,3]. A series of challenges will need to be addressed in order to meet those demands, including low grade ore, more complex minerals, more stringent environmental regulations, to name just a few. To face these challenges, tools are needed to help understand, improve, and facilitate the development of more effective solutions. The use of modeling of various types and levels will undoubtedly be required. The advantages include not only the possibility of cutting the times and costs of experimentation but also the study of phenomena where experimentation is difficult or impossible to employ. On the other hand, a common feature in the processing of mining resources is the presence of multiphase systems. A multiphase system is defined as one in which two or more different phases (i.e., gas, liquid, or solid) are present, including systems with the same type of phases (e.g., liquid–liquid). As such, a series of phenomena associated with processes such as flotation, grinding, magnetic separation, and thickening are related to multiphase systems. With these antecedents, in considering the importance of modeling activities and multiphase systems, we have developed this Special Issue dedicated to the modeling, design, and optimization of multiphase systems in mineral processing to promote discussion, analysis, and cooperation between research groups. The Special Issue contains a review article and eleven articles that cover different methodologies of modeling, design, optimization, and analysis in problems of adsorption, leaching, flotation, and magnetic separation, among others.

Multiphase systems are analyzed at different time and size scales in the review article [4] because the modeling and post-modeling activities depend on those scales (see Figure 1a). For example, molecular modeling is necessary to understand the phenomena that occur at the atomic or molecular level, such as the adsorption of chemical agents on the surface of minerals, while computational fluid dynamics is a suitable tool at the fluid level. The application of molecular modeling is recent in the area of study and has been used to complement experimental studies. Given that the type of information that it delivers cannot be determined experimentally, and the software currently available, numerous new applications are expected. Simulations using computational fluid dynamics codes can give comprehensive information about fluid flow and mass transfer in mineral processing processes and devices, and this can increase the understanding of a given process. The capacity and scope of computational fluid dynamics methodologies have been considerably expanded, and this type of simulation has been utilized to help in understanding a given process and in conducting new process developments. The modeling of experimental results using response surface methodology is also analyzed, given its wide use in mineral processing. Response surface methodology is based on the result of the design of experiment which intends to explain and represent the variation of output variables under conditions that are assumed to reflect the variation. The most commonly used experimental designs in mineral processing are central composite and Box–Behnken designs. One of the limitations of the response surface methodology is the use of second-order polynomials, a behavior rarely observed in multiphasic phenomena. Several applications give reasonable results because the range of the input variables is small. However, a significant amount of work using this modeling strategy has unacceptable or questionable adjustment levels. To solve this problem, new modeling strategies must be proposed, possibly based on artificial intelligence. Precisely, several applications of artificial intelligence in the design, optimization, and modeling of multiphase systems were analyzed in the review [4], including artificial neural networks and support vector machines. In fact, there has been an exponential growth in research associated with artificial intelligence; in 1990, there were 29 publications that included artificial neural networks in their title, while last year this figure was 1430 in Web of Science. Similar behavior was observed in other subjects. Publications that include support vector machine in the title grew over 2600% from 2000 to 2019. In the coming years, with the advancement of these techniques and hardware improvements, many more applications are expected. One of the strategies currently used to understand and model systems is multiscale modeling [5–7]. We have to promote this type of simulation to be able to combine different phenomena that occur at different scales in multiphase systems. The integration of computational fluid dynamic modeling and discrete element simulation, which integrates phenomena at the particle and fluid level, has been an example of an approach that has produced very satisfactory results in terms of its ability to improve the current understanding of the complexity of mineral processing phenomena. Nevertheless, greater efforts are needed in the integration of meso-, micro-, and macro-scales modeling in order to understand and improve multiphase systems, as has been observed in other areas [8]. Uncertainty, both epistemic and stochastic, is an important issue in mineral processing because several phenomena are not well known or difficult to measure, and because several variables (e.g., metal price, particle size, mineral grade) have random variations [9]. Therefore, modeling tools such as uncertainty analysis and global sensitivity analysis were included in the review. Both tools have been shown to be good approaches for considering uncertainties [10–12]. These and other topics are included in the review paper, and readers are recommended to read this review if they are interested in the topic or as an introduction to reading the other articles that cover specific themes.

Published articles can be analyzed following the same scale logic. Molecular modeling has allowed for an improved understanding of the mechanisms of interaction between minerals, the aqueous medium, and flotation reagents [13–15]. For example, studies on the behavior and molecular mechanism of adsorption of the collector sodium oleate were carried out by density functional theory and experimental techniques [13]. A similar study, but of the adsorption of flotation collector N-(carboxymethyl)-N-tetradecylglycine on a fluorapatite surface, was investigated (Figure 1b) [14]. Density functional theory is one of the most used methods in quantum calculations of the electronic structure of matter. Usually, functional density theory is combined with experimental studies; for example, molecular modeling helps identify the most stable structure of ionic species and identify active sites, while experimental techniques such as electrospray ionization-mass spectrometry and ultraviolet-visible spectroscopy allow the existence of molecules or complexes to be validated qualitatively and quantitatively [15]. These three manuscripts are good examples of the useful tool that molecular modeling can be for understanding the performance and development of new reagents.

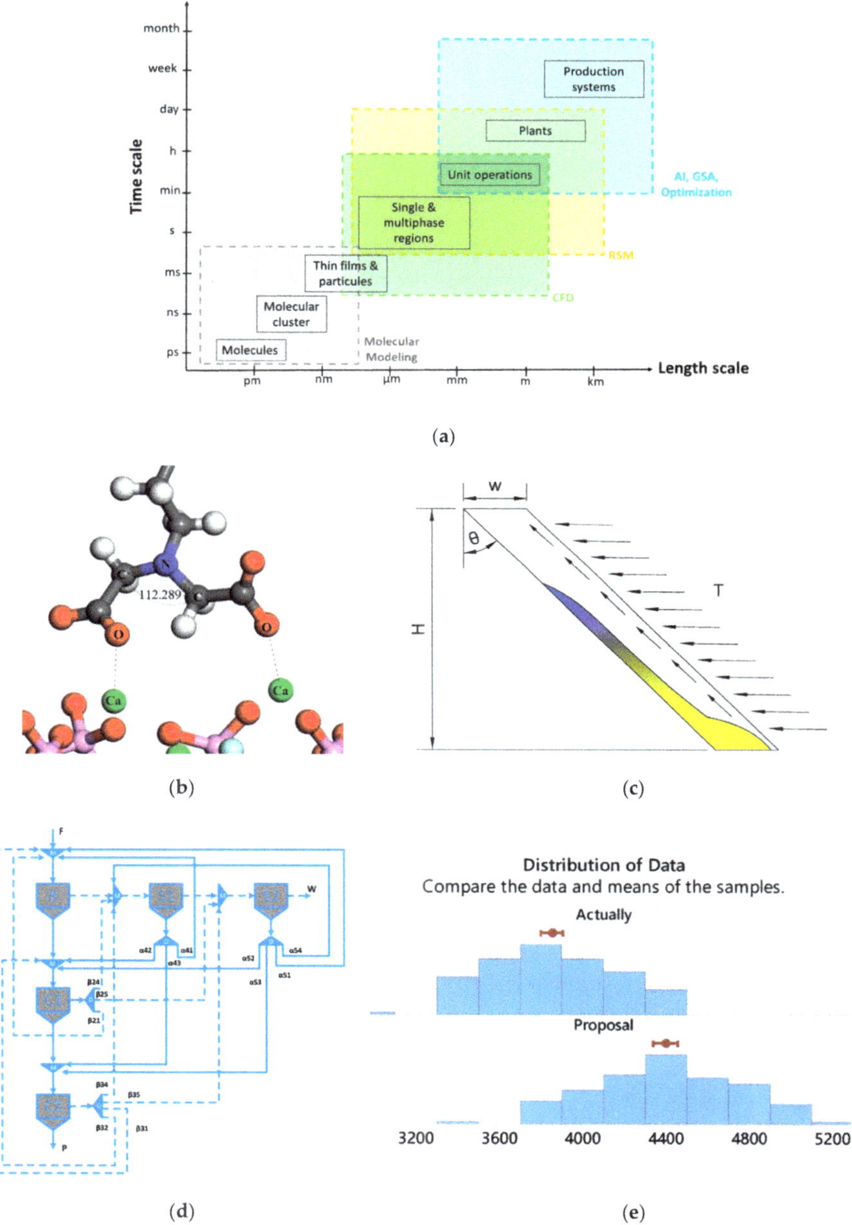

Figure 1. Figures from the special issue. (**a**) Levels of length and time alongside the modeling and optimization tools [4]; (**b**) Adsorption configuration of collector N-(carboxymethyl)-N-tetradecylglycine on fluorapatite. (Ca—green; phosphorus—purple; O—red; H—white; fluorine—light blue; N—dark blue) [14]; (**c**) Scheme of the inclined settler [16]; (**d**) Superstructure for flotation circuit design [17]; (**e**) Production comparison between strategies that do and do not consider changes in the manner of operation [18].

Two manuscripts were published on the numerical simulation of equipment [16,19]. In the first, different turbulent models were compared using computational fluid dynamics in the simulation of cyclonic fields, which are important in cyclonic static microbubble flotation columns. The comparison with experimental values provides important information about which model is most suitable for modeling the different variables in these systems [19]. The second article shows how simulation can be used in the development or improvement of new equipment. Two-dimensional numerical simulations were used to analyze the possibility of improving the separation of particles in inclined settlers [16]. The inclined settler, whose scheme is shown in Figure 1c, has one of its walls exposed to heating. Results show that heating one wall has a significant effect on the particle settling velocity and can help the sedimentation of small particles of the order of 10 μm. These effects can be explained by the change of properties within the settler produced by the temperature profiles. Simulation of kinetic phenomena in multi-phase reactions are also present in this special issue [20,21], although the simulation of the phenomena in these papers have followed traditional methodologies using unreacted shrinking core and progressive conversion models, which have been shown to be unsuitable in several cases [22]. In this sense, it is necessary to move towards multiscale simulation using mesoscale simulation techniques to describe, for example, diffusion and reactive molecular dynamics [23] tools to describe the processes occurring within the interface in order to generate a procedure that can be used to increase our understanding of the heterogeneous gas–solid, liquid–solid, or other multiphase reactions in mineral processing. At the plant level, several articles are included. The use of the tabu search algorithm was applied to determine the optimal flotation circuit within a set of possibilities represented by a superstructure, as shown in Figure 1d [17]. The tabu search algorithm is a method of mathematical optimization classified as a metaheuristic algorithm, which in this work showed a tendency to give better results than the exact methods. The design and optimization at the separation circuit level is an active area in multiphase separation in mineral processing, and several reviews are available in the literature [24,25]. As such, it is not surprising that another study analyzes this same problem [26] but uses analytical methods that significantly simplify the problem, although that can lead to important errors or omissions [27]. A larger time scale problem was considered in the manuscript presented by a multidisciplinary research team [18]. The discrete-event simulation combination with analytical models of leaching processes was used to optimize mineral extraction processes. The methodology helps the planning process by incorporating different possibilities of operation according to the mineralogical changes of the feed. Thus, by simulating a discrete sequence of events over time it is possible to consider the stochastic uncertainties that naturally occur in the mineral. This simulation at the plant level, together with models at the unit operation level, allows for the integration of phenomena that occur at the level of weeks with problems at the level of months or years of operation, giving flexibility to the value chain by adjusting the mineral recovery to the mineralogical variation. This strategy allows for production to be improved compared to strategies that do not consider changes in the manner of operation (Figure 1e).

This Special Issue considered different problems in several areas of multiphase systems in mineral processing. Thus, various strategies and tools were presented to solve or face those problems. On the whole, I hope that this Special Issue will contribute to a superior understanding of multiphase phenomena and will promote future research in the modeling, design, and optimization of multiphase systems in minerals processing.

Authors' contributions from China, Chile, Canada, Germany, Iran, Mexico, and Spain were received. I thank all of them for their contributions that helped in the development of this Special Issue. Finally, I would like to thank the referees and editorial staff of *Minerals* for their valuable efforts that contributed to the success of this initiative.

Acknowledgments: The author is thankful for financial support from the Chilean National Commission for Science and Technology (Fondecyt 1180826) and MINEDUCUA project (code ANT1856).

Conflicts of Interest: The author declares no conflict of interest.

References

1. Mote, C.D.; Dowling, D.A.; Zhou, J. The power of an idea: The international impacts of the grand challenges for engineering. *Engineering* **2016**, *2*, 4–7. [CrossRef]
2. Schlör, H.; Venghaus, S.; Zapp, P.; Marx, J.; Schreiber, A.; Hake, J.F. The energy-mineral-society nexus—A social LCA model. *Appl. Energy* **2018**, *228*, 999–1008. [CrossRef]
3. Cai, M.; Brown, E.T. Challenges in the mining and utilization of deep mineral resources. *Engineering* **2017**, *3*, 432–433. [CrossRef]
4. Cisternas, L.A.; Lucay, F.A.; Botero, Y.L. Trends in modeling, design, and optimization of multiphase systems in minerals processing. *Minerals* **2019**, *10*, 22. [CrossRef]
5. Shu, S.; Vidal, D.; Bertrand, F.; Chaouki, J. Multiscale multiphase phenomena in bubble column reactors: A review. *Renew. Energy* **2019**, *141*, 613–631. [CrossRef]
6. Engquist, B.; Li, X.; Ren, W.; Vanden-Eijnden, E. Heterogeneous multiscale methods: A review. *Commun. Comput. Phys.* **2007**, *2*, 367–450.
7. Liu, T.Y.; Schwarz, M.P. CFD-based multiscale modelling of bubble–particle collision efficiency in a turbulent flotation cell. *Chem. Eng. Sci.* **2009**, *64*, 5287–5301. [CrossRef]
8. Ludwig, A.; Kharicha, A.; Wu, M. Modeling of multiscale and multiphase phenomena in materials processing. *Metall. Mater. Trans. B* **2014**, *45*, 36–43. [CrossRef]
9. Jamett, N.; Cisternas, L.A.; Vielma, J.P. Solution strategies to the stochastic design of mineral flotation plants. *Chem. Eng. Sci.* **2015**, *134*, 850–860. [CrossRef]
10. Sepúlveda, F.D.; Cisternas, L.A.; Gálvez, E.D. The use of global sensitivity analysis for improving processes: Applications to mineral processing. *Comput. Chem. Eng.* **2014**, *66*, 221–232. [CrossRef]
11. Lucay, F.A.; Gálvez, E.D.; Salez-Cruz, M.; Cisternas, L.A. Improving milling operation using uncertainty and global sensitivity analyses. *Miner. Eng.* **2019**, *131*, 249–261. [CrossRef]
12. Lucay, F.; Cisternas, L.A.; Gálvez, E.D. Global sensitivity analysis for identifying critical process design decisions. *Chem. Eng. Res. Des.* **2015**, *103*, 74–83. [CrossRef]
13. Zhang, C.; Xu, Z.; Hu, Y.; He, J.; Tian, M.; Zhou, J.; Zhou, Q.; Chen, S.; Chen, D.; Chen, P.; et al. Novel insights into the hydroxylation behaviors of α-quartz (101) surface and its effects on the adsorption of sodium oleate. *Minerals* **2019**, *9*, 450. [CrossRef]
14. Nan, N.; Zhu, Y.; Han, Y.; Liu, J. Molecular modeling of interactions between N-(Carboxymethyl)-N-tetradecylglycine and fluorapatite. *Minerals* **2019**, *9*, 278. [CrossRef]
15. He, J.; Han, H.; Zhang, C.; Hu, Y.; Yuan, D.; Tian, M.; Chen, D.; Sun, W. New insights into the configurations of lead(II)-benzohydroxamic acid coordination compounds in aqueous solution: A combined experimental and computational study. *Minerals* **2018**, *8*, 368. [CrossRef]
16. Reyes, C.; Ihle, C.F.; Apaz, F.; Cisternas, L.A. Heat-assisted batch settling of mineral suspensions in inclined containers. *Minerals* **2019**, *9*, 228. [CrossRef]
17. Lucay, F.; Gálvez, E.; Cisternas, L. Design of flotation circuits using tabu-search algorithms: Multispecies, equipment design, and profitability parameters. *Minerals* **2019**, *9*, 181. [CrossRef]
18. Saldaña, M.; Toro, N.; Castillo, J.; Hernández, P.; Navarra, A. Optimization of the heap leaching process through changes in modes of operation and discrete event simulation. *Minerals* **2019**, *9*, 421. [CrossRef]
19. Meng, S.; Li, X.; Yan, X.; Wang, L.; Zhang, H.; Cao, Y. Turbulence models for single phase flow simulation of cyclonic flotation columns. *Minerals* **2019**, *9*, 464. [CrossRef]
20. Juárez Tapia, J.; Patiño Cardona, F.; Roca Vallmajor, A.; Teja Ruiz, A.; Reyes Domínguez, I.; Reyes Pérez, M.; Pérez Labra, M.; Flores Guerrero, M. Determination of dissolution rates of Ag contained in metallurgical and mining residues in the $S_2O_3{}^{2-}$-O_2-Cu^{2+} system: Kinetic analysis. *Minerals* **2018**, *8*, 309. [CrossRef]
21. Liu, S.; Han, C.; Liu, J. Study of K-feldspar and lime hydrothermal REACTION: Phase and mechanism with reaction temperature and increasing Ca/Si ratio. *Minerals* **2019**, *9*, 46. [CrossRef]
22. Liddell, K.C. Shrinking core models in hydrometallurgy: What students are not being told about the pseudo-steady approximation. *Hydrometallurgy* **2005**, *79*, 62–68. [CrossRef]
23. Meuwly, M. Reactive molecular dynamics: From small molecules to proteins. *Wiley Interdiscip. Rev. Comput. Mol. Sci.* **2019**, *9*, e1386. [CrossRef]
24. Cisternas, L.A.; Lucay, F.A.; Acosta-Flores, R.; Gálvez, E.D. A quasi-review of conceptual flotation design methods based on computational optimization. *Miner. Eng.* **2018**, *117*, 24–33. [CrossRef]

25. Noble, A.; Luttrell, G.H.; Amini, S.H. Linear circuit analysis: A tool for addressing challenges and identifying opportunities in process circuit design. *Mining, Metall. Explor.* **2019**, *36*, 159–171. [CrossRef]
26. Radmehr, V.; Shafaei, S.; Noaparast, M.; Abdollahi, H. Optimizing flotation circuit recovery by effective stage arrangements: A case study. *Minerals* **2018**, *8*, 417. [CrossRef]
27. Cisternas, L.A.; Acosta-Flores, R.; Gálvez, E.D. Some limitations and disadvantages of linear circuit analysis. In Proceedings of the 7th International Computational Modelling Symposium (Computational Modelling '19), Falmouth, UK, 11–12 June 2019.

Review

Trends in Modeling, Design, and Optimization of Multiphase Systems in Minerals Processing

Luis A. Cisternas [1,*], **Freddy A. Lucay** [2] **and Yesica L. Botero** [1]

[1] Departamento de Ingeniería Química y Procesos de Minerales, Universidad de Antofagasta, Antofagasta 1240000, Chile; yesica.botero@uantof.cl

[2] Escuela de Ingeniería Química, Pontificia Universidad Católica de Valparaíso, Valparaíso 2340000, Chile; freddy.lucay@pucv.cl

* Correspondence: luis.cisternas@uantof.cl; Tel.: +56-552-637-323

Received: 5 November 2019; Accepted: 21 December 2019; Published: 25 December 2019

Abstract: Multiphase systems are important in minerals processing, and usually include solid–solid and solid–fluid systems, such as in wet grinding, flotation, dewatering, and magnetic separation, among several other unit operations. In this paper, the current trends in the process system engineering tasks of modeling, design, and optimization in multiphase systems, are analyzed. Different scales of size and time are included, and therefore, the analysis includes modeling at the molecular level (molecular dynamic modeling) and unit operation level (e.g., computational fluid dynamic, CFD), and the application of optimization for the design of a plant. New strategies for the modeling, design, and optimization of multiphase systems are also included, with a strong focus on the application of artificial intelligence (AI) and the combination of experimentation and modeling with response surface methodology (RSM). The integration of different modeling techniques such as CFD with discrete element simulation (DEM) and response surface methodology (RSM) with artificial neural networks (ANN) is included. The paper finishes with tools to study the uncertainty, both epistemic and stochastic, based on uncertainty and global sensitivity analyses, which is present in all mineral processing operations. It is shown that all of these areas are very active and can help in the understanding, operation, design, and optimization of mineral processing that involves multiphase systems. Future needs, such as meso-scale modeling, are highlighted.

Keywords: computational fluid dynamic; molecular dynamics; density functional theory; discrete element simulation; smoothed particle hydrodynamics; flotation; grinding; response surface methodology; machine learning; artificial neural networks; support vector machine; hydrocyclone; global sensitivity analysis; uncertainty analysis

---

1. Introduction

Multiphase systems are common in mineral processing because most of the process includes the presence of particles, which are usually multiphase mineral particles, and fluids. Examples of operations in mineral processing that include solid–liquid phases are wet grinding, filtration, hydrocyclone, and thickening. An example that includes solid–gas phases is cyclone, examples that include solid–solid phases are magnetic and electrostatic separations, and an example that includes solid–liquid–gas phases is flotation. These operations are generally difficult to study because they are opaque and challenging to measure. Therefore, the modeling of these systems, like other systems, is important, because it allows us to understand their behavior, which allows us to modify them. For example, these models are applied to optimize and design unit operations or plants that depend on multiphase systems. In addition, these models can facilitate the development of new technologies such as new reagents and unit operations.

There are a growing number of tools and methods for the modeling, optimization, and design of these multiphase systems. These increases in the numbers of tools and methods are promoted by the increase in computing power and new algorithms available in the literature. On the other hand, reliable models are needed for the development of new reagents, equipment, and processes. Also, these models are necessary for the optimization of operational conditions. The lack of models increases the dependency on the experience of experts, and also increases the time and cost of scaling up from laboratory- to full-scale. Because the behavior of these systems depends on physical and chemical phenomena that occur at different time and length scales, different tools are available based on these scales. Small scales, e.g., quantum mechanical length scales of 10^{-13} m with time scales of 10^{-16} s, are of significant interest in understanding the interaction of minerals with reagents. Large scales, e.g., plants length scales of 10^3 m with time scales of 10^6 s, are important in terms of plant integration and environmental impact.

This manuscript reviews the main tools and methods for the modeling, design, and optimization of multiphase systems in mineral processing. The idea is not to produce an encyclopedic review, because there are too many tools and methods, but to highlight the most commonly used tools with greater projection. Figure 1, which is based on the work of Grossmann and Westerberg [1], shows different levels of length and time alongside the tools and methods that will be reviewed in this manuscript. First, molecular mechanics and quantum mechanics are analyzed for the purpose of understanding different mineralogical systems. Computational fluid dynamics (CFD), which consists of numerically solving equations of multiphase fluid motion, allows for quantitative predictions and analyses of multiphase fluid flow phenomena. CFD has been applied to mineral processing for both parametric studies and flow-physics investigations. Process design is analyzed next, showing the methods available, with most of them used for flotation processes. Artificial intelligence (AI) is one area with great projection and amount of research, and therefore, is analyzed from the point of view of multiphase systems in mineral processing. Most of the research on mineral processing involves experimental studies, and therefore, experimental design with response surface methodology (RSM) is an important tool to report. Uncertainty, both epistemic and stochastic, must be considered when multiphase systems are studied. The two most important methods for considering uncertainty, uncertainty analysis (UA) and global sensitivity analysis (GSA), are analyzed at the end. Finally, some conclusions and comments are presented to close this report.

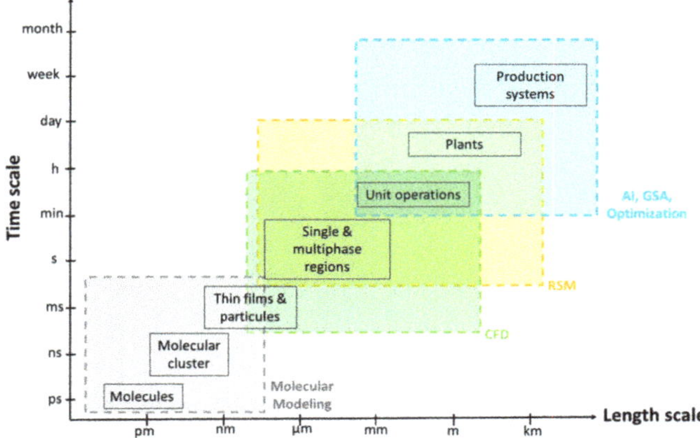

Figure 1. Levels of length and time alongside the modeling and optimization tools analyzed in this manuscript (CFD—computational fluid dynamics; RSM—response surface methodology; AI—artificial intelligence; GSA—global sensitivity analysis).

2. Molecular Dynamic Modeling

The inherent heterogeneous nature and complexity of minerals mineralogy often make the connection between observation and theory very complicated. Additionally, industrial development promotes more and more ore deposit investigation and subsequently, transformation through mineral processing, which adds more phenomena that must be understood. All this complexity from mineralogy and geochemistry requires molecular modeling tools to understand the fundamental properties and mechanisms that control the thermodynamics and kinetics of materials. In this sense, molecular models are often used to supplement experimental observations, providing a powerful complementary tool to the researcher [2,3]. In 1998, De Villiers [4] from Miltek analyzed the potential of molecular modeling to improve mineral processes, using the South African industry as an example. He identified several potential studies including new reagents, the development of new materials, and a theoretical understanding of surface interactions.

According to the abovementioned reference, this tool can be used to understand all microscopic effects (atomic level) that occur on mineral surfaces in different field applications. For example, in the solid–fluid interactions in the flotation process (hydrophobicity and hydrophilicity), and in thickening (water absorption, hydrate minerals, layered double hydroxides, mineral interlayers, clay minerals), among other applications. All these applications have made molecular simulation an accepted approach to solve a number of mineralogical and geochemical problems in multiphase systems [5].

Molecular modeling tools consist of calculating the total energy of the molecular (isolated cluster) or periodic system (crystalline or amorphous structure) under investigation. Two fundamental approaches are typically used: molecular mechanics and quantum mechanics. Figure 2 shows a diagram of molecular mechanics and quantum mechanics methods. Both methods are related and are used to examine the structure and energy of a molecule or periodic system [2].

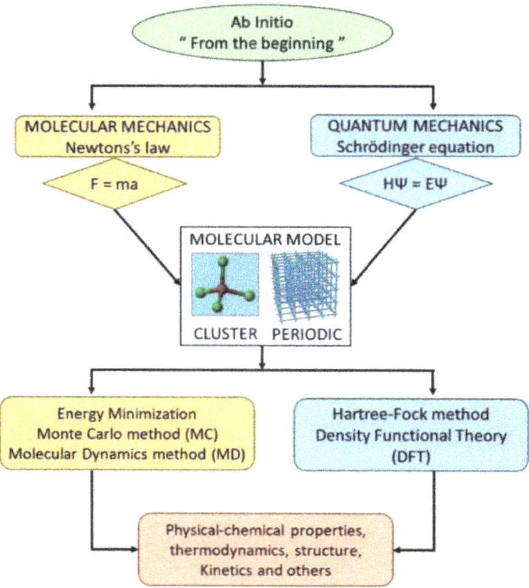

Figure 2. Diagram of molecular mechanics and quantum mechanics methods.

To better understand this diagram, it is necessary to know some concepts regarding how molecular modeling works. According to this, firstly, ab initio refers to the quantum approach for obtaining the electronic properties of a molecule based on the Schrödinger equation ($H\psi = E\psi$), which describes the wave function or state function of a quantum-mechanical system. Secondly, the molecular mechanism relies on the use of analytical expressions that have been parameterized through either experimental observation or quantum calculations using an energy forcefield, based on Newtonian physics ($F = ma$, classical mechanics) to evaluate the interaction energies for the given structure or configuration. In contrast, in quantum mechanics, the analog of Newton's law is Schrödinger's equation, which does not use empirical parameters to evaluate the energy system. In this sense, in a molecular mechanics simulation, the most important requirement is the forcefield used to describe the potential energy of the system. It is essential to have an accurate energy forcefield to achieve a successful energy minimization. The energy of interaction for an assemblage of atoms in either a molecular or crystalline configuration is described by the interatomic potential, generated by the forcefield. This interatomic potential—named potential energy—can be obtained as a function of geometric variables, such as angle, distance, and other geometric measurements [2].

Therefore, it is possible to describe the potential energy for a complex multibody system by the summation of all energy interactions in the system. The energy components are the following: the coulombic energy (electrostatic energy) and the Van der Waals energy (short-range energy associated with atomic interactions), which represent the non-bonded energy components, and the bond stretching (bond energy associated with length changes), angle bending, and torsion, which represent the bonded energy components [6]. From all of these energy components, the total potential energy of a system can be calculated. These types of energies will not be explained in detail because this work provides a general overview of molecular modeling.

Thirdly, energy minimization is another concept that must be understood. This concept also refers to the geometry optimization for obtaining a stable configuration for a molecule or periodic system. This energy involves the repeated measurement of the potential energy on the surface until the minimum potential energy is obtained, which corresponds to the configuration where the forces between atoms are equal to zero. Finally, there are two molecular mechanism approaches—the Monte Carlo (MC) method and the molecular dynamics (MD) simulation—to analyze all the energies and chemical systems on mineral surfaces. The MC method is a stochastic analysis that consists of random sampling of the potential energy surface to obtain a selection of possible equilibrium configurations. The MD simulation is a deterministic molecular modeling tool that involves the calculations of forces based on Newtonian physics used to make a mathematical prediction to evaluate the time evolution of a system on the time scale of pico- and nano-seconds [2,5]. Examples of molecular modeling applications using MD and MC will be presented later, where a detailed discussion will be presented on the use of these techniques for various minerals and mineral surfaces.

The quantum mechanism is a method that evaluates the electronic structure and energy of molecular systems using the Schrödinger equation, which is based on the quantized nature of electronic configurations in atoms and molecules. This technique permits the obtainment of a detailed description of reaction mechanisms, properties of molecular and crystalline structures, electrostatic potentials, thermodynamics properties, and other phenomena that occur in a multiphasic system. The application of this method in the mineralogical and geochemical field is the most challenging task for today's computational modeling.

Quantum chemistry methods can be divided into different classes, where the most used are the Hartree–Fock method and density functional theory (DFT). The Hartree–Fock method uses an antisymmetric determinant of one-electron orbitals to define the total wavefunction. A trial wavefunction is iteratively improved until self-consistency is attained. On the other hand, DTF is a method in which the total energy is expressed as a function of the electron density, and in which all correlation contributions are based on the Schrödinger equation for an electron gas.

Finally, a variety of molecular modeling methods have been implemented by a fair number of research works to study all the interactions between reagents and mineral surfaces, such as adsorption/desorption of reagents on mineral surfaces (collectors, depressors, frothers) in the flotation process, the interaction of water and solute species with mineral surfaces and their behavior in mineral interlayers, and the impact of clay minerals on the dewatering of coal slurry. Next, we focus on an overview of the use of molecular modeling and simulation in the last three years to address specific applications associated with mineralogical and geochemical problems [2,3,5]. Molecular modeling examples are shown following the study of solid–liquid interactions to obtain a good structural model for the material.

2.1. Collector/Depressor Adsorption on Different Mineral Surfaces in the Flotation Process

Leal Filho et al. [7] used MD to demonstrate the ability of two polysaccharides to promote the selective depression of calcite from apatite. They made a good selection of the interaction to represent the chemical and physico-chemical processes during the depression of calcite. The mineral surface of calcite and hydroxyapatite were modeled together with the corn starch and ethyl-cellulose. Firstly, measurements of the unit cell parameters were realized to study the crystal structure of calcite and hydroxy-apatite by X-ray diffraction. Later, the crystallographic orientations of particles of hydroxyapatite and particles of calcite were characterized by optical microscopy and scanning electron microscopy, respectively. The planes predominant for calcite were (101), (401), and (021), and for hydroxy-apatite it was (001). It was observed that calcium species were common active sites at the calcite/water and hydroxy-apatite/water interface, and those sites interacted with starch molecules via the hydroxyl groups existing along with the polymer structure. However, depending on partition planes (hkl), it was demonstrated that the major steric compatibility was in the calcite/starch system. The total fitting number Ft (parameter to define steric compatibility between reagents and mineral orientation) for calcite was: plane (101) Ft = 51.5, plane (401) Ft = 20.1, and plane (021) Ft = 30.3, and for hydroxy-apatite it was: plane (001) Ft = 8.5. Therefore, from these results, it was concluded that the larger the Ft, the greater the expected steric compatibility between reagent structure and crystallographic orientation [7]. Then, these results were compared with micro-flotation experiments of calcite and hydroxy-apatite with sodium oleate in the presence of starch, and it was proven, by calculating recoveries, that the Ft was calculated accurately because the recovery was less with the increase in starch concentration on the calcite surface. Finally, molecular modeling provides appropriate theoretical representations to understand the depressing ability of starch and ethyl cellulose on the mineral surface.

Similar studies using MD simulation were developed by Zhang et al. [8]. The adsorption of collectors on a coal surface was studied. The findings showed that the collector oil absorbed on the coal surface decreases the number of hydrogen bonds between the modified coal surface and contacting water molecules. This can be attributed to the improvement of coal surface hydrophobicity. The hydrophobicity occurs due to the interaction force weakening between water molecules and the coal surface [8]. The same methodology was used by Zhang et al. [9], but this time studying the adsorption behavior of methyl laurate and dodecane on the coal surface. It was determined that methyl laurate is a more successful collector to improve the hydrophobicity of the modified coal surface because the water molecule mobility in methyl laurate was greater than in dodecane. Finally, Nan et al. [10], using the DFT calculation, studied a flotation collector, N-(carboxymethyl)-N-tetradecylglycine (NCNT), in order to understand the adsorption ability of the collector on a fluorapatite (001) surface. They confirmed that the NCNT collector could be used in the fluorapatite flotation process.

2.2. Interaction of Clay Minerals, Water, and Interlayer Structures

Clay minerals such as kaolinite, montmorillonite, smectites, and others are very common in soils, sediments, and sedimentary rocks. In this sense, their properties and behavior have received considerable industrial importance. The interaction of clay minerals with water promotes the water adsorption in the interlayer structure on the clay surface, which generates complex systems. In this sense, avoiding water absorption becomes a difficult task. Hence, computational studies of clay minerals are required to understand the swelling, interlayer structure, and dynamics of water distribution on a clay surface. Today, several studies have been developed to obtain significant dynamical information about these systems. In this section, we show some of the most recent works in this area.

Ma et al. [11] studied the impact of clay minerals (kaolinite and montmorillonite) on the dewatering of coal slurry using a molecular-simulation study, followed by an experimental section to corroborate data accuracy. The molecular simulation results show different adsorptions of water on the side surfaces of kaolinite and montmorillonite. Water molecules could scarcely diffuse into kaolinite from the edge but could easily propagate into the montmorillonite layers from the edge surface because of the existence of a hydrated cation in montmorillonite and a weak interlayer connection. This means that a small amount of montmorillonite caused a major decrease in the filtration velocity and a huge rise in the moisture of the filter cake. Therefore, the efficiency of the dewatering process has a strong dependency on the interaction between kaolinite/montmorillonite and water. Figure 3 shows an equilibrium snapshot from an MD simulation of water adsorption on the side surfaces of kaolinite and montmorillonite. Another study of water absorption on a mineral surface was conducted by Wang et al. [12]. They evaluated the water adsorption on the β-dicalcium silicate (cement) surface from DFT simulations. This work studied how to improve the hydration rate on the cement surface. Then, they studied the adsorption mechanics of the water/cement system. The cement hydration is a crucial step that controls the final properties of cement materials. However, the industrial production of cement produces a large amount of CO_2 emissions and energy consumption. For this reason, understanding cement hydration mechanisms was the main motivation of this study to provide an academic basis for the design of new environmentally friendly cement. Finally, Kubicki et al. [13] studied the vibrational spectra on clays by DFT approaches. Herein, they presented an overview of quantum mechanical calculations to predict vibrational frequencies of molecules and materials such as clays and silicates. For creating a realistic model, the vibrational frequencies were calculated by two analytical methods, Raman and infrared intensities.

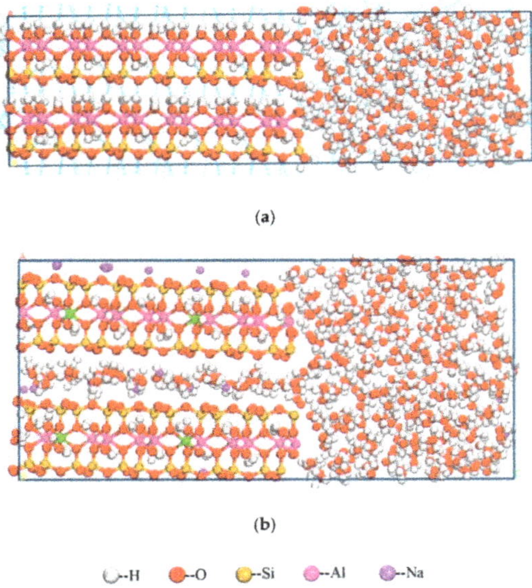

Figure 3. Example of an equilibrium snapshot from a molecular dynamics (MD) simulation of water adsorption on the side surfaces of (**a**) kaolinite and (**b**) montmorillonite at 298 K and 1 bar [11].

The analytical methods combined with computational molecular-scale modeling studies reviewed in this paper illustrate how these methods can provide otherwise unobtainable structural, dynamical, and energetic information about mineral-fluid systems. Using modern supercomputers, molecular modeling can readily model geo-chemically relevant systems containing up to millions of atoms for times up to milliseconds. Thus, these methods can provide dynamic information at frequencies of the order of and greater than the gigahertz range. This approach will continue to play an important role in understanding different mineralogical systems. Future applications in wet grinding can help to better understand and improve this operation as, currently, it is used in nanoscale grinding [14,15]. The development of experimental methods from computational modeling can be an appropriate route for future research in this field.

3. Computational Fluid Dynamics (CFD) in Multiphase Systems

Traditional modeling in mineral processing is strongly based on empirical or semi-empirical models [16–20]. Typically, these models work well under the condition of the experimental data used in the fitting stage but are not reliable for new operational conditions. For new operational conditions or new equipment, new equations or parameters must be determined based on additional experimental data. Several papers have been published that review modeling in multiphase systems including the flotation process [21], either for classical mathematical models [22] or a soft computers approach [23], ball mill [24], and hydrocyclone [25]. Some of the papers are more specific, such as the review of the modeling of bubble-particle detachment [26] or entrainment [27] and water recovery [28] in flotation. These days, engineers are increasingly using CFD to analyze flow and performance in the design of new equipment and processes [29]. The secret behind the success of CFD is its ability to simulate flows, similar to those observed in practical conditions—in terms of tackling real, three-dimensional, irregular flow geometries and phenomena involving complex physics [30]. This is made possible by resorting to a numerical solution of the equations' governing fluid flow rather than seeking an analytical solution. Usually, the equations describing the flow of fluids consist of mathematical statements of conservation of such fundamental quantities as mass, momentum, and energy during fluid flow and allied phenomena. The variables in these equations are three velocity components, pressure, and temperature of the fluid. In a typical case, each of these varies with location and time within the flow domain. Their variation is governed and determined by the conservation equations, which take the form of non-linear partial differential equations. CFD deals with the numerical solution of these equations [30]. For this, a region of space is discretized by creating what is known as a spatial mesh, dividing a region of space into small volumes of control. Then, the discretized conservation equations are solved iteratively in each of them until the residue is sufficiently small. Therefore, a CFD solution requires a large number of arithmetic computations on real numbers; hence, its rise coincided with the advent of computers and the rapid expansion of computer power that ensued in the subsequent decades. In fact, in several cases, even with simplified equations, only approximate results can be obtained. Figure 4 shows examples of CFD modeling.

Multiphase flows are usually modeled using the Euler–Lagrange (E–L) model, the Euler–Euler (E–E) model, and the mixture model. In E–L modeling, the fluid phase is modeled as a continuum, while for the dispersed phase, a large number of individual particles are modeled. The dispersed phase can exchange momentum, mass, and energy with the fluid phase. Since the particle or droplet trajectories are computed for each particle or for a bundle of particles that are assumed to follow the same trajectory, the approach is limited to systems with a low volume fraction of the dispersed phase. Typical applications are dissolved air flotation and air classification. In E–E models, the different phases are all treated as continuous phases, and momentum and continuity equations are solved for each phase. The E–E method can become computationally expensive as the number of equations increases with the number of phases present in the system. The E–E model can handle very complex flows but does not always give the best results since empirical information is needed for the momentum equations. Typical applications are flotation cells and magnetic separators. Another E–E model is the volume-of-fluid (VOF) model, whereby the interface between the different phases is tracked. This model is suitable for hydrocyclone separators. Since the interface between the fluids must be resolved, it is not applicable to a system with many small drops or bubbles. The mixture phase model shortens the E–E method, considering a single momentum equation for all the phases, assuming that they are components of a mixture. In this model, the viscosity is estimated for the mixture. The velocities of the different phases are subsequently calculated from buoyancy, drag, and other forces, giving the relative velocities in comparison with the mean velocity of the mixture [29]. Typical applications are bubble columns, fine particle suspensions, and stirred-tank reactors.

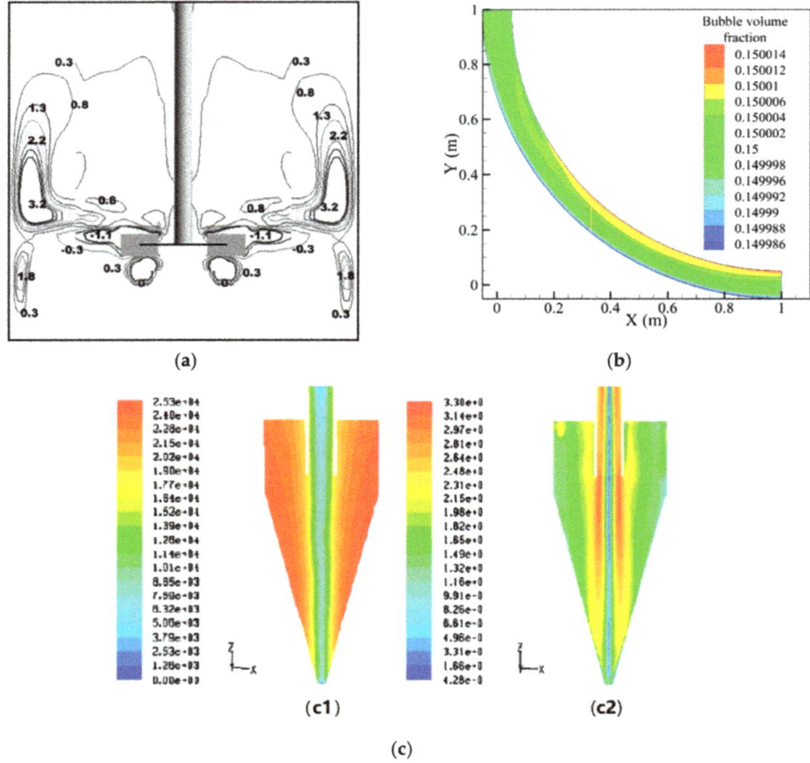

Figure 4. Examples of computational fluid dynamics (CFD) multiphase modeling in mineral processing. (**a**) CFD-predicted net attachment rates after flotation time in the stirred cell [31]; (**b**) Bubble volume fraction (unit in vol %) distribution in a pipe for a backfill material [32]; (**c**) Predicted contours of (**c1**) pressure and (**c2**) tangential velocities in Renner's cyclone [33].

Several factors affect the selection of the most appropriate multiphase model, and the physics of the system must be analyzed and understood. For example, it must be considered whether the phases are separated or dispersed and if the particles follow the continuous phase, among several other factors.

Examples of applications of CFD in mineral processing are given below. CFD was used to improve the understanding of the influence of the geometric design of the classifier on the cut size and the resulting particle size distribution in a centrifugal air classification [34]. The E–L approach was used to investigate how the internal airflow in the second stage of the air classifier affects the classification efficiency. The simulation results show that the classification results are affected by the airflow velocity, particle shape, particle size, the geometry of the air classifier, and turbulence in the airflow. The performance of a wet, high-intensity, magnetic separator was analyzed using CFD [35]. The behavior of these systems relies on the interaction between magnetic, hydrodynamic, gravitational, and interparticle forces. These forces are controlled by the process as well as design parameters. A three-dimensional E–E approach was developed to predict the flow profile as well as the concentration profile of solid particles between two parallel plates. Three phases, i.e., one liquid and two solid phases, were considered. Simulation results agree with the results observed experimentally. Another application of CFD is the study of flow behavior in a hydrocyclone, which is a highly swirling and turbulent multiphase structure. Narasimha et al. [33] developed a multiphase CFD model to understand the particle size segregation inside a six inches hydrocyclone. The predictions were validated against experimental data, and were shown to be in good agreement. An application, outside of separators, is the study of the complex flow behavior in the pulp lifter of autogenous and semi-autogenous grinding mills as it controls the throughput, performance, and efficiency of mills. CFD modeling—the VOF approach—was used to study the efficient and effective removal of pulp/slurry from the mill by a pulp lifter design [36]. Comparison with experimental data shows that CFD can be a useful tool to understand and improve complex flow behavior. In the same direction, a CFD model, a mixture phases model, was developed to study a three-dimensional backfill pipeline transport of three-phase foam slurry backfill (TFSB) [32]. The simulation results indicate that TFSB can maintain a steady state during pipeline transport, experience a markedly reduced pipeline transport resistance, and exhibit better liquidity than conventional cement slurry. Last but not least, the flotation process is one of the most studied systems using CFD, and a comprehensive review of the published literature regarding the CFD modeling of the flotation process was presented [37]. The advances made in the modeling and simulations of the equipment were critically analyzed, and specific emphasis was given to the bubble–particle interactions and the effect of turbulence on these interactions. The simulation of flow behavior of flotation cells has been studied using multiphase E–E [38,39], mixture phase [40], and E–L [41,42] approaches. Mostly, the finite volume approach has been utilized in the reported studies, wherein local values of the flow properties are calculated by solving the governing continuity and momentum equation for each phase [37].

The combination of macroscale CFD simulation with microscale simulation can be a powerful tool in predicting complex phenomena in multiphase systems [43]. Liu and Schwarz [44,45] proposed an integrated CFD-based scheme for the prediction of bubble–particle collision efficiency in turbulent flow from a multiscale modeling perspective. The proposed model can account for changes at the macroscale in the flotation cell geometry and structure, inlet and exit configurations, impeller structure and tip speed, air nozzle structure and airflow rate, and at the microscale in turbulence and collision mechanisms. Similarly, CFD modeling can be combined with discrete element simulation (DEM) to understand the behavior of individual particles. For example, Lichter et al. [46] combined CFD with DEM to analyze the effect of cell size and inflow rate on the retention time distribution in flotation cells. Ji et al. [47] developed two numerical models to model the multiphase flow in hydrocyclones: one is a combined approach of the VOF model and DEM with the concept of the coarse-grained (CG) particle, which can be applicable to a relatively dilute flow, and the other is a combined approach of the mixture model and DEM model with the CG concept, which can be quantitatively applicable to both dilute and dense flows. Finally, Chu et al. [48] studied the coal-medium flow in a dense medium cyclone using DEM to model the motion of coal particles, while the flow of the medium was modeled using the VOF model.

Modeling of wet grinding, including autogenous grinding, semi-autogenous grinding, ball or stirred mills, has been developed using DEM because it is an adequate method to represent the movement and collision of particles. Essentially, the Newton's equation of motion is solved together with a collision/contact law to resolve inter-particle forces. Its application has included the design, optimization, and operation of grinding devices. A complete review on DEM application to comminution, including autogenous/semi-autogenous grinding and mills, is available [49]. However, DEM is computationally intensive, which limits the number of particles that can be considered. Therefore, for a large number of small particles, the method may not be appropriate. In addition, the aqueous phase, which is added to improve the solid transportation and suppress the dust, must be considered in the simulation to obtain a realistic or complete description of the behavior of a wet grinding device. The usual approach to predict the transportation of solid particles interacting with the slurry flow is a coupled DEM and smoothed particle hydrodynamics (SPH) method [50]. SPH is a type of particle-based method, known as mesh-free methods, that employ a set of finite numbers of discrete particles to represent the state and evolution of a flow system [51]. This coupled DEM–SPH method has been successfully applied to autogenous grinding [52,53]. Figure 5 shows on example of how DEM and SPH are coupled in this type of simulation. DEM–CFD (e.g., the k–ε-turbulence model) methods have also been applied to wet grinding, such as stirred media and planetary ball mills [54]. Similar to DEM–SPH methods, here, CFD takes the fluid flow into account and DEM is used for modeling of the grinding media.

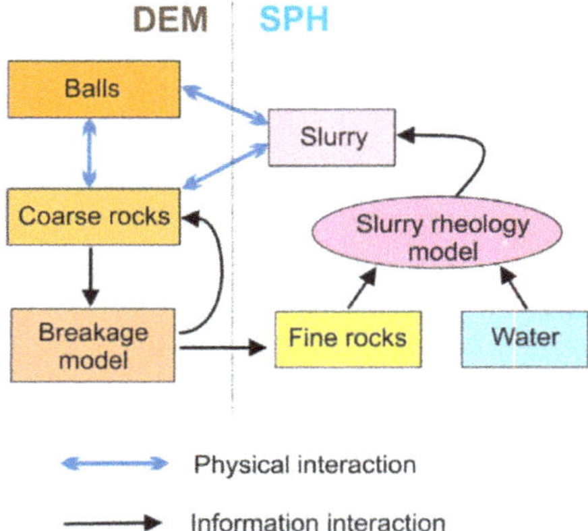

Figure 5. Example of the physical interactions, the model components, and the data flows in discrete element simulation–smoothed particle hydrodynamics (DEM–SPH) modeling of wet grinding [53].

Since the beginning of numerical modeling in mineral processing, in the past two decades, significant advances have been made to simulate multiphase flow behavior. However, it is still far from complete due to the multiscale nature of the problem, which requires integration of the complex interplay between the molecular level and system hydrodynamics. One of the important challenges is the development of methodologies and theories for an adequate representation of the meso-scales. The meso-scales are important in linking micro and macro behaviors and in showing the complexity and diversity of phenomena that occur [55]. The meso-scales can be decisive in the modeling and understanding of multiphase systems where the behavior of the system is strongly influenced by the phenomena at that level [56]. For example, Figure 6 shows the relationship between macro and micro scales with the meso-scale of bubble behavior in flotation.

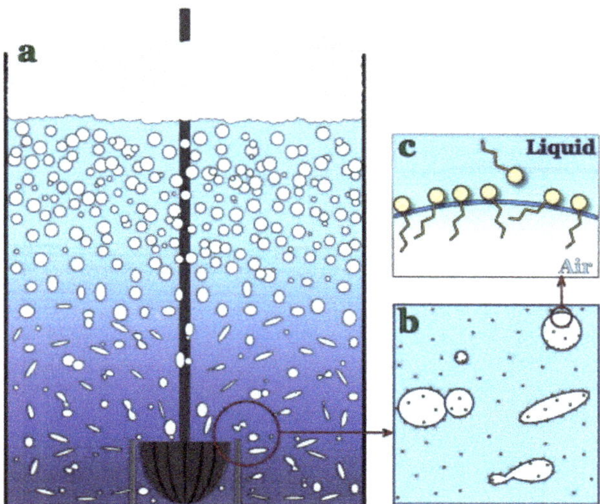

Figure 6. Relationship between scales in bubble behavior in flotation (**a**) macro-scale: cell level, (**b**) meso-scale: bubble coalescence and breakup studies, and (**c**) micro/nano-scale: molecular interactions [57].

4. Design and Optimization

The development of systematic methods for process design in multiphase mineral processing has been active, but to our knowledge, has still not been applied in the industry. The development of these methods has been motived by the search to increase productivity, reduce costs, reduce the adverse environmental impact of waste, and to develop simpler, more economical processes [58]. In general, there are three methods for process design: heuristic-based methods, hybrid methods, and rigorous methods. The heuristic-based method uses rules-of-thumb to help identify process alternatives. Hybrid methods combine first principles with the insight of the designer to obtain a feasible process design. Rigorous methods use a mathematical model to represent a set of alternatives and an optimization algorithm to search for optimal solutions. As they move from heuristic to rigorous methods, the mathematical complexity of the problem increases, the probability of getting better designs increases, the importance of the designer's experience decreases, and the design process goes from being closer to art to being closer to a science.

Most of the work published in the literature is related to the design of flotation circuits. Few works have been published based on heuristic design methods [59,60]. Chan and Price [60] presented a method to design a process for non-sharp separations based on heuristics. The process design is built up unit by unit, stopping when further addition does not increase the profit. The method was applied to flotation circuits. Because heuristic design methods do not guarantee the finding of optimal solutions, this approach to solve the design problem has lost interest from the scientific community.

The most important hybrid method for designing mineral processing facilities is linear circuit analysis (LCA). This technique provides fundamental insights into how unit operations interact and respond when arranged in multistage processing circuits [61]. The method, proposed by Meloy [62] and then developed by Meloy, Williams, and Fuerstanou over several years [63–66], consists of representing the separation yield of a process unit by a transfer function, and then expressing recoveries of the concentrate and tailing as a function of this transfer function. This mass balance approach is extended to the circuit by expressing the global recovery of the circuit as a function of the transfer function of each process unit. Figure 7 shows an example of LCA formulation.

LCA has successfully been utilized to improve the operating performance of industrial processing circuits including magnetic separators [67] and spirals separators [68,69]. More recently, advanced versions of this tool have also been developed to include techno-economic objective functions [70] and circuit uncertainty analysis [71]—a complete review, written by Noble et al. [61], is available. Despite its applications, LCA has several disadvantages, mainly caused by the common practice of assuming that all transfer functions are equal. This simplification does not allow researchers to examine the whole behavior of a concentration circuit because it reduces a multidimensional function of the transfer functions of all the units to one dimension [72]. For example, differences of up to 10% have been observed in the overall circuit recovery for two- and three-stage circuits in the case of identical and non-identical stage recovery [73].

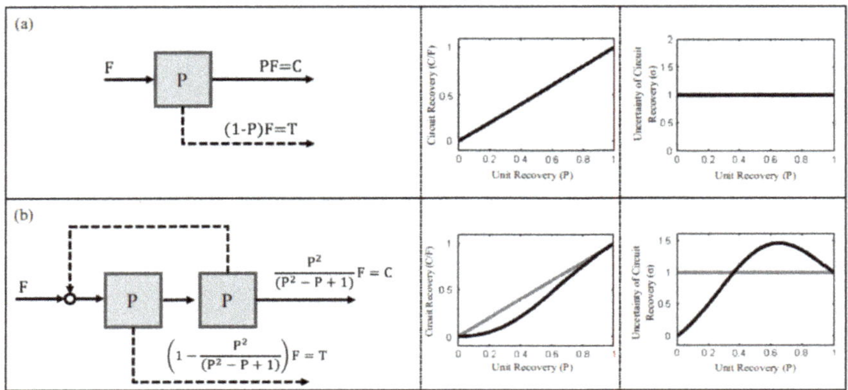

Figure 7. Linear circuit analysis (LCA) formulation example, including uncertainty analysis: (**a**) individual unit, (**b**) two-stage circuit [71].

Rigorous methods are based on optimization procedures. The methodologies consist of developing a superstructure that represents a set of alternatives in which to search for the optimal solution. A mathematical model based on mass balance and kinetics expressions of each operation unit is developed to represent the superstructure, and then, using an objective function, it is solved to obtain the optimal solution. The mathematical model results in a mixed-integer nonlinear programming model (MINLP), which is difficult to solve due to the nonconvex nature. Most of the methodologies proposed are for flotation circuit design, but one methodology has been proposed for a dewatering system [74]. Several reviews on flotation circuit design are available [75–79], and therefore, a brief description is given here. Table 1 shows a list of methodologies that use optimization for the design of flotation circuits. It can be observed that most of the works use few components and/or few process units because the problem is difficult to solve. Also, some simplification of the problem has been applied so that the model is linear programming (LP), nonlinear programming (NLP), or mixed-integer linear programming (MILP). Only in the last few years can it be observed that methodologies are applied to real size plants with at least six species and five process units. The application to real size plants has been possible due to the advances in computer power, optimization algorithm improvements, and the fact that the stage recovery (unit transfer function) uncertainty has a low effect on the optimal circuit structure [80]. By now, this type of methodology can generate a set of optimal alternatives that can be subject to further study by the designer. Also, the case studies analyzed generated new knowledge that could be difficult to obtain from plant experience.

The design of concentration circuits using rigorous methods requires further development to incorporate regrinding and equipment selection, which can affect the circuit performance. A few works have considered regrinding in the design of flotation circuits [81–83]; however, they have used simplified models or the grinding stage was not considered in the decision problem. Therefore, strategies to incorporate grinding in the design of these systems must be considered. A major challenge is the modification of the way the design process occurs in the organizations, which is based on designer experience. Consequently, the usefulness of this type of methodology must be highlighted and adapted in the innovation of the mining sector. For example, breakthrough circuit design may not be the best option because of high capex and small research and development activities in the mining industry. Then, in-process changes of the circuit design, where one or two units or structures are changed, may be a better option.

Table 1. Flotation circuit design methodologies (adapted from Reference [84]) (LP linear programming; NLP nonlinear programming; MILP, mixed-integer linear programming; MINLP, mixed-integer nonlinear programming).

Reference	Model Type	Cell or Bank Model	Entrainment Model	Froth Recovery Model	Algorithm Used	Maximum Number of Species	Maximum Number of Cell or Bank
Mehrotra and Kapur [85]	NLP	Bank	no	no	Mathematical programming	3	4
Reuter et al. [86]	LP	Bank	no	no	Mathematical programming	3	4
Reuter and Van Deventer [87]	LP	Bank	no	no	Mathematical programming	3	5
Schena et al. [88]	MINLP	Bank	no	no	Mathematical programming	2	4
Schena et al. [83]	MINLP	Bank	no	no	Mathematical programming	2	6
Guria et al. [89]	NLP	Cell	no	no	Genetic Algorithm	3	4
Guria et al. [90]	NLP	Cell	no	no	Genetic Algorithm	2	2
Cisternas et al. [81]	MINLP	Bank	no	no	Mathematical programming	3	4
Méndez et al. [82]	MINLP	Bank	no	no	Mathematical programming	3	3
Ghobadi et al. [91]	MINLP	Bank	yes	no	Genetic Algorithm	3	2
Maldonado et al. [92]	NLP	Bank	no	no	Mathematical programming	2	6
Hu et al. [93]	MINLP	Cell	yes	yes	Genetic Algorithm	2	8
Cisternas et al. [94]	MINLP	Bank	no	no	Mathematical programming	3	5
Pirouzan et al. [95]	NLP	Bank	no	no	Genetic Algorithm	2	4
Calisaya et al. [96]	MILP MINLP	Bank	no	no	Mathematical programming	5	7
Acosta-Flores et al. [84]	MILP MINLP	Bank Cell	no	yes	Mathematical programming	15	3 8
Lucay et al. [97]	MINLP	Bank	no	no	Tabu-search	7	5

5. Artificial Intelligence (AI) Applied to Multiphase Systems

The term AI appeared in 1955 [98]. AI is a branch of computer science dedicated to the development of computer algorithms to accomplish tasks traditionally associated with human intelligence. In recent years, the interest from the mining industry in utilizing AI techniques in areas such as geology and minerals processing has increased. This trend is repeated in the ambit of scientific research [23,99–101]. Among these techniques, soft computing is highlighted, which has been used in the modeling, design, and optimization of mining processes.

Soft computing is defined as the group of methodologies and tools that can assist in the design, development, and operation of intelligent systems that are capable of adaptation, learning, and operating autonomously in an environment of uncertainty and imprecision [102]. Soft computing can be divided into two groups: probability reasoning, and functional approximation and randomized search. The first group, in turn, can be divided into probabilistic models and fuzzy logic. The second group, in turn, can be divided into evolutionary computing (EC), swarm optimization (SO), and machine learning [103–105]. The developed tools in each group mentioned earlier are shown in Figure 8. Note that ML includes a broad set of methods used to extract useful models from empirical data. Machine learning tools are focused on endowing programs with the ability to "learn" and adapt [106]. These tools need training algorithms to learn, some of which are shown in Figure 8 (SO and EC). These algorithms, as seen later, can be used for optimizing and designing metallurgical processes [93,97].

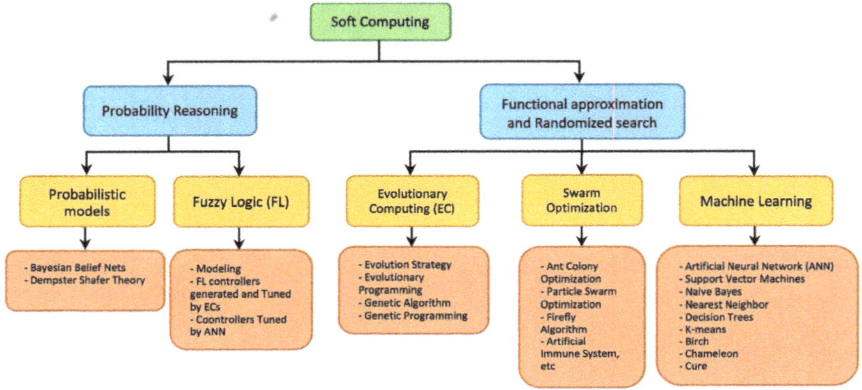

Figure 8. Methodologies and tools considered in soft computing.

Modeling can be divided into data-driven, fault detection and/or diagnosis, and machine vision. The first considers building models for complementing or replacing physically-based models. The second involves a statistical model based on data that are considered representative of the normal operating condition (NOC) of the process. Any observations that exceed a certain limit in this NOC model are considered as faults [107]. The third considers a type of data-driven modeling that uses images or video, rather than process measurements.

Data-based modeling uses information extracted from experimental, simulated, or industrial data. At an industrial scale, these methods are applied as soft sensors for the prediction of measurements that are difficult to measure. Some applications of these methods include the modeling of metallurgical responses or subprocesses involved in integral processes: grinding [108–114], thickening [115], flotation [116–121], and hydrocyclones [122], among other processes. For example, Estrada-Ruiz and Pérez-Garibay [118] used multilayer perceptron, which is a type of neural network, for estimating the mean bubble diameter and bubble size distribution on the mineralized froth surface. Meanwhile, Jahedsaravani et al. [120] used multilayer perceptron for predicting the copper recovery, copper concentrate grade, mass recovery of the concentrate, and water recovery in the concentrate obtained through batch flotation. Saravani et al. [121] developed a fuzzy model for estimating the performance of an industrial flotation column. Núñez et al. [114] developed a fuzzy model for predicting the future weight of a semi-autogenous grinding (SAG) mill. Artificial neural networks (ANNs), support vector machine (SVM), and fuzzy models substantially reduce the computational cost involved in simulation, and uncertainty and sensitivity analyses [123].

Fault detection is commonly carried out using principal component analysis (PCA) or its extensions/modifications [107], and it has been used in flotation systems [124,125] and milling circuits [126,127]. Fault diagnosis, i.e., the identification of variables associated with faulty conditions, is usually achieved via the use of approaches based on PCA. Some applications of these methods include flotation [128–130] and grinding [126,127]. For example, Wakefield et al. [127] simulated a milling circuit for investigating faults related to particle size estimates and mill liners. They applied statistical tools (PCA) for detecting faults, in conjunction with process topology data-driven techniques (Granger causality) for root cause analysis. These authors reported that the statistical monitoring method took slightly longer to detect the mill liner fault, due to the incipient nature of the fault. However, this method is significantly faster than what has been achieved by monitoring only the economic performance of the circuit. The fault diagnosis identified the mill power as the root cause of the fault.

Machine vision is the study of techniques for extracting meaningful information from high-dimensional images, and it has been used almost exclusively in flotation [131–133]. Developed models were used for classifying flotation froth images, and commonly, these were based on SVM, ANNs, and decision trees [23]. For example, Zhu and Yu [132] proposed an ANN model based on features extracted from digital froth images at a hematite flotation plant. This model was used to help identify flotation conditions and to adjust the reagent's quantity. Zhao et al. [134] estimated the bubble size distribution using image processing techniques based on decision trees.

Many of the developed models were used to optimize the process, for example, Curilem et al. [113] used ANNs and SVM models for online optimization of the energy consumption in SAG. Zhu and Yu [132] used the developed model for optimizing the reagent's dosage. Saravani et al. [121] used a fuzzy model for optimizing and stabilizing the industrial flotation column. Note that ANN and SVM, among other tools included in machine learning, require training algorithms, which can be divided into exact and approximate algorithms. This last group considers genetic algorithms, particles swarm optimization, and differential evolution, among others, including their hybridizations. According to the related literature, these algorithms have been used for tuning the parameters of the ANN, SVM, and fuzzy models [135], and for minimizing/maximizing the objective function in optimization problems and process design.

Process optimization via approximate algorithms has been reported by several authors, for example, Tandon et al. [136] developed an ANN for predicting cutting forces in a milling process, which, in turn, was used to optimize both feed and speed through particle swarm optimization. Massinaei et al. [137] used ANN and gravitational search algorithms to model and optimize the metallurgical performance of a flotation column. Shunmugam et al. [138] used genetic algorithms to optimize the minimum production cost in a face milling operation. Here, the trend is using hybrid algorithms to explore the search space efficiently and find a global optimal solution [101,139,140].

Process design via approximate algorithms has been performed almost exclusively in froth flotation, specifically in flotation circuit design. The latter considers three ingredients: first, a superstructure for representing the alternatives for design, second, a mathematical model for modeling the alternatives for design, included goals, constraints, and objective function, and third, an optimization algorithm [76]. Lucay et al. [97] considered a stage superstructure composed of five stages of flotation, which were modeled using a bank model. Here, the single-objective function was of the economic type, and the Tabu-Search algorithm was used for solving the design problem. Hu et al. [93] used a superstructure of eight cells, which were modeled using a cell model. They used a single-objective function of the economic type and genetic algorithm for solving the problem. Ghobahi et al. [91] used genetic algorithms, a superstructure of stages, and single-objective functions of the technical type. Pirouzan et al. [95] also applied genetic algorithms but considered a multi-objective function of the technical type. Due to the multi-objective nature of the problem, the authors used the Pareto method to obtain a set of solutions. The superstructure used involved three and four flotation stages, out of all possible combinations, which were modeled using a bank model. These authors applied their methodology to improve the design of a flotation circuit processing coal. Figure 9a shows the initial flotation circuit design, and Figure 9b shows the new design of the flotation circuit. They reported that the new design provided a recovery of ash that was 6.7% higher than that of the initial design. In addition, to consider designs of four stages would increase the recovery by 3.8%, and the ash grade would be 11.2%, which is within the acceptable quality level. Here, the common factor is the use of objective functions of the economic type because objective technical functions are difficult to define, including approaches using the Pareto method to address multi-objective problems [76].

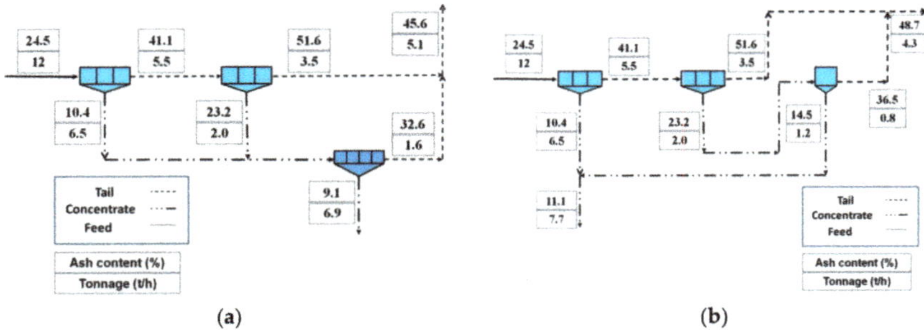

Figure 9. Comparison of flotation circuit design: (**a**) design and ash contents of the initial circuit, (**b**) design and ash contents of the new circuit, adapted from Pirouzan et al. [95].

Obtaining data can be very expensive, so one challenge in data-based modeling is developing methodologies that allow for the attainment of robust models using a small amount of data. In addition, at an industrial scale, the data frequently exhibit as being high-dimensional, non-normally distributed, and nonstationary, with nonlinear relationships, including noise and outliers, which makes it even more difficult to develop a model capturing the true relationships between the input variables. These comments are also valid for fault detection and diagnosis and for machine vision.

6. Response Surface Methodology (RSM)

RSM is used for modeling and optimization processes. RSM involves the following three steps [141]: first, a design of experiments (DoE) for driving the experiments, second, the response surface is modeled based on empirical models, and third, the optimization of the responses is carried out using the empirical model. According to Garud et al. [142], DoE can be divided into broad families, i.e., classical and modern design of experiments. The first is based on laboratory experiments. This includes approaches such as full- and half-factorial design, central composite design, Plackett–Burman design, and Box–Behnken design, among others [143]. The second is based on computer simulations. This includes approaches such as full-factorial design [144], fractional factorial design, central composite design [145], Latin hypercube sampling [146], and symmetric Latin hypercube sampling [147], among others.

One advantage of the classical RSM is that it needs a smaller number of experiments, which means it is cheaper and requires less time. These characteristics explain the large number of applications, including flotation [148,149], grinding [150–152], and thickening [153], among other processes.

The related literature shows that classical RSM is commonly applied using a second-order polynomial as a prediction model [143]. For example, optimal conditions of rotation speed, solid concentration, and grinding time were obtained for wet grinding in a ball mill using central composite design with a second-order polynomial [152]. In fact, an excellent determination coefficient ($R^2 =$ 0.9989) was obtained, which indicates a good agreement with experimental values. Similar good results were observed using a Box–Bhenken design in copper sulphide ore grinding in a ball mill [150]. However, several processes do not follow a second-order polynomial behavior and, consequently, a poor adjustment of the model is obtained (see Figure 10). The immediate consequence is incorrect optimization. The related literature proposes different approaches in the modeling of surface response instead of polynomial models. For example, regression of Gaussian processes has been proposed, since these models can model complex functions [141,154]. Also, the use of SVM regression as a prediction model has been proposed [155]. However, the most popular alternative has been ANNs [156].

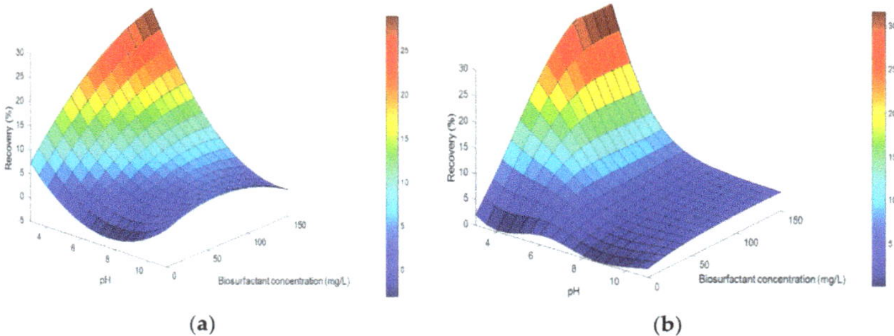

(a)   (b)

Figure 10. (**a**) Quartz recovery using a second-order polynomial as a prediction model ($R^2 = 0.931$), (**b**) quartz recovery using an artificial neural network as a prediction model ($R^2 = 0.982$) [156].

On the other hand, according to Garud et al. [142], the chemical and process system engineering community has exclusively employed modern DoE techniques in the context of surrogate approximation and surrogate-assisted optimization. Modern RSM is also called response surface surrogate (RSS). Surrogate modeling techniques are grouped by some authors into two broad families, which are statistical or empirical data-driven models that emulate the high-fidelity model response, and lower-fidelity physically-based surrogates, which are simplified models of the original system [123].

Data-driven surrogates involve empirical approximations of the complex model output calibrated in a set of inputs and outputs of the complex model. Some approximate techniques proposed in the related literature are: polynomial, kriging (Gaussian process), k nearest neighbors, proper orthogonal decomposition, radial basis functions, support vector machines, multivariate adaptive regression splines, high-dimensional model representation, treed Gaussian processes, Gaussian emulator, smoothing splines analysis of variance (ANOVA) models, polynomial chaos expansions, genetic programming, Bayesian networks, and ANNs [123,157].

This approach has been applied in flotation [158], thickening [159,160], and comminution [161], among others. Usually, these works are based on CFD models, which consider several complex phenomena involved in the studied process. However, these models are computationally expensive to evaluate. This limits their application in continuous process modeling for dynamic simulation, optimization algorithms, and control purposes. Surrogate model techniques can help to overcome this disadvantage. For example, Rabhi et al. [158] developed surrogate models via a hierarchical polynomial using a dataset obtained through simulations of a CFD model of froth flotation. The surrogate model was used for estimating the bubble–particle collision probability (see Figure 11a). These authors reported that the surrogate models developed were highly accurate with a negligible CPU (central processing unit) time. This accuracy increases with an increasing number of interpolation points (see Figure 11b). Stephens et al. [159] developed surrogate models of a CDF model of flocculant adsorption in an industrial thickener because the latter is impractical for performing sensitivity analysis (SA). These authors used radial basis functions, ANNs, and least squares-support vector machines as surrogate models, and they reported that the radial angle between the flocculant sparge and feed pipe, and the distance from the feed well to the flocculant sparge, are the most important parameters in flocculant loss (output variable).

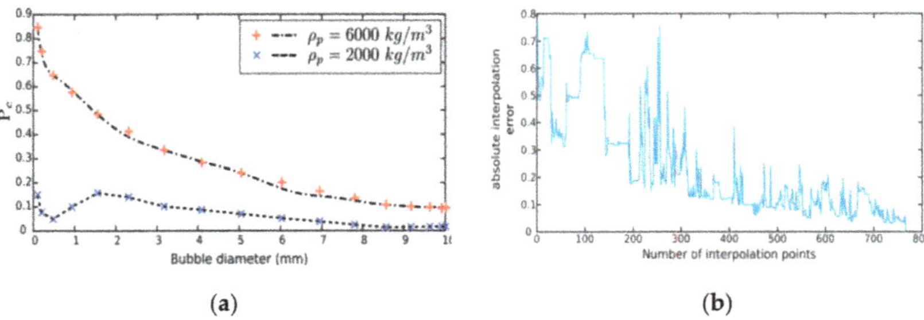

Figure 11. (**a**) Bubble–particle collision probability, P_c, versus bubble diameter, (**b**) prediction absolute error versus number of interpolation points [158].

7. Uncertainty and Sensitivity Analyses

UA corresponds to determining the uncertainty in the output variables as a result of the uncertainty in the input variables. For performing UA, the related literature proposed several theories, such as fuzzy theory and probability theory, among other theories [162]. Meanwhile, SA can be defined as the study of how the uncertainty in the output of a model can be apportioned to different sources of uncertainty in the model input. There are two types of SA: local sensitivity analysis (LSA) and GSA. The second is the most robust because it considers the full range of uncertainty of the input variables.

According to Saltelli et al. [163], SA is an ingredient of modeling. These authors suggested that SA could considerably assist in the use of models, by providing objective criteria of judgment for different phases of the model-building process: model identification and discrimination, model calibration, and model corroboration. In line with this, Lane and Ryan [164] indicate that a well-developed model should include model verification, validation, and uncertainty quantification. Model verification is used for ensuring that the model is behaving properly, for example, the model can be compared with other models or with known analytical solutions. Model validation involves the comparison with experimental data. Uncertainty quantification (UQ) studies the effect of uncertainties on the model. UQ can be performed using uncertainty and sensitivity analyses. These provide a general overview of the effect of uncertainties. Typically, model verification, calibration, and corroboration are not applied in mineral processing, but they must be considered in future model development. The interested readers can see the model developed by Mellado et al. [165] for heap leaching, which has been validated, verified, and corroborated [166–168].

UA and GSA have also been used to identify the operational conditions of a mill system under uncertainty. Lucay et al. [162] applied UA for studying the effect of the distribution and magnitude of the uncertainties of input variables in the responses of the grinding process (see Figure 12a). GSA was utilized to identify influential input variables. Then, the regionalization of the influential input variables was applied to identify the operational regions (see Figure 12b). In other words, the control of the uncertainty of the significant input variables allows for the control of uncertainty in the mill system. GSA has also been applied in the design or optimization of flotation circuits under uncertainty [169–172]. Sepúlveda et al. [169] proposed a methodology for the conceptual design of flotation circuits. The methodology involved three decision levels: level I—the definition of the analysis of the problem, level II—the synthesis and screening of alternatives, and level III—the final design. This last level considers the identification of gaps and opportunities for improvement, among other aspects. Identification was performed using LSA and GSA. Figure 13a shows the flotation circuits designed using this methodology, and Figure 13b shows how the uncertainty on the recovery of each species in the flotation stages affects their global recovery. These authors reported that if the target is increasing the recovery of chalcopyrite, it is recommended that modifications should be made in cleaner 3 (see Figure 13b) because changes at this stage have a significant effect on the global recovery of chalcopyrite and little effect on the global recovery of other species (higher Sobol index values).

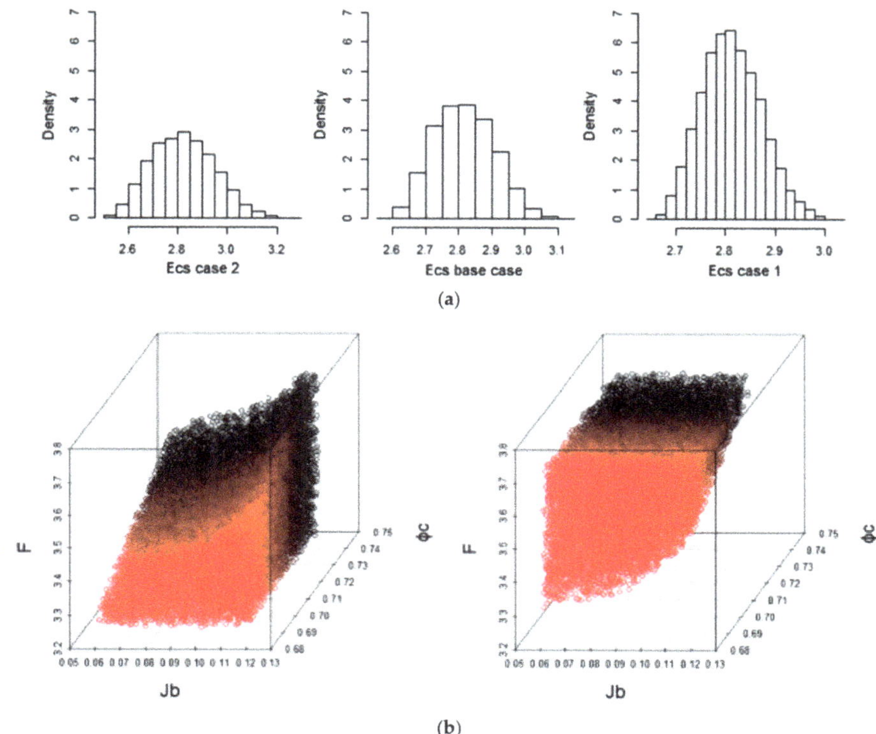

Figure 12. (**a**) The comminution-specific energy histogram of a SAG (semi-autogenous grinding) mill under three uncertainty magnitudes. (**b**) The regionalization of fresh ore flux fed (F), percentage of mill volume occupied by steel balls (J_b), and percentage of critical speed (ϕ_c) [162].

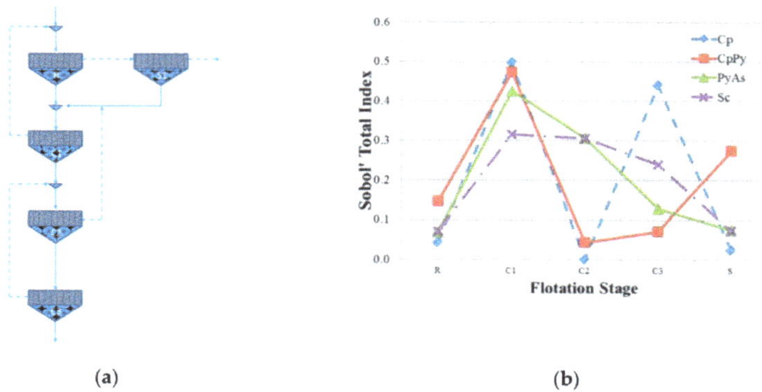

Figure 13. (**a**) Designed circuit using the methodology. (**b**) Sobol total index for each stage and for chalcopyrite (Cp), chalcopyrite–pyrite (CpPy), pyrite–arsenopyrite, and silica (Sc) [169].

Here, the trend is using methods of GSA based on the decomposition of variance due to its versatility [173]. However, this last approach is computationally expensive. This drawback has been overcome in other engineering areas via the development of metamodels or surrogate models [123,157,174], such as ANNs. This approach has not been applied to multiphase mineral processing systems; however, we estimate that this will change due to metamodels that are not only more efficient for performing GSA and UA, but also for carrying complementary analyses, such as data classification.

8. Discussion and Conclusions

Experimentally based research is time demanding and costly, but necessary in multiphase mineral processing systems. These systems include operations and phenomena such as flotation, hydroclyclone, grinding, and magnetic separation. The need for models for these systems is not only necessary to reduce the cost and time associated with research activities but also, because if we do not have a model, we do not understand the system, and if we do not understand the system, we cannot modify it to obtain the desired conditions. The models and tools available to study multiphase systems in mineral processing depend on the length and time scales of the phenomenon that needs to be analyzed. Important advances have been developed in different tools, such as MD at the molecular level, CFD at the fluid level, and mathematical programming at the plant level. RSM can be applied to all levels to model experimental data and numerical experiments. UA and GSA are the most powerful tools to analyze uncertainty. AI can have applications at all levels and, in the future, new applications and developments are expected.

MD has recently emerged in the study of multiphase systems in mineral processing. New studies and applications will undoubtedly show the benefits of understanding phenomena at the molecular level in these systems. There are other challenges that have not been analyzed in detail in the literature, such as integration in multiscale modeling, design, and optimization. Some advances have been observed, for example, the integration of CFD modeling with DEM to integrate particle and fluid phenomena. However, new research on meso-scale modeling integrated with micro- and macro-scale modeling is essential to better describe and optimize multiphase systems. Also, some tools have been combined to increase the capabilities of these methods, for example, ANNs have been combined with RSM to be able to model complex behavior. Examples in design are the integration of process design with control design and molecular modeling with process design. The simultaneous design of process/control or process/molecular can, as a result, produce a better overall design. For example, explicit process control structures can be included in the process design problem, which allows for consideration of the operability in early stages of the design process [175]. A good understanding of molecular phenomena can aid the consideration of the properties in process design, for example, deciding the best location for stream recycling not only based on mineral concentrations, but also considering the final result in properties (such as pH, chemical potential, dissolved oxygen). A technique for property integration based on property clustering can be used for this purpose [176–178]. To achieve these objectives, more research and development efforts in this area are necessary.

Despite the advantages of the use of optimal design tools to identify better process structures, they have not been used in practice in mineral processing, at least to the authors' knowledge. Therefore, the usefulness of this type of methodology must be highlighted and adapted in the innovation of the mining sector. In-process changes of the circuit design, where one or two units or structures are changed, can be explored for this purpose.

Author Contributions: Conceptualization, L.A.C.; methodology and review development, L.A.C., F.A.L. and Y.L.B.; writing—original draft preparation, L.A.C., F.A.L. and Y.L.B.; writing—review and editing, L.A.C. All authors have read and agreed to the published version of the manuscript.

Funding: This research was funded by CONICYT, PIA program grant number ACM 170005 and Fondecyt program grant number 1180826.

Acknowledgments: The authors thank financial support from Chilean National Commission for Science and Technology, CONICYT, through PIA program grant number ACM 170005 and Fondecyt program grant number 1180826. L.A.C. Thanks the supported of MINEDUCUA project, code ANT1856.

Conflicts of Interest: The authors declare no conflict of interest.

References

1. Grossmann, I.E.; Westerberg, A.W. Research challenges in process systems engineering. *AIChE J.* **2000**, *46*, 1700–1703. [CrossRef]
2. Cygan, R.T. Molecular Modeling in Mineralogy and Geochemistry. *Rev. Mineral. Geochem.* **2010**, *42*, 1–35. [CrossRef]
3. Kirkpatrick, R.J.; Kalinichev, A.G.; Bowers, G.M.; Yazaydin, A.Ö.; Krishnan, M.; Saharay, M.; Morrow, C.P. Review. NMR and computational molecular modeling studies of mineral surfaces and interlayer galleries: A review. *Am. Mineral.* **2015**, *100*, 1341–1354. [CrossRef]
4. De Villiers, J.P.R. Minerals Processing in South Africa: Materials Modelling Opportunities. *Mol. Simul.* **1999**, *22*, 81–90. [CrossRef]
5. Creton, B.; Nieto-Draghi, C.; Pannacci, N. Prediction of Surfactants' Properties using Multiscale Molecular Modeling Tools: A Review. *Oil Gas Sci. Technol.* **2012**, *67*, 969–982. [CrossRef]
6. Kirkpatrick, R.J.; Kalinichev, A.G.; Wang, J. Molecular dynamics modelling of hydrated mineral interlayers and surfaces: Structure and dynamics. *Mineral. Mag.* **2005**, *69*, 289–308. [CrossRef]
7. Filho, L.S.L.; Seidl, P.R.; Correia, J.C.G.; Cerqueira, L.A.C.K. Molecular modeling of Reagents for flotation process. *Miner. Eng.* **2000**, *13*, 1495–1503. [CrossRef]
8. Zhang, Z.; Wang, C.; Yan, K. Adsorption of collectors on model surface of Wiser bituminous coal: A molecular dynamics simulation study. *Miner. Eng.* **2015**, *79*, 31–39. [CrossRef]
9. Zhang, H.; Liu, W.; Xu, H.; Zhuo, Q.; Sun, X. Adsorption Behavior of Methyl Laurate and Dodecane on the Sub-Bituminous Coal Surface: Molecular Dynamics Simulation and Experimental Study. *Minerals* **2019**, *9*, 30. [CrossRef]
10. Nan, N.; Zhu, Y.; Han, Y.; Liu, J. Molecular Modeling of Interactions between N-(Carboxymethyl)-N-tetradecylglycine and Fluorapatite. *Minerals* **2019**, *9*, 278. [CrossRef]
11. Ma, X.; Fan, Y.; Dong, X.; Chen, R.; Li, H.; Sun, D.; Yao, S. Impact of Clay Minerals on the Dewatering of Coal Slurry: An Experimental and Molecular-Simulation Study. *Minerals* **2018**, *8*, 400. [CrossRef]
12. Wang, Q.; Manzano, H.; López-Arbeloa, I.; Shen, X. Water Adsorption on the β-Dicalcium Silicate Surface from DFT Simulations. *Minerals* **2018**, *8*, 386. [CrossRef]
13. Kubicki, J.; Watts, H. Quantum Mechanical Modeling of the Vibrational Spectra of Minerals with a Focus on Clays. *Minerals* **2019**, *9*, 141. [CrossRef]
14. Wang, Q.; Fang, Q.; Li, J.; Tian, Y.; Liu, Y. Subsurface damage and material removal of Al–Si bilayers under high-speed grinding using molecular dynamics (MD) simulation. *Appl. Phys. A* **2019**, *125*, 514. [CrossRef]
15. Ren, J.; Hao, M.; Lv, M.; Wang, S.; Zhu, B. Molecular dynamics research on ultra-high-speed grinding mechanism of monocrystalline nickel. *Appl. Surf. Sci.* **2018**, *455*, 629–634. [CrossRef]
16. Safari, M.; Deglon, D. An attachment-detachment kinetic model for the effect of energy input on flotation. *Miner. Eng.* **2018**, *117*, 8–13. [CrossRef]
17. Gorain, B.K.; Franzidis, J.-P.; Manlapig, E.V. The empirical prediction of bubble surface area flux in mechanical flotation cells from cell design and operating data. *Miner. Eng.* **1999**, *12*, 309–322. [CrossRef]
18. Wang, L.; Runge, K.; Peng, Y.; Vos, C. An empirical model for the degree of entrainment in froth flotation based on particle size and density. *Miner. Eng.* **2016**, *98*, 187–193. [CrossRef]
19. Savassi, O.N.; Alexander, D.J.; Franzidis, J.P.; Manlapig, E.V. An empirical model for entrainment in industrial flotation plants. *Miner. Eng.* **1998**, *11*, 243–256. [CrossRef]
20. Kraipech, W.; Chen, W.; Dyakowski, T.; Nowakowski, A. The performance of the empirical models on industrial hydrocyclone design. *Int. J. Miner. Process.* **2006**, *80*, 100–115. [CrossRef]
21. Gharai, M.; Venugopal, R. Modeling of flotation process—An overview of different approaches. *Miner. Process. Extr. Metall. Rev.* **2015**, *37*, 120–133. [CrossRef]
22. Jovanović, I.; Miljanović, I. Modelling of Flotation Processes by Classical Mathematical Met—A Review. *Arch. Min. Sci.* **2015**, *60*, 905–919.

23. Jovanović, I.; Miljanović, I.; Jovanović, T. Soft computing-based modeling of flotation processes—A review. *Miner. Eng.* **2015**, *84*, 34–63. [CrossRef]

24. Tavares, L.M. A Review of Advanced Ball Mill Modelling. *KONA Powder Part. J.* **2017**, *34*, 106–124. [CrossRef]

25. Narasimha, M.; Brennan, M.; Holtham, P.N. A Review of CFD Modelling for Performance Predictions of Hydrocyclone. *Eng. Appl. Comput. Fluid Mech.* **2007**, *1*, 109–125. [CrossRef]

26. Wang, G.; Nguyen, A.V.; Mitra, S.; Joshi, J.B.; Jameson, G.J.; Evans, G.M. A review of the mechanisms and models of bubble-particle detachment in froth flotation. *Sep. Purif. Technol.* **2016**, *170*, 155–172. [CrossRef]

27. Wang, L.; Peng, Y.; Runge, K.; Bradshaw, D. A review of entrainment: Mechanisms, contributing factors and modelling in flotation. *Miner. Eng.* **2015**, *70*, 77–91. [CrossRef]

28. Zheng, X.; Franzidis, J.P.; Johnson, N.W. An evaluation of different models of water recovery in flotation. *Miner. Eng.* **2006**, *19*, 871–882. [CrossRef]

29. Andersson, B.; Andersson, R.; Hakansson, L.; Mortensen, M.; Sudiyo, R.; Van Wachem, B. *Computational Fluid Dynamics for Engineers*; Cambridge University Press: Cambridge, UK, 2012; ISBN 978-1-107-01895-2.

30. Jayanti, S. *Computational Fluid Dynamics for Engineers and Scientists*; Cambridge University Press: Cambridge, UK, 2018; ISBN 9789402412178.

31. Koh, P.T.L.; Schwarz, M.P. CFD modelling of bubble—Particle attachments in flotation cells. *Miner. Eng.* **2006**, *19*, 619–626. [CrossRef]

32. Chen, X.; Zhou, J.; Chen, Q.; Shi, X.; Gou, Y. CFD Simulation of Pipeline Transport Properties of Mine Tailings Three-Phase Foam Slurry Backfill. *Minerals* **2017**, *7*, 149. [CrossRef]

33. Narasimha, M.; Brennan, M.S.; Holtham, P.N. CFD modeling of hydrocyclones: Prediction of particle size segregation. *Miner. Eng.* **2012**, *39*, 173–183. [CrossRef]

34. Johansson, R.; Evertsson, M. CFD simulation of a centrifugal air classifier used in the aggregate industry. *Miner. Eng.* **2014**, *63*, 149–156. [CrossRef]

35. Mohanty, S.; Das, B.; Mishra, B.K. A preliminary investigation into magnetic separation process using CFD. *Miner. Eng.* **2011**, *24*, 1651–1657. [CrossRef]

36. Weerasekara, N.S.; Powell, M.S. Performance characterisation of AG/SAG mill pulp lifters using CFD techniques. *Miner. Eng.* **2014**, *63*, 118–124. [CrossRef]

37. Wang, G.; Ge, L.; Mitra, S.; Evans, G.M.; Joshi, J.B.; Chen, S. A review of CFD modelling studies on the flotation process. *Miner. Eng.* **2018**, *127*, 153–177. [CrossRef]

38. Farzanegan, A.; Khorasanizadeh, N.; Sheikhzadeh, G.A.; Khorasanizadeh, H. Laboratory and CFD investigations of the two-phase flow behavior in flotation columns equipped with vertical baffle. *Int. J. Miner. Process.* **2017**, *166*, 79–88. [CrossRef]

39. Sarhan, A.R.; Naser, J.; Brooks, G. CFD simulation on influence of suspended solid particles on bubbles' coalescence rate in flotation cell. *Int. J. Miner. Process.* **2016**, *146*, 54–64. [CrossRef]

40. Lakghomi, B.; Lawryshyn, Y.; Hofmann, R. A model of particle removal in a dissolved air flotation tank: Importance of stratified flow and bubble size. *Water Res.* **2015**, *68*, 262–272. [CrossRef]

41. Xia, Y.K.; Peng, F.F.; Wolfe, E. CFD simulation of alleviation of fluid back mixing by baffles in bubble column. *Miner. Eng.* **2006**, *19*, 925–937. [CrossRef]

42. Bondelind, M.; Sasic, S.; Kostoglou, M.; Bergdahl, L.; Pettersson, T.J.R. Single- and two-phase numerical models of Dissolved Air Flotation: Comparison of 2D and 3D simulations. *Colloids Surf. A Physicochem. Eng. Asp.* **2010**, *365*, 137–144. [CrossRef]

43. Schwarz, M.P.; Koh, P.T.L.; Verrelli, D.I.; Feng, Y. Sequential multi-scale modelling of mineral processing operations, with application to flotation cells. *Miner. Eng.* **2016**, *90*, 2–16. [CrossRef]

44. Liu, T.Y.; Schwarz, M.P. CFD-based modelling of bubble-particle collision efficiency with mobile bubble surface in a turbulent environment. *Int. J. Miner. Process.* **2009**, *90*, 45–55. [CrossRef]

45. Liu, T.Y.; Schwarz, M.P. CFD-based multiscale modelling of bubble–particle collision efficiency in a turbulent flotation cell. *Chem. Eng. Sci.* **2009**, *64*, 5287–5301. [CrossRef]

46. Lichter, J.; Potapov, A.; Peaker, R. The use of computational fluid dynamics and discrete element modeling to understand the effect of cell size and inflow rate on flotation bank retention time distribution and mechanism performance. In Proceedings of the 39th Annual Canadian Mineral Processors Operators Conference, Ottawa, ON, Canada, 23–25 January 2007.

47. Ji, L.; Chu, K.; Kuang, S.; Chen, J.; Yu, A. Modeling the Multiphase Flow in Hydrocyclones Using the Coarse-Grained Volume of Fluid—Discrete Element Method and Mixture-Discrete Element Method Approaches. *Ind. Eng. Chem. Res.* **2018**, *57*, 9641–9655. [CrossRef]

48. Chu, K.W.; Wang, B.; Yu, A.B.; Vince, A.; Barnett, G.D.; Barnett, P.J. CFD–DEM study of the effect of particle density distribution on the multiphase flow and performance of dense medium cyclone. *Miner. Eng.* **2009**, *22*, 893–909. [CrossRef]

49. Weerasekara, N.S.; Powell, M.S.; Cleary, P.W.; Tavares, L.M.; Evertsson, M.; Morrison, R.D.; Quist, J.; Carvalho, R.M. The contribution of DEM to the science of comminution. *Powder Technol.* **2013**, *248*, 3–24. [CrossRef]

50. Cleary, P.W. Prediction of coupled particle and fluid flows using DEM and SPH. *Miner. Eng.* **2015**, *73*, 85–99. [CrossRef]

51. Ye, T.; Li, Y.A. Comparative Review of Smoothed Particle Hydrodynamics, Dissipative Particle Dynamics and Smoothed Dissipative Particle Dynamics. *Int. J. Comput. Methods* **2018**, *15*, 1850083. [CrossRef]

52. Cleary, P.W.; Sinnott, M.; Morrison, R. Prediction of slurry transport in SAG mills using SPH fluid flow in a dynamic DEM based porous media. *Miner. Eng.* **2006**, *19*, 1517–1527. [CrossRef]

53. Cleary, P.W.; Delaney, G.W.; Sinnott, M.D.; Morrison, R.D. Inclusion of incremental damage breakage of particles and slurry rheology into a particle scale multiphase model of a SAG mill. *Miner. Eng.* **2018**, *128*, 92–105. [CrossRef]

54. Beinert, S.; Fragnière, G.; Schilde, C.; Kwade, A. Analysis and modelling of bead contacts in wet-operating stirred media and planetary ball mills with CFD–DEM simulations. *Chem. Eng. Sci.* **2015**, *134*, 648–662. [CrossRef]

55. Li, J.; Ge, W.; Wang, W.; Yang, N. Focusing on the meso-scales of multi-scale phenomena—In search for a new paradigm in chemical engineering. *Particuology* **2010**, *8*, 634–639. [CrossRef]

56. Li, J.; Kwauk, M. *Particle-Fluid Two-Phase Flow—The Energy-Minimization Multi-Scale Method*; Metallurgical Industry Press: Beijing, China, 1994.

57. Jávor, Z.; Schreithofer, N.; Heiskanen, K. Multi-scale analysis of the effect of surfactants on bubble properties. *Miner. Eng.* **2016**, *99*, 170–178. [CrossRef]

58. Cisternas, L.A. On the synthesis of inorganic chemical and metallurgical processes, review and extension. *Miner. Eng.* **1999**, *12*, 15–41. [CrossRef]

59. Loveday, B.K.; Brouckaert, C.J. An analysis of flotation circuit design principles. *Chem. Eng. J. Biochem. Eng. J.* **1995**, *59*, 15–21. [CrossRef]

60. Chan, W.K.; Prince, R.G.H. Heuristic evolutionary synthesis with non-sharp separators. *Comput. Chem. Eng.* **1989**, *13*, 1207–1219. [CrossRef]

61. Noble, A.; Luttrell, G.H.; Amini, S.H. Linear Circuit Analysis: A Tool for Addressing Challenges and Identifying Opportunities in Process Circuit Design. *Min. Metall. Explor.* **2019**, *36*, 159–171. [CrossRef]

62. Meloy, T.P. Analysis and optimization of mineral processing and coal-cleaning circuits—Circuit analysis. *Int. J. Miner. Process.* **1983**, *10*, 61–80. [CrossRef]

63. Williams, M.C.; Meloy, T.P. Dynamic model of flotation cell banks—Circuit analysis. *Int. J. Miner. Process.* **1983**, *10*, 141–160. [CrossRef]

64. Williams, M.C.; Fuerstenau, D.W.; Meloy, T.P. Circuit analysis—General product equations for multifeed, multistage circuits containing variable selectivity functions. *Int. J. Miner. Process.* **1986**, *17*, 99–111. [CrossRef]

65. Williams, M.C.; Meloy, T.P. Feasible designs for separation networks: A selection technique. *Int. J. Miner. Process.* **1991**, *32*, 161–174. [CrossRef]

66. Williams, M.C.; Fuerstenau, D.W.; Meloy, T.P. A graph-theoretic approach to process plant design. *Int. J. Miner. Process.* **1992**, *36*, 1–8. [CrossRef]

67. Luttrell, G.H.; Kohmuench, J.; Mankosa, M. Optimization of magnetic separator circuit configurations. In Proceedings of the SME Annual Conference and Expo, Englewood, CO, USA, 23–25 February 2004.

68. Luttrell, G.H.; Kohmuench, J.N.; Stanley, F.A.L.; Trump, G.D. Improving spiral performance using circuit analysis. *Miner. Metall. Process.* **1998**, *15*, 16–21. [CrossRef]

69. Mckeon, T.; Luttrel, G.H. Optimization of multistage circuits for gravity concentration of heavy mineral sands. *Miner. Metall. Process.* **2012**, *29*, 1–5. [CrossRef]

70. Noble, A.; Luttrell, G.H. Micro-pricing optimization: Value based partition curve analysis with applications to coal separation. In Proceedings of the 2015 SME Annual Conference and Expo and CMA 117th National Western Mining Conference—Mining: Navigating the Global Waters, Denver, CO, USA, 15–18 February 2015.

71. Amini, S.H.; Noble, A. Application of linear circuit analysis in the evaluation of mineral processing circuit design under uncertainty. *Miner. Eng.* **2017**, *102*, 18–29. [CrossRef]

72. Cisternas, L.A.; Acosta-Flores, R.; Gálvez, E.D. Some Limitations and Disadvantages of Linear Circuit Analysis. In Proceedings of the 7th International Computational Modelling Symposium (Computational Modelling '19, Falmouth, UK, 11–12 June 2019.

73. Radmehr, V.; Shafaei, S.; Noaparast, M.; Abdollahi, H. Optimizing Flotation Circuit Recovery by Effective Stage Arrangements: A Case Study. *Minerals* **2018**, *8*, 417. [CrossRef]

74. Cruz, R.; Cisternas, L.A.; Gálvez, E.D. Optimal Design of a Solid-Liquid Separation System. *Comput. Aided Chem. Eng.* **2013**, *32*, 907–912.

75. Mehrotra, S.P. Design of optimal flotation circuits—A review. *Miner. Metall. Process.* **1988**, *5*, 142–152. [CrossRef]

76. Cisternas, L.A.; Lucay, F.A.; Acosta-Flores, R.; Gálvez, E.D. A quasi-review of conceptual flotation design methods based on computational optimization. *Miner. Eng.* **2018**, *117*. [CrossRef]

77. Mendez, D.A.; Gálvez, E.D.; Cisternas, L.A. State of the art in the conceptual design of flotation circuits. *Int. J. Miner. Process.* **2009**, *90*, 1–15. [CrossRef]

78. Saito, F.; Baron, M.; Dodds, J.; Muramatsu, A. *Morphology Control of Materials and Nanoparticles: Chapter I and II*; Waseda, Y., Muramatsu, A., Eds.; Springer: Berlin/Heidelberg, Germany, 2004; ISBN 978-3-642-05671-0 (Print) 978-3-662-08863-0 (Online).

79. Yingling, J.C. Parameter and configuration optimization of flotation circuits, part I. A review of prior work. *Int. J. Miner. Process.* **1993**, *38*, 21–40. [CrossRef]

80. Cisternas, L.A.; Jamett, N.; Gálvez, E.D. Approximate recovery values for each stage are sufficient to select the concentration circuit structures. *Miner. Eng.* **2015**, *83*, 175–184. [CrossRef]

81. Cisternas, L.A.; Méndez, D.A.; Gálvez, E.D.; Jorquera, R.E. A MILP model for design of flotation circuits with bank/column and regrind/no regrind selection. *Int. J. Miner. Process.* **2006**, *79*, 253–263. [CrossRef]

82. Méndez, D.A.; Gálvez, E.D.; Cisternas, L.A. Modeling of grinding and classification circuits as applied to the design of flotation processes. *Comput. Chem. Eng.* **2009**, *33*, 97–111. [CrossRef]

83. Schena, G.D.; Zanin, M.; Chiarandini, A. Procedures for the automatic design of flotation networks. *Int. J. Miner. Process.* **1997**, *52*, 137–160. [CrossRef]

84. Acosta-Flores, R.; Lucay, F.A.; Cisternas, L.A.; Gálvez, E.D. Two phases optimization methodology for the design of mineral flotation plants including multi-species, bank or cell models. *Miner. Met. Process. J.* **2018**, *35*, 24–34.

85. Mehrotra, S.P.; Kapur, P.C. Optimal-Suboptimal Synthesis and Design of Flotation Circuits. *Sep. Sci.* **1974**, *9*, 167–184. [CrossRef]

86. Reuter, M.A.; van Deventer, J.S.J.; Green, J.C.A.; Sinclair, M. Optimal design of mineral separation circuits by use of linear programming. *Chem. Eng. Sci.* **1988**, *43*, 1039–1049. [CrossRef]

87. Reuter, M.A.; Van Deventer, J.S.J. The use of linear programming in the optimal design of flotation circuits incorporating regrind mills. *Int. J. Miner. Process.* **1990**, *28*, 15–43. [CrossRef]

88. Schena, G.; Villeneuve, J.; Noël, Y. A method for a financially efficient design of cell-based flotation circuits. *Int. J. Miner. Process.* **1996**, *46*, 1–20. [CrossRef]

89. Guria, C.; Verma, M.; Mehrotra, S.P.; Gupta, S.K. Multi-objective optimal synthesis and design of froth flotation circuits for mineral processing, using the jumping gene adaptation of genetic algorithm. *Ind. Eng. Chem. Res.* **2005**, *44*, 2621–2633. [CrossRef]

90. Guria, C.; Verma, M.; Gupta, S.K.; Mehrotra, S.P. Simultaneous optimization of the performance of flotation circuits and their simplification using the jumping gene adaptations of genetic algorithm. *Int. J. Miner. Process.* **2005**, *77*, 165–185. [CrossRef]

91. Ghobadi, P.; Yahyaei, M.; Banisi, S. Optimization of the performance of flotation circuits using a genetic algorithm oriented by process-based rules. *Int. J. Miner. Process.* **2011**. [CrossRef]

92. Maldonado, M.; Araya, R.; Finch, J. Optimizing flotation bank performance by recovery profiling. *Miner. Eng.* **2011**, *24*, 939–943. [CrossRef]

93. Hu, W.; Hadler, K.; Neethling, S.J.; Cilliers, J.J. Determining flotation circuit layout using genetic algorithms with pulp and froth models. *Chem. Eng. Sci.* **2013**, *102*, 32–41. [CrossRef]
94. Cisternas, L.A.; Lucay, F.; Gálvez, E.D. Effect of the objective function in the design of concentration plants. *Miner. Eng.* **2014**, *63*, 16–24. [CrossRef]
95. Pirouzan, D.; Yahyaei, M.; Banisi, S. Pareto based optimization of flotation cells configuration using an oriented genetic algorithm. *Int. J. Miner. Process.* **2014**, *126*, 107–116. [CrossRef]
96. Calisaya, D.A.; López-Valdivieso, A.; de la Cruz, M.H.; Gálvez, E.E.; Cisternas, L.A. A strategy for the identification of optimal flotation circuits. *Miner. Eng.* **2016**, *96–97*, 157–167. [CrossRef]
97. Lucay, F.; Gálvez, E.; Cisternas, L. Design of Flotation Circuits Using Tabu-Search Algorithms: Multispecies, Equipment Design, and Profitability Parameters. *Minerals* **2019**, *9*, 181. [CrossRef]
98. McCarthy, J.; Minsky, M.; Rochester, N.; Shannon, C. A proposal for the Dartmouth summer research project on artificial intelligence. *AI Mag.* **2012**, *27*, 12–14.
99. McCoy, J.T.; Auret, L. Machine learning applications in minerals processing: A review. *Miner. Eng.* **2019**, *132*, 95–109. [CrossRef]
100. Bergh, L. Artificial Intelligence in Mineral Processing Plants: An Overview. In Proceedings of the 2016 International Conference on Artificial Intelligence: Technologies and Applications, Bangkok, Thailand, 24–25 January 2016.
101. Hoseinian, F.S.; Rezai, B.; Kowsari, E.; Safari, M. A hybrid neural network/genetic algorithm to predict Zn(II) removal by ion flotation. *Sep. Sci. Technol.* **2019**, *2019*, 1–10. [CrossRef]
102. Zadeh, L.A. Soft Computing and Fuzzy Logic. *IEEE Softw.* **1994**, *11*, 48–56. [CrossRef]
103. Thyagharajan, K.K.; Vignesh, T. Soft Computing Techniques for Land Use and Land Cover Monitoring with Multispectral Remote Sensing Images: A Review. *Arch. Comput. Methods Eng.* **2019**, *26*, 275–301. [CrossRef]
104. Das, S.K.; Kumar, A.; Das, B.; Burnwal, B. On Soft Computing Techniques in Various Areas. *Int. J. Inform. Technol. Comput. Sci.* **2013**, *3*, 59–68. [CrossRef]
105. Sharma, D.; Chandra, P. A comparative analysis of soft computing techniques in software fault prediction model development. *Int. J. Inf. Technol.* **2018**, *11*, 1–10. [CrossRef]
106. Jeffers, J.; Reinders, J.; Sodani, A. Machine learning. In *Intel Xeon Phi Processor High Performance Programming*; Elsevier: Amsterdam, The Netherlands, 2016; pp. 527–548. ISBN 9780128091944.
107. Qin, S.J. Statistical process monitoring: Basics and beyond. *J. Chemom.* **2003**, *17*, 480–502. [CrossRef]
108. Umucu, Y.; Çağlar, M.F.; Gündüz, L.; Bozkurt, V.; Deniz, V. Modeling of grinding process by artificial neural network for calcite mineral. In Proceedings of the INISTA 2011 International Symposium on INnovations in Intelligent SysTems and Applications, Istanbul, Turkey, 15–18 June 2011.
109. Umucu, Y.; Deniz, V.; Bozkurt, V.; Çağlard, M.F. The evaluation of grinding process using artificial neural network. *Int. J. Miner. Process.* **2016**, *146*, 46–53. [CrossRef]
110. Warren Liao, T.; Chen, L.J. A neural network approach for grinding processes: Modelling and optimization. *Int. J. Mach. Tools Manuf.* **1994**, *34*, 919–937. [CrossRef]
111. Makokha, A.B.; Moys, M.H. Multivariate approach to on-line prediction of in-mill slurry density and ball load volume based on direct ball and slurry sensor data. *Miner. Eng.* **2012**. [CrossRef]
112. Mitra, K.; Ghivari, M. Modeling of an industrial wet grinding operation using data-driven techniques. *Comput. Chem. Eng.* **2006**, *30*, 508–520. [CrossRef]
113. Curilem, M.; Acuña, G.; Cubillos, F.; Vyhmeister, E. Neural Networks and Support Vector Machine models applied to energy consumption optimization in semiautogeneous grinding. *Chem. Eng. Trans.* **2011**, *25*, 761–766.
114. Núñez, F.; Silva, D.; Cipriano, A. Characterization and modeling of semi-autogenous mill performance under ore size distribution disturbances. *IFAC Proc. Vol.* **2011**, *44*, 9941–9946. [CrossRef]
115. Vini, M.H. Using Artificial Neural Networks to predict rolling force and real exit thickness of steel strips. *J. Mod. Process. Manuf. Prod.* **2014**, *3*, 53–60.
116. Nakhaei, F.; Irannajad, M. Comparison between neural networks and multiple regression methods in metallurgical performance modeling of flotation column. *Physicochem. Probl. Miner. Process.* **2013**, *49*, 255–266. [CrossRef]
117. Hosseini, M.R.; Shirazi, H.H.A.; Massinaei, M.; Mehrshad, N. Modeling the Relationship between Froth Bubble Size and Flotation Performance Using Image Analysis and Neural Networks. *Chem. Eng. Commun.* **2015**, *202*, 911–919. [CrossRef]

118. Estrada-Ruiz, R.H.; Peérez-Garibay, R. Neural networks to estimate bubble diameter and bubble size distribution of flotation froth surfaces. *J. S. Afr. Inst. Min. Metall.* **2009**, *109*, 441–446.

119. Chelgani, S.C.; Shahbazi, B.; Rezai, B. Estimation of froth flotation recovery and collision probability based on operational parameters using an artificial neural network. *Int. J. Miner. Metall. Mater.* **2010**, *17*, 526–534. [CrossRef]

120. Jahedsaravani, A.; Marhaban, M.H.; Massinaei, M. Prediction of the metallurgical performances of a batch flotation system by image analysis and neural networks. *Miner. Eng.* **2014**, *69*, 137–145. [CrossRef]

121. Saravani, A.J.; Mehrshad, N.; Massinaei, M. Fuzzy-Based Modeling and Control of an Industrial Flotation Column. *Chem. Eng. Commun.* **2014**, *201*, 896–908. [CrossRef]

122. Karimi, M.; Dehghani, A.; Nezamalhosseini, A.; Talebi, S.H. Prediction of hydrocyclone performance using artificial neural networks. *J. S. Afr. Inst. Min. Metall.* **2010**, *110*, 207–212.

123. Razavi, S.; Tolson, B.A.; Burn, D.H. Review of surrogate modeling in water resources. *Water Resour. Res.* **2012**, *48*. [CrossRef]

124. Aldrich, C.; Jemwa, G.T.; Krishnannair, S. Multiscale process monitoring with singular spectrum analysis. *IFAC Proc. Vol.* **2007**, *12*, 167–172. [CrossRef]

125. Bergh, L.G.; Yianatos, J.B.; León, A. Multivariate projection methods applied to flotation columns. *Miner. Eng.* **2005**, *18*, 721–723. [CrossRef]

126. Groenewald, J.W.D.V.; Coetzer, L.P.; Aldrich, C. Statistical monitoring of a grinding circuit: An industrial case study. *Miner. Eng.* **2006**, *19*, 1138–1148. [CrossRef]

127. Wakefield, B.J.; Lindner, B.S.; McCoy, J.T.; Auret, L. Monitoring of a simulated milling circuit: Fault diagnosis and economic impact. *Miner. Eng.* **2018**, *120*, 132–151. [CrossRef]

128. Groenewald, J.W.D.; Aldrich, C. Root cause analysis of process fault conditions on an industrial concentrator circuit by use of causality maps and extreme learning machines. *Miner. Eng.* **2015**, *74*, 30–40. [CrossRef]

129. Lindner, B.; Auret, L. Application of data-based process topology and feature extraction for fault diagnosis of an industrial platinum group metals concentrator plant. *IFAC-PapersOnLine* **2015**, *48*, 102–107. [CrossRef]

130. Cao, B.; Xie, Y.; Gui, W.; Wei, L.; Yang, C. Integrated prediction model of bauxite concentrate grade based on distributed machine vision. *Miner. Eng.* **2013**, *53*, 31–38. [CrossRef]

131. Rughooputh, H.C.S.; Rughooputh, S.D.D.V. Neural network process vision systems for flotation process. *Kybernetes* **2002**, *31*, 529–535. [CrossRef]

132. Zhu, J.; Yu, K.W. Application of image recognition system in flotation process. In Proceedings of the World Congress on Intelligent Control and Automation (WCICA), Chongqing, China, 25–27 June 2008.

133. Sharad, M.; Augustine, C.; Panagopoulos, G.; Roy, K. Ultra low energy analog image processing using spin based neurons. In Proceedings of the 2012 IEEE/ACM International Symposium on Nanoscale Architectures (NANOARCH), Amsterdam, The Netherlands, 4–6 July 2012.

134. Zhao, L.; Peng, T.; Xie, Y.; Yang, C.; Gui, W. Recognition of flooding and sinking conditions in flotation process using soft measurement of froth surface level and QTA. *Chemom. Intell. Lab. Syst.* **2017**, *169*, 45–52. [CrossRef]

135. Çinar, A. A Method for Local Tuning of Fuzzy Membership Functions. In Proceedings of the ICCS 2010: "Celebrating 10 years of Advancing Computational Thinking", Amsterdam, The Netherlands, 31 May–2 June 2010.

136. Tandon, V.; El-Mounayri, H.; Kishawy, H. NC end milling optimization using evolutionary computation. *Int. J. Mach. Tools Manuf.* **2002**, *42*, 595–605. [CrossRef]

137. Massinaei, M.; Falaghi, H.; Izadi, H. Optimisation of metallurgical performance of industrial flotation column using neural network and gravitational search algorithm. *Can. Metall. Q.* **2013**, *52*, 115–122. [CrossRef]

138. Shunmugam, M.S.; Bhaskara Reddy, S.V.; Narendran, T.T. Selection of optimal conditions in multi-pass face-milling using a genetic algorithm. *Int. J. Mach. Tools Manuf.* **2000**, *40*, 401–414. [CrossRef]

139. Moscato, P. On evolution, search, optimization, genetic algorithms and martial arts: Towards memetic algorithms. In *Caltech Concurrent Computation. Program, C3P Report*; California Institute of Technology, (Caltech): Pasadena, CA, USA, 1989.

140. Ojha, V.K.; Abraham, A.; Snásel, V. Metaheuristic Design of Feedforward Neural Networks: A Review of Two Decades of Research. *Eng. Appl. Artif. Intell.* **2017**, *60*, 97–116. [CrossRef]

141. Tang, Q.; Lau, Y.B.; Hu, S.; Yan, W.; Yang, Y.; Chen, T. Response surface methodology using Gaussian processes: Towards optimizing the trans-stilbene epoxidation over Co^{2+}—NaX catalysts. *Chem. Eng. J.* **2010**, *156*, 423–431. [CrossRef]

142. Garud, S.S.; Karimi, I.A.; Kraft, M. Design of computer experiments: A review. *Comput. Chem. Eng.* **2017**, *106*, 71–95. [CrossRef]

143. Bezerra, M.A.; Santelli, R.E.; Oliveira, E.P.; Villar, L.S.; Escaleira, L.A. Response surface methodology (RSM) as a tool for optimization in analytical chemistry. *Talanta* **2008**, *76*, 965–977. [CrossRef]

144. Gutmann, H.M. A Radial Basis Function Method for Global Optimization. *J. Glob. Optim.* **2001**, *19*, 201–227. [CrossRef]

145. Montgomery, D.C. *Design and Analysis of Experiments*, 8th ed.; Wiley: Hoboken, NJ, USA, 2012.

146. Gan, Y.; Duan, Q.; Gong, W.; Tong, C.; Sun, Y.; Chu, W.; Ye, A.; Miao, C.; Di, Z. A comprehensive evaluation of various sensitivity analysis methods: A case study with a hydrological model. *Environ. Model. Softw.* **2014**, *51*, 269–285. [CrossRef]

147. Ye, K.Q.; Li, W.; Sudjianto, A. Algorithmic construction of optimal symmetric Latin hypercube designs. *J. Stat. Plan. Inference* **2000**, *90*, 145–159. [CrossRef]

148. Shahreza, S.; Shafaei, S.; Noaparast, M.; Sarvi, M. Optimization of Galena Flotation Process of Irankouh Complex Ore Using A Statistical Design of Experiments. *Curr. World Environ.* **2015**, *10*, 626–636. [CrossRef]

149. Vieceli, N.; Durão, F.O.; Guimarães, C.; Nogueira, C.A.; Pereira, M.F.C.; Margarido, F. Grade-recovery modelling and optimization of the froth flotation process of a lepidolite ore. *Int. J. Miner. Process.* **2016**, *157*, 184–194. [CrossRef]

150. Ebadnejad, A.; Karimi, G.R.; Dehghani, H. Application of response surface methodology for modeling of ball mills in copper sulphide ore grinding. *Powder Technol.* **2013**, *245*, 292–296. [CrossRef]

151. Krajnik, P.; Kopac, J.; Sluga, A. Design of grinding factors based on response surface methodology. *J. Mater. Process. Technol.* **2005**, *162*, 629–636. [CrossRef]

152. Chen, G.; Peng, J.H.; Chen, J. Optimizing conditions for wet grinding of synthetic rutile using response surface methodology. *Miner. Metall. Process.* **2011**, *28*, 44–48. [CrossRef]

153. Alireza, A.S.; Ataallah, S.G.; Majid, E.G.; Amir, S.; Mohammad, R.; Hadi, A. Application of response surface methodology and central composite rotatable design for modeling the influence of some operating variables of the lab scale thickener performance. *Int. J. Min. Sci. Technol.* **2013**, *23*, 717–724. [CrossRef]

154. Costa, N.R.; Lourenço, J. Gaussian Process Model—An Exploratory Study in the Response Surface Methodology. *Qual. Reliab. Eng. Int.* **2016**, *32*. [CrossRef]

155. Charte, F.; Romero, I.; Pérez-Godoy, M.D.; Rivera, A.J.; Castro, E. Comparative analysis of data mining and response surface methodology predictive models for enzymatic hydrolysis of pretreated olive tree biomass. *Comput. Chem. Eng.* **2017**, *101*, 23–30. [CrossRef]

156. Gutiérrez, A.; Olivera, C.A.C.; Hacha, R.R.; Torem, M.L.; Santos, B.F. dos Optimization of hematite and quartz BIOFLOTATION by AN artificial neural network (ANN). *J. Mater. Res. Technol.* **2019**, *8*, 3076–3087.

157. Asher, M.J.; Croke, B.F.W.; Jakeman, A.J.; Peeters, L.J.M. A review of surrogate models and their application to groundwater modeling. *Water Resour. Res.* **2015**, *51*, 5957–5973. [CrossRef]

158. Rabhi, A.; Chkifa, A.; Benjelloun, S.; Latifi, A. Surrogate-based modeling in flotation processes. *Comput. Aided Chem. Eng.* **2018**, *43*, 229–234.

159. Stephens, D.W.; Gorissen, D.; Crombecq, K.; Dhaene, T. Surrogate based sensitivity analysis of process equipment. *Appl. Math. Model.* **2011**, *35*, 1676–1687. [CrossRef]

160. Stephens, D.; Fawell, P. Optimization of Process Equipment Using Global Surrogate Models. In Proceedings of the Conference on CFD in the Minerals and Process Industries CSIRO, Melbourne, Australia, 10–12 December 2012.

161. Metta, N.; Ramachandran, R.; Ierapetritou, M. A Computationally Efficient Surrogate-Based Reduction of a Multiscale Comill Process Model. *J. Pharm. Innov.* **2019**, *2019*, 1–21. [CrossRef]

162. Lucay, F.A.; Gálvez, E.D.; Salez-Cruz, M.; Cisternas, L.A. Improving milling operation using uncertainty and global sensitivity analyses. *Miner. Eng.* **2019**, *131*, 249–261. [CrossRef]

163. Saltelli, A.; Tarantola, S.; Campolongo, F. Sensitivity analysis as an ingredient of modeling. *Stat. Sci.* **2000**, *15*, 377–395.

164. Lane, W.A.; Ryan, E.M. Verification, validation, and uncertainty quantification of a sub-grid model for heat transfer in gas-particle flows with immersed horizontal cylinders. *Chem. Eng. Sci.* **2018**, *176*, 409–420. [CrossRef]
165. Mellado, M.E.; Cisternas, L.A.; Gálvez, E.D. An analytical model approach to heap leaching. *Hydrometallurgy* **2009**, *95*, 33–38. [CrossRef]
166. Mellado, M.; Lucay, F.; Cisternas, L.; Gálvez, E.; Sepúlveda, F. A Posteriori Analysis of Analytical Models for Heap Leaching Using Uncertainty and Global Sensitivity Analyses. *Minerals* **2018**, *8*, 44. [CrossRef]
167. Mellado, M.E.; Casanova, M.P.; Cisternas, L.A.; Gálvez, E.D. On scalable analytical models for heap leaching. *Comput. Chem. Eng.* **2011**, *35*, 220–225. [CrossRef]
168. Mellado, M.E.; Gálvez, E.D.; Cisternas, L.A. Stochastic analysis of heap leaching process via analytical models. *Miner. Eng.* **2012**, *33*, 93–98. [CrossRef]
169. Sepúlveda, F.D.; Lucay, F.; González, J.F.; Cisternas, L.A.; Gálvez, E.D. A methodology for the conceptual design of flotation circuits by combining group contribution, local/global sensitivity analysis, and reverse simulation. *Int. J. Miner. Process.* **2017**, *164*, 56–66. [CrossRef]
170. Lucay, F.; Cisternas, L.A.; Gálvez, E.D. Global sensitivity analysis for identifying critical process design decisions. *Chem. Eng. Res. Des.* **2015**, *103*, 74–83. [CrossRef]
171. Lucay, F.; Cisternas, L.A.; Gálvez, E.D. Retrofitting of Concentration Plants Using Global Sensitivity Analysis. *Comput. Aided Chem. Eng.* **2015**, *37*, 311–316.
172. Sepúlveda, F.D.; Cisternas, L.A.; Gálvez, E.D. The use of global sensitivity analysis for improving processes: Applications to mineral processing. *Comput. Chem. Eng.* **2014**, *66*, 221–232. [CrossRef]
173. Saltelli, A.; Annoni, P.; Azzini, I.; Campolongo, F.; Ratto, M.; Tarantola, S. Variance based sensitivity analysis of model output. Design and estimator for the total sensitivity index. *Comput. Phys. Commun.* **2010**, *181*, 259–270. [CrossRef]
174. Li, S.; Yang, B.; Qi, F. Accelerate global sensitivity analysis using artificial neural network algorithm: Case studies for combustion kinetic model. *Combust. Flame* **2016**, *168*, 53–64. [CrossRef]
175. Sakizlis, V.; Perkins, J.D.; Pistikopoulos, E.N. Recent advances in optimization-based simultaneous process and control design. *Comput. Chem. Eng.* **2004**, *28*, 2069–2086. [CrossRef]
176. Eljack, F.T.; Solvason, C.C.; Chemmangattuvalappil, N.; Eden, M.R. A Property Based Approach for Simultaneous Process and Molecular Design. *Chin. J. Chem. Eng.* **2008**, *16*, 424–434. [CrossRef]
177. Kazantzi, V.; Qin, X.; El-Halwagi, M.; Eljack, F.; Eden, M. Simultaneous Process and Molecular Design through Property Clustering Techniques: A Visualization Tool. *Ind. Eng. Chem. Res.* **2007**, *46*, 3400–3409. [CrossRef]
178. Eljack, F.T.; Eden, M.R.; Kazantzi, V.; Qin, X.; El-Halwagi, M.M. Simultaneous process and molecular design—A property based approach. *AIChE J.* **2007**, *53*, 1232–1239. [CrossRef]

Article

Novel Insights into the Hydroxylation Behaviors of α-Quartz (101) Surface and its Effects on the Adsorption of Sodium Oleate

Chenyang Zhang [1,2], Zhijie Xu [1], Yuehua Hu [1], Jianyong He [1], Mengjie Tian [1], Jiahui Zhou [1], Qiqi Zhou [1], Shengda Chen [1,3], Daixiong Chen [2,*], Pan Chen [1,*] and Wei Sun [1,*]

[1] Key Laboratory of Hunan Province for Clean and Efficient Utilization of Strategic Calcium-containing Mineral Resources, School of Minerals Processing and Bioengineering, Central South University, Changsha 410083, China

[2] Key Laboratory of Hunan Province for Comprehensive Utilization of Complex Copper-Lead Zinc Associated Metal Resources, Hunan Research Institute for Nonferrous Metals, Changsha 410100, China

[3] Institute of Theoretical and Computational Chemistry, School of Chemistry and Chemical Engineering, Nanjing University, Nanjing 210023, China

[*] Correspondence: sunmenghu@csu.edu.cn (W.S.); chendaixiong888@163.com (D.C.); panchen@csu.edu.cn (P.C.)

Received: 9 May 2019; Accepted: 18 July 2019; Published: 19 July 2019

Abstract: A scientific and rigorous study on the adsorption behavior and molecular mechanism of collector sodium oleate (NaOL) on a Ca^{2+}-activated hydroxylated α-quartz surface was performed through experiments and density functional theory (DFT) simulations. The rarely reported hydroxylation behaviors of water molecules on the α-quartz (101) surface were first innovatively and systematically studied by DFT calculations. Both experimental and computational results consistently demonstrated that the adsorbed calcium species onto the hydroxylated structure can significantly enhance the adsorption of oleate ions, resulting in a higher quartz recovery. The calculated adsorption energies confirmed that the adsorbed hydrated Ca^{2+} in the form of $Ca(H_2O)_3(OH)^+$ can greatly promote the adsorption of OL^- on hydroxylated quartz (101). In addition, Mulliken population analysis together with electron density difference analysis intuitively illustrated the process of electron transfer and the Ca-bridge phenomenon between the hydroxylated surface and OL^- ions. This work may offer new insights into the interaction mechanisms existing among oxidized minerals, aqueous medium, and flotation reagents.

Keywords: quartz; DFT calculation; hydroxylation; adsorption; flotation

---

1. Introduction

Quartz is one of the main gangue minerals in iron ores [1], and it is also the common gangue mineral for most metal oxidized ores, non-metal oxidized ores, sulfide ores, silicate minerals, phosphate minerals [2,3]. There are many varieties of quartz in nature, such as α-quartz, β-quartz, coesite, and stishovite, among which α-quartz is the most widely distributed, and is the main rock-forming minerals of magmatic, sedimentary, and metamorphic rocks [4]. In addition, as an important industrial raw material, quartz has been widely used in electronic devices, optical instruments, glass raw materials, abrasive materials, refractories, and other aspects [5–8]. Therefore, the separation of quartz from other minerals makes great sense for mineral processing and relevant industries.

Flotation, one of the most efficient mineral processing methods which selectively concentrates target minerals based on their physicochemical properties and differences resulting from the intrinsic properties of ores and modification of flotation reagents, has been widely employed in the separation of iron ores [9–11]. It is generally acknowledged that cationic/anionic reverse flotation is one of the most efficient technologies for the removal of quartz and enrichment of iron-containing minerals from iron ores [12]. The collectors of quartz reverse flotation could be classified into cationic collectors and anion collectors, among which cationic collectors are sensitive to slime and cause large foam viscosity while anion collectors have good selectivity and adaptability to ores [13]. Therefore, anion collectors (e.g., sodium oleate) are usually used in practical production to reduce SiO_2 content to obtain high-grade iron concentrate [14]. However, low recovery of quartz is gained only when sodium oleate is used. In most cases, before the addition of sodium oleate, metal ion activators should be previously added to the pulp and adsorbed onto the quartz's surface as active sites. Actually, the activation mechanisms of metal ions on mineral surfaces have been research hotspots in the field of flotation [15,16]. Divalent ions, such as Cu^{2+}, Pb^{2+}, Ca^{2+}, etc., are widely used to activate specific mineral surfaces [15,17–19]. Calcium ion possesses more widespread applications for separating quartz from other valuable minerals compared with most other metallic ions [17,20]. The activation of calcium ion on quartz has been previously studied by other scholars through various characterization methods and theory calculations, but no consistent conclusion has been reached. For instance, Shi et al. [21] considered that $Ca(OH)_2$ precipitation was the main activation component of quartz; Guo et al. [2] concluded that the activation of calcium ions on quartz was due to the preferential chemical adsorption of Ca^{2+} on oxygen sites of quartz surface; and Gong et al. [22] believed that the adsorption of sodium oleate on the quartz surface was mainly attributed to the activation of $Ca(OH)^+$.

In recent years, with the rapid developments of theoretical and computational chemistry, more effective methods can be adopted to visually investigate the interaction mechanisms between reagents and minerals at a microscopic level. Density functional theory (DFT) calculation is a kind of simulation method whose application in mineral processing is relatively mature [12,23] and it has great potential to provide novel and microcosmic insights into the flotation process [24,25] which cannot be obtained in conventional experimental studies. For example, Zhu et al. [12] intensively studied the interaction mechanism between collector α-Bromolauric acid and Ca^{2+}-activated quartz (101) surface, and concluded that the essence of activation and flotation of quartz was that $Ca(OH)^+$ ions served as a bridge between the collector and the mineral surface. Rath et al. [26] found that magnetite could form the most stable surface complexes with oleate through comparing the interaction of oleate with hematite, magnetite, and goethite based on DFT calculation. Zhao et al. [27] investigated the adsorption behaviors of $Ca(OH)^+$ on pyrite, marcasite, and pyrrhotite surfaces, and discussed in depth the corresponding bonding mechanism and electron transfer using DFT simulation. Long et al. [28] confirmed by researching the effects of the three typical thiol collectors on galena and sphalerite in the presence of water on the basis of the first principles, that there existed distinct differences in the electron distribution, the atoms' activity on the mineral's surface, and the interactions between the collectors and the minerals. It is worth mentioning that metal ions in aqueous systems will experience complex behaviors to generate hydrated metal ions cluster [18,29–31], which has profound impacts on the interactions between collectors and the mineral surface. Wang et al. [32] conducted an in-depth simulations study on the activation mechanism of calcium ions on quartz and found that the major activation component of calcium ions adsorbed on the surface of quartz was $(Ca(H_2O)_4)^{2+}$ which transformed into $(Ca(H_2O)_3(OH))^+$ after the adsorption process. Hu et al. [15], by studying the activation mechanism of calcium ion on a sericite surface, concluded that adsorption onto a sericite (001) surface of hydrated calcium ions in the form of $(Ca(H_2O)_3(OH))^+$ was the most favorable. According to the latest and thorough studies mentioned above regarding the activation mechanism of calcium ions, it is believed that the $Ca(H_2O)_3(OH)^+$ cluster acts as the main active components in alkaline solution. Moreover, water molecules play an indispensable role in the flotation process [28,33] and quartz possesses strong hydrophilicity [34,35]. De Leeuw's first principles calculations show that the existence of hydration behavior has a very important influence on mineral surface structures and reactivities, and a solvent effect must be considered in the DFT simulations of the mineral flotation process in order to accurately predict the affinity between flotation reagents and mineral surface [36]. Pradip et al. [37] believe that flotation reagents and mineral surfaces need to match each other in spatial structure and properties to display productive effects. Therefore, only when the hydroxylation behaviors of water molecules on quartz surfaces are firstly fully considered in DFT calculations can the adsorption and flotation mechanisms of flotation reagents onto quartz surfaces be reasonably and accurately investigated.

Unfortunately, although many in-depth studies on the activation or flotation of quartz have been carried out through experiments and theoretical simulations, there are two serious shortcomings as follows [12,14,22,32]: There is a lack of rigorous and systematic studies on the hydroxylation behaviors of quartz in aqueous solution, as well as a lack of further investigations on the microscopic mechanisms between flotation agents and pre-hydroxylated quartz. Therefore, the main objective of this study was to firstly systematically study the rarely reported strong hydroxylation behaviors of water molecules on quartz surfaces and further to rigorously research the adsorption and flotation mechanisms of flotation reagents onto hydroxylated quartz structures. The adopted experimental characterization methods included X-ray diffraction (XRD) spectrums, micro-flotation tests, zeta potential tests and Fourier transform infrared (FTIR) spectrums. In addition, the interaction mechanism at micro aspects were further investigated by first principles DFT calculations. This work may provide novel insights into the interaction mechanisms existing among aqueous medium, α-quartz surfaces, calcium ions, and anionic collectors.

2. Experimental and Computational Details

2.1. Materials and Reagents

The high-purity massive quartz crystal sample was smashed into small pieces with a maximum particle size of 30 mm by a hammer and sent to the JC6 jaw crusher for crushing to prepare smaller particles with a size below 2 mm. The crushed minerals were ground with a porcelain mortar and then sieved, among which a partial fraction (-38 µm) was further ground using an agate mortar to obtain a particle size less than 5 µm for various analyses and the sample within the range of 38–74 µm was used for micro-flotation tests [38]. The X-ray diffraction (XRD) spectrum of the sample is shown in Figure 1. The chemical element analysis results indicated that the content of SiO_2 in the quartz samples was as high as 97.63%, and the sample contained very small amounts of impurities (0.035% Ca and 0.025% Fe), which was pure enough for the following tests. In the experiments, chemical pure sodium oleate (NaOL), analytical pure calcium chloride ($CaCl_2$), sodium hydroxide (NaOH), hydrochloric acid (HCl), potassium nitrate (KNO_3), and deionized water were used.

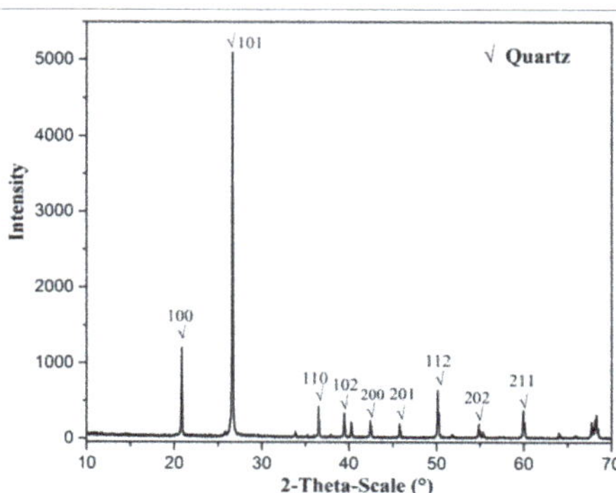

Figure 1. The XRD spectrum of the used quartz pure mineral sample.

2.2. Methodology

2.2.1. Micro-Flotation

All flotation tests were carried out using an XFG flotation machine with a 40 mL cell operated at 1650 rpm. In three parallel tests, 35 mL deionized water and 2 g pure mineral sample were mixed and stirred [15]. Firstly, pH regulators were added to adjust the flotation pulp to the specified pH value in the subsequent 2 min. Secondly, the pulp was conditioned with $CaCl_2$ for 3 min. Finally, a certain amount of NaOL was added into the cell with a 3 min conditioning time. The scraping operation lasted for 4 min and the concentrate was scraped out in the form of mineralized froth every 5 s. It should be mentioned that the addition of flotation agents would make the pH of the slurry fluctuate, so the pH regulators were discontinuously added to keep the pH of the slurry stable until 1 min before scraping. The products were filtrated, dried, and weighed, and the recovery was calculated based on the solid weight distribution among the two products [38]. The specific process of flotation tests is presented in Figure 2.

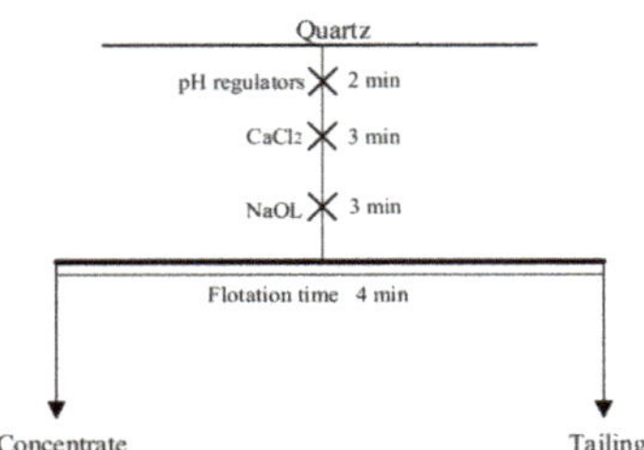

Figure 2. The flowsheet and conditions of the micro-flotation tests.

2.2.2. Zeta Potential Measurements

Zeta potentials were measured by Malvern ZETASIZER Nano-Z instrument at 25 °C. The suspension with a mass concentration of 1% (40 mg −5 µm quartz: 40 mL aqueous solution) containing 1×10^{-2} mol/L KNO_3, serving as a background electrolyte, was agitated for 5 min to make the quartz particles fully disperse [35]. The desired reagent(s) was added in accordance with the actual flotation test conditions. Similar to the actual micro-flotation process, pH regulators were discontinuously added to adjust the slurry to the specified pH range until 5 min before stirring stops (note: 1 min before scraping operation, plus 4 min during bubble scraping process) during the zeta potential measurements. After 5 min standing time, a small amount of supernatant was slowly injected into a test cell with a syringe. Each sample result presented is the average of three independent measurements with a measurement tolerance of ±3 mV.

2.2.3. FTIR Spectroscopy

Characteristic peaks in the FTIR spectrums can be used to analyze the adsorption behaviors between reagents and the mineral surface [35,39]. The pure mineral was first ground to −2 µm in an agate mortar and then treated according to the procedures described in the flotation flowsheet, and finally measured by Fourier transform infrared spectroscopy (FTIR) Vertex 80v manufactured by Bruker, Germany. This equipment can create a vacuum environment to guarantee the test results are accurate enough for FTIR analysis. The infrared spectra of samples were recorded by the FTIR spectrometer at a resolution of 4 cm^{-1} in the range of 400 cm^{-1} to 4000 cm^{-1} at room temperature [33].

2.2.4. Computational Details

In this work, all periodic calculations were performed by utilizing the Cambridge Serial Total Energy Package (CASTEP) module [40] in Materials Studio 2017. A periodic model of the quartz bulk phase was built from the XRD crystal structure obtained from the American Mineralogist Crystal Structure Database [41]. Firstly, the quartz unit-cell underwent crystal optimization convergence tests for exchange-correlation functions and the k-point set mesh and cutoff energy, respectively [12]. Afterwards, the optimized quartz lattice parameters were compared with the obtained experimental values to ascertain the optimum parameters. The Fast Fourier Transformation (FFT) grid quality was set to a fine level. For self-consistent iteration, the density mixing electronic minimizer was used with a self-consistent field (SCF) tolerance of 2.0×10^{-6} eV/atom. The convergence tolerance during the entire simulation work set for the energy, maximum force, and maximum displacement tolerance was 2.0×10^{-5} eV/atom, 0.05 eV/Å, and 0.002 Å, respectively. All the optimal parameters confirmed in the unit-cell geometry optimization tests remained constant in all subsequent simulation jobs unless otherwise mentioned.

Surface energy is defined as the work needed to generate a unit's new surface [25]. For higher surface energy, the corresponding cleavage surface is hard to create and thermodynamically unstable, which can serve as a good measurement for judging the stability of a specified surface [25,42]. According to previous literature and aforementioned XRD results, we found that the quartz (101) surface was the most stable cleavage plane [12,32,43]. Therefore, our subsequent simulation was based on the quartz (101) surface. The surface slab was first created by cleaving the optimized quartz bulk unit-cell at the (101) cleavage plane; then, the symmetric Si_9O_{18} surface slab was obtained by adjusting the top position and the thickness position. The final periodic structure, whose cell formula was $Si_{54}O_{108}$, was obtained by using ($2 \times 3 \times 1$) super-cell with 20 Å vacuum thickness, which is sufficient for the prevention of interactions among slabs [32,43]. According to a previous study [35], the contact angle of quartz in distilled water is lower than 25°, which implies that the surface of quartz is highly hydrophilic. Owing to its strong hydrophilicity, it is obligatory to innovatively study the rarely reported hydroxylation behaviors between quartz (101) surface and water molecules in DFT simulations. Usually, the adsorption energy of a spontaneous chemical reaction is subtractive, and the more negative the adsorption energy, the more stable the hydroxylation configuration obtained [25].

According to the work of Wang et al. [32], the hollow site of the same Si center and the top sites of O atoms are implied as the major adsorption sites, thus we put adsorbates in the vacancy above the Si atom (this Si atom was attached to three O atoms and at the center of the O atoms) on the hydroxylated quartz (101) surface to obtain a relatively stable adsorption configuration in each simulation. The interaction strength of different adsorbates on quartz surface (101) can also be measured by the adsorption energy, which is regarded as a good measure to assess the relative interaction strength [32,44]. It is worth mentioning that the reasonable initial oleate ion structure was obtained from the National Center for Biotechnology Information [45]. Under the comprehensive consideration of acceptable calculation accuracy and affordable calculation cost, the $Si_{18}O_{36}$ layer at the bottom of the quartz periodic model and counterions (Na^+, Cl^-) were constrained, while the rest of the parts were allowed structural relaxation within constrained volume during all optimization calculations. The k-point set was changed to gamma point and energy-cutoff was set as 381.0 eV for the subsequent structural optimization calculations.

3. Results and Discussion

3.1. Micro-Flotation Tests

The flotation recovery of quartz as a function of pH is presented in Figure 3a. When $CaCl_2$ was absent, the recovery of quartz increased slightly in the whole range of pH tests. The maximum recovery of quartz at pH 12 was only 38%, which is consistent with the flotation phenomena of Wang [46]. This indicates that NaOL has a poor collecting ability for quartz. According to research by Tian et al. [47], it was found that the dosing methods of benzohydroxamic acid and lead nitrate have a great influence on the flotation behavior of cassiterite. Therefore, it is worthwhile to study the effects of mixed-dosing and sequential-dosing methods of NaOL and $CaCl_2$ on quartz. Obviously, when the $CaCl_2$ concentration is selected as 25 mg/L, the recovery trend of quartz in the two methods (sequential dosing and mixed dosing) is consistent and there is basically no difference in the value, with the pH increasing. However, when the pH value is lower than 10, the recovery of quartz increases slowly with increasing pH, and the recovery is lower than that without $CaCl_2$. The reason behind this phenomenon can be explained as the generation of $Ca(IO)_2$, which consumes a lot of oleate ions. As pH continues to increase, the recovery of quartz increases sharply and reaches a balance after the pH increases to

11.5. At this point, the recovery was around 93%, which was much higher than that without $CaCl_2$. This means that $CaCl_2$ can greatly improve the collecting ability of NaOL for quartz in a reasonable pH range. Therefore, the optimal pH range of 11.5–12.0 was chosen for the succeeding flotation tests.

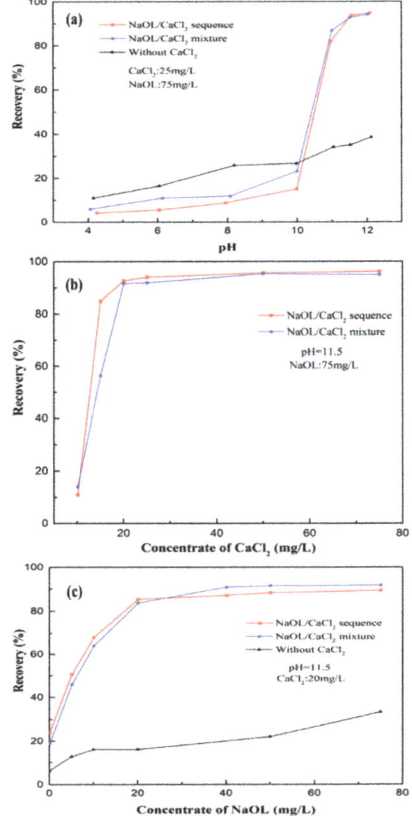

Figure 3. Quartz recovery as a function of pH (**a**), $CaCl_2$ concentration (**b**), and NaOL concentration (**c**).

Figure 3b shows the recovery of quartz as a function of $CaCl_2$ concentration. When the pH value was 11.5 and NaOL concentration was fixed at 75 mg/L, with an increase in $CaCl_2$ concentration from 10 mg/L to 75 mg/L, the quartz recovery first increased sharply to the maximum value of about 92% and then reached equilibrium. The difference in dosing methods had little effect on the final experimental results. The proper $CaCl_2$ dosage was determined as 20 mg/L.

Figure 3c illustrates the recovery of quartz as a function of NaOL concentration. When the pH value of the slurry was fixed at 11.5, for the condition without $CaCl_2$ activation, as the NaOL concentration increased from 0 mg/L to 75 mg/L, the recovery of quartz gradually and consistently increased to a maximum value about 33% at 75 mg/L. When $CaCl_2$ existed, the trend in quartz recovery using the two dosing methods both increased rapidly and then remained stable with NaOL concentration increasing, and the recovery of the same dosage was basically the same. The equilibrium recovery increased by 70% compared with that of NaOL alone at 40 mg/L, which indicates that the addition of $CaCl_2$ significantly improves the flotation performance of NaOL on quartz. The most reasonable NaOL dosage can be determined as 40 mg/L.

3.2. Solution Species Distribution Analysis

In aqueous solution, calcium ion may possess various forms such as Ca^{2+}, $Ca(OH)^+$, $Ca(OH)_2$ molecule and $Ca(OH)_2$ precipitate [48–50]. Figure 4A displays the distribution diagram of calcium species in aqueous solution at calcium (II) ions concentration of 1.8×10^{-4} mol/L ($CaCl_2$ 20 mg/L). It could be easily found that positively Ca^{2+} and $Ca(OH)^+$ are the dominant species when pH is lower than 12.6 while $Ca(OH)_2$ molecule becomes the major component when pH value is higher than 12.6. Figure 4B shows the distribution diagram of oleate ions species with total concentration of 1.8×10^{-4} mol/L (NaOL 40 mg/L) [48,51]. Obviously, oleic acid molecules and OL^- are the predominant components when pH is below 8.33; with pH increasing, oleate ions (OL^- and $(OL)_2^{2-}$) become the main components. Theoretically speaking, these species can absorb on the quartz surface due to the coexistence of electrostatic attraction and chemisorption [38]. However, according to aforesaid experimental results, the flotation recovery of quartz in the presence of calcium ion reaches highest values when the pulp pH ranges from 11 to 12. Thus, calcium species (Ca^{2+} and $Ca(OH)^+$) and collector ions were considered as the possible components in the adsorption process for this study [32]. In consideration of the far-reaching influence of the hydration structures formed by metal ions in aqueous solution on the interactions of flotation reagents and mineral surfaces along with the latest research results of Wang et al. [32] and Hu et al. [15], it can be found that calcium ions are usually absorbed on the quartz surface in the form of a $(Ca(OH)(H_2O)_3)^+$ cluster. Therefore, $(Ca(OH)(H_2O)_3)^+$ and OL^- were adopted as the effective activator and collector components in the flotation system of quartz for the following DFT calculations.

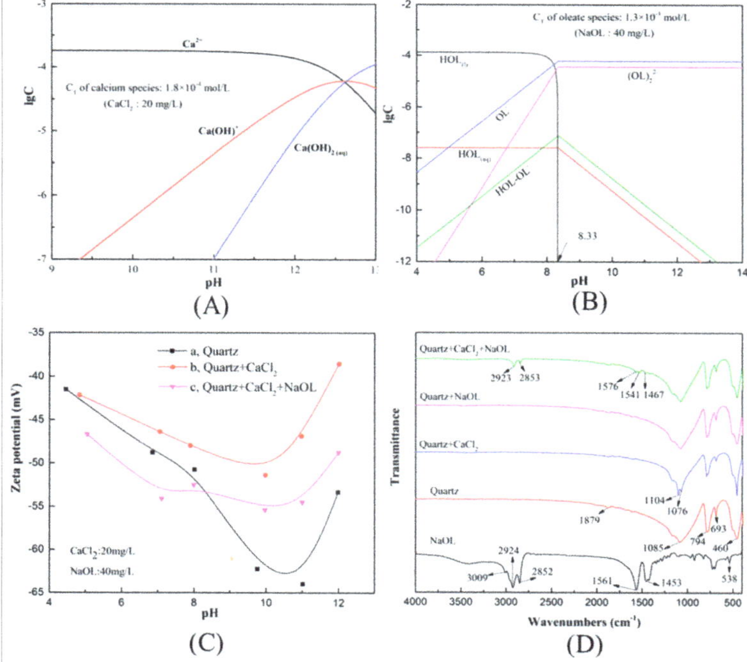

Figure 4. (**A**) Species distribution diagram of calcium ion as a function of pH; (**B**) species distribution diagram of oleate ion as a function of pH; (**C**) zeta potentials of quartz as a function of pH in the absence or presence of flotation reagents; (**D**) FTIR spectra of the quartz test sample with different conditions.

3.3. Zeta Potential Analysis

It is generally acknowledged that the zeta potential is useful for understanding the adsorption behaviors of flotation reagents on minerals' surface, which can largely explain the differences in flotation performance and flotation efficiency [33]. Figure 4C presents the zeta potential of quartz as a function of pH in the absence or presence of reagents. "a" in Figure 4C displays the variation trend in zeta potential of the quartz sample with increasing pH. Obviously, the zeta potential of the original quartz sample always had a negative value under the tested pH range of 4.5–12. During the stage with a pH 4.5–10.5, more OH^- ions adsorbed on the surface of quartz with increasing pH resulting in a decrease in the zeta potential [52]. However, when the pH was around 10.5, the zeta potential of quartz reached a turning point, which was attributed to the saturation adsorption of negatively charge OH^- ions [35]. As pH continues increasing, the zeta potential will rise due to the shielding effect of negatively charged ions and the compression of double layers [35]. The zeta potential of quartz treated by $CaCl_2$, shown in "b" of Figure 4C, was always higher than that of pure quartz in the whole tested range of pH, and it became more evident with the increase of pH, suggesting the increased adsorption quantity of calcium species on the surface of the quartz. When NaOL was added to the quartz suspension treated with $CaCl_2$, the zeta potential change process illustrated in "c" of Figure 4C can be divided into two parts. When pH was lower than 8.3, the zeta potential was lower than that of the original quartz. In terms of Figure 4B and "b" of Figure 4C, it's clear that the decrease in zeta potential was mainly attributed to a small amount of adsorption of oleate ions on quartz's surface. Moreover, the presence of large amounts of free Ca^{2+} led to the decrease in oleate ions, which is consistent with the conclusion from the micro-flotation tests that the recovery of quartz with the $CaCl_2$ and NaOL treatments was lower than that of NaOL alone in this range. With the continuous increase in pH (pH > 8.33), the adsorption of oleate ions onto the quartz surface caused the decrease in zeta potential compared with "b" of Figure 4C and the increase in zeta potential compared with "a" of Figure 4C. This implies that the presence of calcium species enhanced the positive adsorption of oleate ions on the surface of the quartz, which can be confirmed by better flotation performance, as presented in Figure 3.

3.4. FTIR Tests

The FTIR spectra of NaOL and quartz test samples are presented in Figure 4D for proposing a possible adsorption mechanism of reagents on the surface of quartz. In the FTIR spectrum of NaOL, the peaks at 2924 and 2852 cm^{-1} resulted from the symmetric C–H vibration of $-CH_2-$ and $-CH_3-$, respectively [53]. The peaks at 1561 and 1453 cm^{-1} corresponded to the asymmetric and symmetric stretching vibration of $-COO-$, respectively [53,54]. The quartz spectrum displayed several peaks in the region of 2000–400 cm^{-1}. The peaks at 1085 and 794 cm^{-1} can be attributed to the asymmetric and symmetric stretching vibration of Si–O, and the peaks at 693 and 460 cm^{-1} were associated with the symmetrical and asymmetrical bending vibration [55,56]. No new peaks were formed in the FTIR spectrum of quartz treated by NaOL alone, and one can consider NaOL as the physical adsorption or non-adsorption on the quartz surface combined with the experimental results of micro-flotation. Interestingly, the peak of 1085 cm^{-1} disappeared in the FTIR spectrum of quartz treated with $CaCl_2$ alone, and the peaks of 1104 and 1076 cm^{-1} appeared nearby, indicating that the adsorption of calcium species will change the surface structure of quartz to some extent. When quartz was treated in the order of $CaCl_2$ and NaOL, it was observed that five distinct new peaks appeared in the FTIR spectrum, namely, 2923 cm^{-1}, 2853 cm^{-1}, 1576 cm^{-1}, 1541 cm^{-1}, and 1467 cm^{-1}, respectively, corresponding to the stretching vibration of $-CH_2-$ and $-CH_3-$, and $-COO-$. Therefore, it can be concluded that oleate ions were indirectly chemisorbed on the quartz surface via the action of calcium species under appropriate pH conditions. Two possible effects from the calcium species exist: one is that the adsorption on the surface of the quartz increased the action sites of the oleate ions, and the other is that the generated calcium oleate complexes (such as $Ca(OL)^+$, $Ca(OL)_2$, $Ca(OH)(OL)$, etc.) assembled on the quartz surface to increase its hydrophobicity and float up.

3.5. Computational Results

3.5.1. Quartz Bulk Cell Optimization

Firstly, the influence of various exchange-correlation functionals on the rationality of α-quartz bulk cells was studied under the cut-off energy of 489.8 eV and a k-point set of $3 \times 3 \times 2$. At present, generalized gradient approximation (GGA) and local density approximation (LDA) were the two favorite types in practice [12,25]. Taking the limitation of LDA into consideration, GGA is more appropriate for this simulation's calculation [12,57]. Thus, several functionals such as Perdew–Wang's 1991 (PW91) [58], the revised Perdew–Burke–Ernzerhof (RPBE) [59], Perdew–Burke–Ernzerhof solids (PBESOL) [60], Perdew–Burke–Ernzerhof (PBE) [61], and Wu–Cohen (WC) [62] were investigated based on the GGA framework. The experimental values and optimized quartz lattice parameters are tabulated in Table 1, and the optimal functional was determined by comparing the Difference (%) between the calculated values and the experimental values. The formulas for calculating "Difference/%" are attached below:

$$\text{Total Difference (Å)} = (a_m - a) + (b_m - b) + (c_m - c)$$

$$\text{Difference (\%)} = \text{Total Difference} / (a + b + c)$$

where a, b, and c represent the experimental lattice parameter values obtained from the American Mineralogist Crystal Structure Database (AMCSD) website; a_m, b_m, and c_m refer to the simulated lattice parameter values based on the identical α-quartz crystal structure combined with different functionals.

Table 1. Comparison of the computed quartz parameters from different generalized gradient approximation (GGA) functionals with experimental values under the condition of a cut-off energy of 489.8 eV and k-point mesh $3 \times 3 \times 2$.

Data Sources	Functionals	a/Å	b/Å	c/Å	Total Difference/Å	Difference/%
AMCSD	-	4.914	4.914	5.405	-	-
Calculated Values	GGA-WC	5.054	5.054	5.524	0.399	2.62
	GGA-PBE	5.088	5.088	5.574	0.518	3.40
	GGA-RPBE	5.142	5.142	5.631	0.684	4.49
	GGA-PBESOL	5.035	5.035	5.526	0.363	2.38
	GGA-PW91	5.123	5.123	5.584	0.597	3.92

It is obvious from Table 1 that the calculated lattice parameters on account of the five GGA functionals coincided well with the experimental values from AMCSD and that the total lattice constant differences did not exceed 5%. Moreover, when the functional was determined as GGA-PBESOL, the minimal lattice parameter difference of 2.38% could be obtained [12]. Therefore, GGA-PBESOL was selected for the subsequent convergence tests.

Figure 5a shows that under the conditions of a GGA-PBESOL functional and a cut-off energy of 489.8 eV, when the k-point set increased from $2 \times 2 \times 2$ to $4 \times 4 \times 6$, the variation trends of the total energy and lattice constants difference were similar, both changing considerably and then tending to become stable. It was obvious that the total energy and lattice difference of the system converged to relatively low constant values at k-point $3 \times 3 \times 2$. Thus, the proper k-point set was determined as $3 \times 3 \times 2$.

Figure 5b presents the total energy and lattice constants difference as a function of cut-off energy. As illustrated in Figure 5b, the total energy and lattice constants differences both kept decreasing to an equilibrium value with the increase of cut-off energy. Then, the decrease in total energy was lower than 0.05 eV and the reduction in the lattice parameters difference did not exceed 0.003% when the cut-off energy was beyond 571.4 eV. Therefore, the appropriate cutoff energy was selected as 571.4 eV.

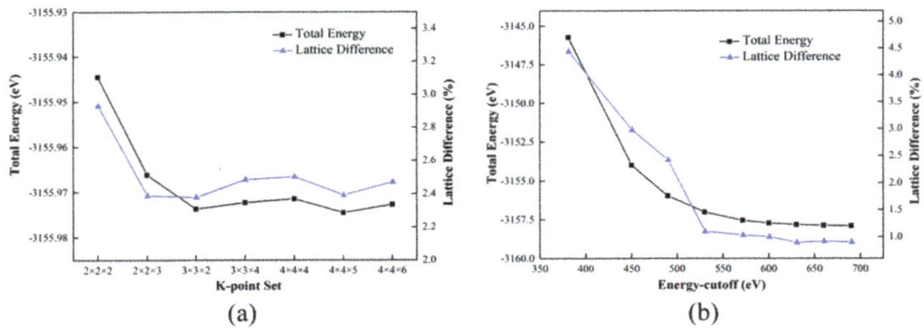

Figure 5. Total energy and lattice parameter difference as a function of k-point set with a cutoff energy of 489.8 eV and a GGA-PBESOL functional (**a**), as a function of cutoff energy with the k-point set to $3 \times 3 \times 2$ and with a GGA-PBESOL functional (**b**).

In conclusion, the rational parameters for quartz bulk cell optimization were determined as a GGA-PBESOL functional, k-point set $3 \times 3 \times 2$, and a cutoff energy of 571.4 eV, which could generate a sufficiently accurate structure at an acceptable cost. The experimental conclusions basically tally with previous research results [12]. The optimized quartz bulk cell structure on the strength of optimum parameters is shown in Figure 6.

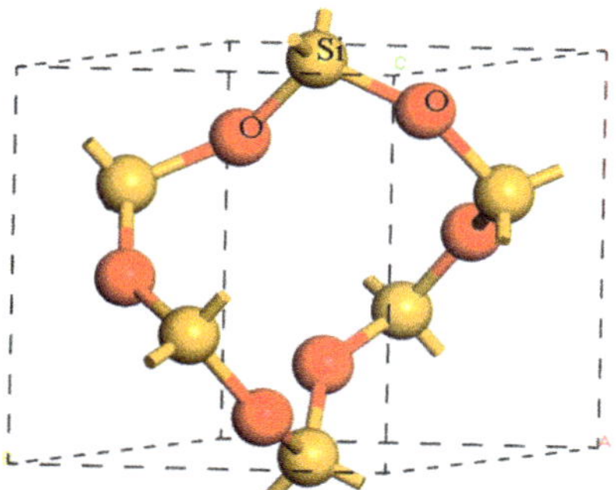

Figure 6. The optimized quartz bulk unit-cell with the optimum parameters of a GGA-PBESOL functional, $3 \times 3 \times 2$ k-point set, 571.4 eV cut-off energy (color codes: yellow—Si, red—O).

3.5.2. Hydroxylation Model of Quartz (101) Surface

The initial cleaved (101) surface structure of the quartz is shown in Figure 7a, and its optimized structure at the GGA-PBESOL level is presented in Figure 7b. It can be seen from Figure 7 that the newly generated (101) surface structure will undergo a slight reconstruction to obtain a stable surface. Each Si atom was coordinated with four O atoms to form a tetrahedral structure, where the bond lengths of the Si–O bond formed by O atoms connected up and down with a Si atom extend from 1.539 Å to 1.546 Å and 1.677 Å to 1.704 Å, respectively, and the bond lengths of the Si–O bond formed by O connected left and right, with Si shortened from 1.688 Å to 1.680 Å and 1.742 Å to 1.727 Å, respectively. Before simulating the adsorptions of various adsorbates onto the quartz surface, the first consideration was to investigate the rarely reported hydroxylation behaviors of water molecules on the α-quartz (101) surface to more accurately describe the subsequent interaction mechanisms [36]. The hydroxylation reaction of the quartz (101) surface can be evaluated by the adsorption energy (E_{ads}) of water on the quartz's surface. The equation for adsorption energy (E_{ads}) can be defined as the following [15]: $E_{ads} = E_{system} - (E_{quartz(101)} + E_{H2O})$, where E_{system} represents the total energy of the optimized hydroxylated configuration of the H_2O-absorbed quartz (101) surface, $E_{quartz(101)}$ and E_{H2O} refer to the total energy of the optimized quartz (101) surface and optimized H_2O molecules in the periodic structure, respectively. Firstly, individual water molecules are placed above unsaturated Si atoms on the quartz (101) surface. Due to the strong hydrophilicity of the quartz surface, it was found that H_2O would spontaneously crack into H and OH, and separately combine with O atoms and Si atoms onto the quartz (101) surface after optimization. At the same time, one can easily find that two new Si–OH silanol groups were generated accompanied by the formation of O_w–H_2···O_q hydrogen bonds with distance and bond angles of 2.598 Å and 169.2°, as illustrated in Figure 8a (note that O_q and Si_q belong to the surface atoms of quartz; H_1, H_2, and O_w belong to the water molecule). The adsorption energy of a single water molecule on the quartz (101) surface was −501.4 kJ/mol, which indicates that the hydroxylation process was thermodynamically favorable. In consideration of the existence of six unsaturated Si atoms on the cleaved quartz (101) surface, six H_2O molecules were placed directly above six different unsaturated Si atoms to explore the hydroxylation behaviors of water molecules on quartz. Figure 8b presents the optimized complete hydroxylation models of the quartz (101) surface with six water molecules adsorbed at the same parameter settings. The configuration A in Figure 8b is the original optimized hydroxylation model, and the other three possible configurations can be obtained in consideration of the symmetry of the system. Clearly, the bond lengths of the newly generated O_q–H and Si_q–OH in the four hydroxylation models show little difference, but there is a significant diversity in the orientation of the bonds. The adsorption energy of the four different hydroxylation configurations of the quartz (101) surface is listed in Table 2. When all the O–H bonds on the surface of the quartz intersect with each other, the obtained configuration (i.e., configuration D in Figure 8b) has the lowest energy about −2517.8 kJ/mol, indicating that configuration D was the most stable in terms of thermodynamics.

Table 2. The adsorption energy of the hydroxylation models of the quartz (101) surface with water molecules adsorbed.

Configuration	a	A	B	C	D
Adsorption energy (kJ/mol)	−501.4	−2483.8	−2489.3	−2467.9	−2517.8

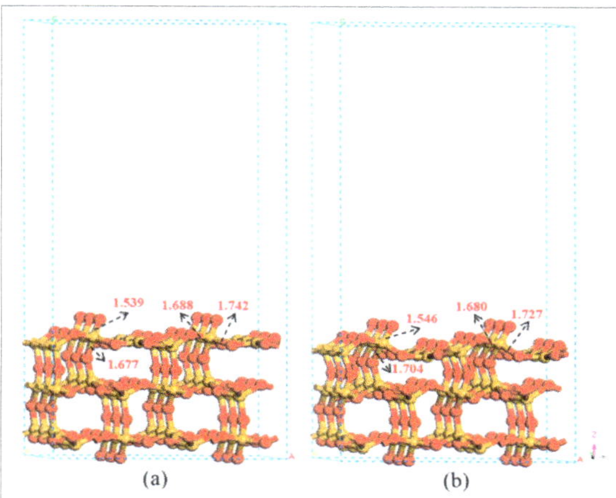

Figure 7. Quartz (101) surface structures: (**a**) original configuration and (**b**) optimized configuration (color codes: red—O, yellow—Si).

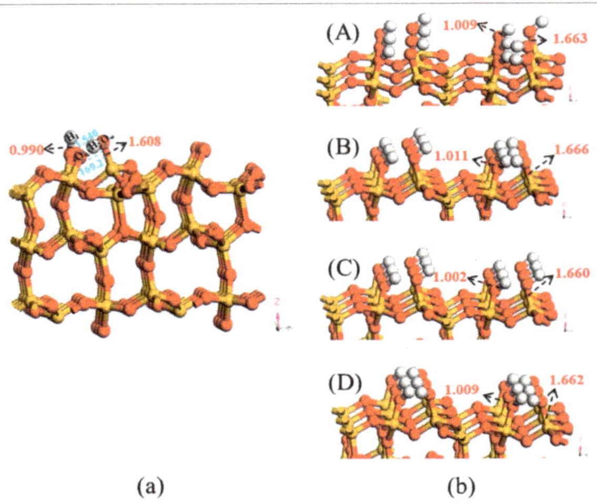

Figure 8. The hydroxylation model of the quartz (101) surface with one water molecule adsorbed (**a**), four possible complete hydroxylation models of the quartz (101) surface with six water molecules adsorbed (**b**) (some structural fragments were cut out for the convenience of comparing structural differences) (color codes: gray—H, red—O, yellow—Si).

3.5.3. Adsorbates on the Hydroxylated Quartz (101) Surface

Based on the results of the flotation tests, solution chemistry calculations, and previous research work [15,29,32], it can be concluded that the main adsorbates in the pulp include H_2O molecules, OH^- ions, $Ca(H_2O)_3(OH)^+$ ions, OL^- (oleate ions). As the sequential-dosing method and the mixed-dosing method for NaOL and $CaCl_2$ had no effect on the recovery of quartz flotation, the sequential-dosing method was selected to gradually study the activation and flotation mechanisms in DFT simulations to thoroughly elucidate the flotation process of quartz. The optimized adsorption structures of various adsorbates on hydroxylated quartz (101) surface are displayed in Figure 9 and the corresponding adsorption energy is listed in Table 3. Figure 9a shows the adsorption of water molecules. The closest distances between the two hydrogen atoms of the water molecule and the hydroxylated surface were 1.927 Å and 2.824 Å, respectively, and the $O_w–H_1\cdots O_q$ hydrogen bond with the $O_w\cdots O$ distance of 2.879 Å and bond angle of 158.6° can be observed. The corresponding interaction energy was −20.9 kJ/mol, suggesting that a weaker interaction existed between the water molecule and the hydroxylated structure and that other adsorbates can easily expel the water molecular layers to act on the hydroxylated surface. Figure 9b depicts the adsorption of the OH^- ion. The distances between the O atom in the OH^- ion and H atoms in hydroxylated structure were 1.232 Å and 1.162 Å, respectively, and the interaction energy was −291.6 kJ/mol. The bond lengths of the covalent O–H bonds of the involved Si–OH silanol groups were 1.161 Å and 1.258 Å, respectively. It can be inferred that the O atom in the OH^- ion forms covalent bonds with the H atoms of the hydroxylated surface. Figure 9c describes the adsorption of the OL^- ion. The shorter distances between the oxygen atoms of oleate ions and the H atoms on the hydroxylated surface were 1.284 Å and 1.675 Å, respectively, and the strong hydrogen bond ($O_q–H\cdots O_1$ hydrogen bond: $O_q\cdots O_1$ distance, 2.651 Å, $O_q–H\cdots O_1$ angle, 159.7°) was formed [63]. The corresponding interaction energy was −203.6 kJ/mol, manifesting that pure OL^- had a certain collection capacity for quartz by hydrogen bonds and Van der Waals forces, which is consistent with the results of flotation tests. Figure 9d represents the adsorption of $Ca(H_2O)_3(OH)^+$, and the bond length of the newly formed $Ca–O_w$ bond was 2.343 Å (note that the O_w atom belongs to water molecules involved in hydroxylation). The interaction energy of $Ca(H_2O)_3(OH)^+$ on the hydroxylated surface of approximately −254.3 kJ/mol was 50.7 kJ/mol lower than that of pure OL^-

on the quartz surface and 37.3 kJ/mol higher than that of OH^- on the quartz surface, which indicates that $Ca(H_2O)_3(OH)^+$ was easier to adsorb on the hydroxylated surface than OL^- and could activate the hydroxylated quartz (101) surface. Through careful analysis of the optimized structures, one can observe that $Ca(H_2O)_3(OH)^+$ adsorbed on the quartz surface via O_w atom sites on the hydroxylated structure, which is different from OH^- and OL^- through H atom sites, so the adsorption of OH^- and OL^- had little effect on the adsorption of $Ca(H_2O)_3(OH)^+$. Figure 9e presents the optimized system of hydroxylated Quartz–$Ca(H_2O)_3(OH)$–OL, formed by OL^- coordinating with the Ca atom based on the previous optimized structure (Figure 9d). We found that the O atoms in the OL^- and Ca atoms in the optimized structure (Figure 9d) formed new O–Ca bonds with bond lengths of 2.343 Å and 2.369 Å, respectively. Simultaneously, a water molecule originally bound with a Ca atom was spontaneously dissociated into free H_2O due to the stable six-coordinated structure of calcium ion in the solution [29]. Besides, the bond length of the previous O_w–Ca was shortened from 2.343 Å to 2.253 Å. The adsorption energy of OL^- on the activated surface was −460.7 kJ/mol, which decreased by 257.1 kJ/mol compared with that of OL^- without activation. This indicates that activation of $Ca(H_2O)_3(OH)^+$ markedly promoted the adsorption of OL^- on the hydroxylated quartz (101) surface, which enhanced the hydrophobicity of the surface and greatly improved the recovery of quartz.

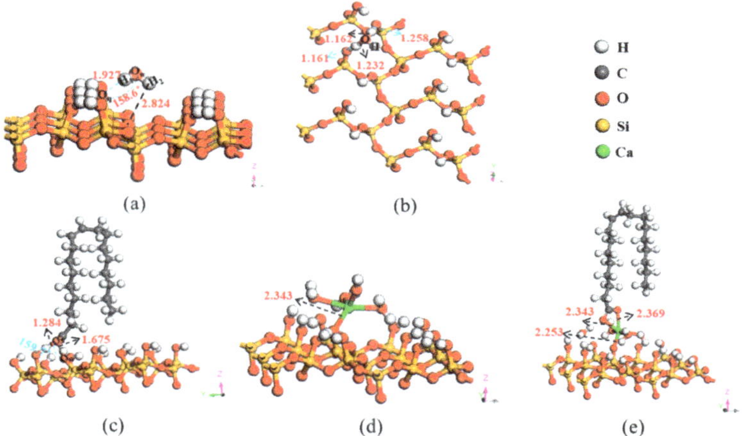

Figure 9. The optimized adsorption model of a water molecule on the hydroxylated quartz (101) surface (**a**), OH^- ions on the hydroxylated quartz (101) surface (**b**), OL^- on the hydroxylated quartz (101) surface (**c**), $Ca(H_2O)_3(OH)^+$ ions on the hydroxylated quartz (101) surface (**d**), OL^- on the previously activated structure (**e**).

Table 3. The calculated adsorption energy of various adsorbates on the hydroxylated quartz (101) surface; unit is kJ/mol.

Adsorbate	H_2O	OH^-	OL^-	$Ca(H_2O)_3(OH)^+$	OL^- on the $Ca(H_2O)_3(OH)^+$-Activated Surface
Interaction energy	−20.9	−291.6	−203.6	−254.3	−460.7

3.5.4. Electronic Properties Analyses

The Mulliken population analysis can explain the process of electron transfer, and the magnitude of bond populations was used as a handy assessment of the bond covalency [25]. The whole adsorption process can be divided into two stages: the first stage is the adsorption of $Ca(H_2O)_3(OH)^+$ on the hydroxylated quartz (101) surface through the interaction with O_w atoms, followed by the adsorption of OL$^-$ ions on the activated structure by calcium cluster. The involved atoms are labeled as presented in Figure 10 and the Mulliken populations for pertinent atoms and bonds are tabulated in Tables 4 and 5, respectively. Obviously, during the adsorption process, the Mulliken charges of Ca atoms increased from +1.33 to +1.43, while the charges of O_w and O_1 decreased from −1.06 to −1.13 and −0.55 to −0.70, respectively. This indicates that the electrons of the calcium cluster were transferred to the oxygen atom on the hydroxylated surface and to the oxygen atoms of OL$^-$. The bond population of Ca–O_w increased from 0.08 to 0.13, revealing the strengthened covalent character of the Ca–O_w bonds in the adsorption process of OL$^-$, which is consistent with shorter Ca–O_w bond length. The bond population of Ca–O_1 changed from 0.00 to 0.07, corresponding to the formation of new bonds (note that the O_1 atom belonged to the OL$^-$ ion). In addition, the bond populations of Ca–O_{w1} and Ca–O_{w2} in the calcium cluster structure decreased from 0.05 to 0.00 and 0.32 to 0.04, respectively, implying that the adsorption of OL$^-$ resulted in a certain degree of weakening of the calcium cluster structure.

Figure 10 represents the electron density difference of OL$^-$ ion on Ca^{2+}-activated hydroxylated α-quartz (101) surface. Obviously, the electron density of calcium atom decreased and the electron density of O_w and O_1 atoms increased. The electron density of the overlap area between Ca and O atoms was very low, indicating a weaker covalency of Ca–O bonds, which accords with the lower bond population of Ca–O bonds. In terms of population analysis and electron density difference, it can be concluded that Ca atoms interact with the hydroxylated surface and OL$^-$ by ionic Ca–O bonds during the whole adsorption process, where Ca atoms serve as a bridge between the hydroxylated quartz (101) surface and NaOL collector [12].

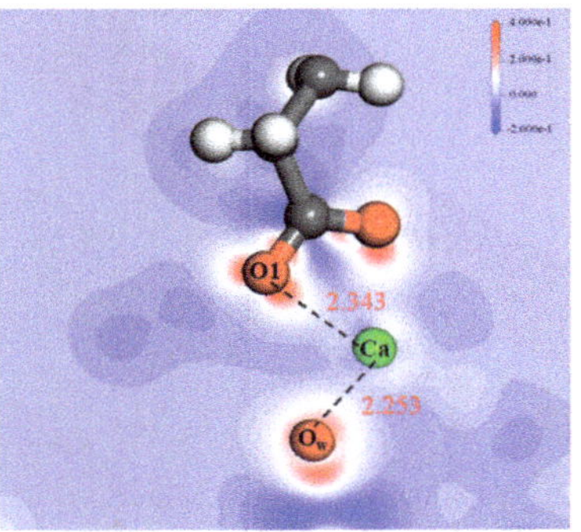

Figure 10. The electron density difference of OL$^-$ ions adsorbed on the Ca^{2+}-activated hydroxylated α-quartz (101) surface after optimization (blue contour represents decreased electron density and red contour indicates increased electron density).

Table 4. Mulliken populations of NaOL on the $Ca(H_2O)_3(OH)^+$-activated hydroxylated α-quartz (101) surface.

Atom	In Sequence	s	p	d	Total	Charge (e)
Ca	Before	2.09	5.99	0.58	8.67	1.33
	After	2.10	6.00	0.47	8.57	1.43
O_w	Before	1.84	5.21	0.00	7.06	−1.06
	After	1.87	5.25	0.00	7.13	−1.13
O_1	Before	1.85	4.70	0.00	6.55	−0.55
	After	1.81	4.89	0.00	6.70	−0.70

The energy levels corresponding to the angular quantum number 0, 1, 2 are marked by the spectroscopy symbols s, p, d, respectively.

Table 5. The bond populations of NaOL on the $Ca(H_2O)_3(OH)^+$-activated hydroxylated α-quartz (101) surface.

Bond	Mulliken Populations			
	Before	Length/Å	After	Length/Å
$Ca-O_w$	0.08	2.343	0.13	2.253
$Ca-O_{w1}$	0.05	2.553	0.00	-
$Ca-O_{w2}$	0.32	2.036	0.04	2.511
$Ca-O_1$	0.00	-	0.07	2.343

4. Conclusions

In this work, the rarely reported hydroxylation behaviors of water molecules on the α-quartz (101) surface were innovatively and systematically studied for the first time by DFT calculations, and the activation and flotation mechanisms of Ca (II) species and OL^- ions onto hydroxylated quartz structures were further scientifically and rigorously investigated by micro-flotation tests, zeta potential measurements, solution species distribution analysis, FTIR analysis, and first principles calculations.

The flotation results showed that the equilibrium recovery of quartz exceeded 87% in the presence of $CaCl_2$ and NaOL in the optimal flotation pH range, which was up 70% compared with NaOL alone. Zeta potential analyses manifested calcium species can adsorb on the quartz surface with pH increasing and enhance the positive adsorption of oleate ions; The FTIR spectra of the quartz samples showed pure oleate ions exist physical adsorption or non-adsorption on quartz surface and oleate ions are mainly indirectly chemisorbed on the surface of quartz via calcium species. Owing to the strong hydrophilicity of quartz, the hydroxylation behaviors of H_2O molecules on the α-quartz (101) surface were spontaneous and intense, and the most stable hydroxylated structure can be obtained when all the O–H bonds of the quartz surface intersect with each other in pairs.

According to the adsorption energies of various adsorbates on the pre-hydroxylated α-quartz (101) surface, the interaction strengths of the H_2O, OH^- ions, $Ca(H_2O)_3(OH)^+$ ions, and OL^- ions are in the order of $H_2O < OL^- < Ca(H_2O)_3(OH)^+ < OH^-$. However, due to the different action sites, the adsorptions of OH^- and OL^- have little effect on the adsorption of $Ca(H_2O)_3(OH)^+$ on the hydroxylated surface. The interaction strength of OL^- on the activated structure was much higher than that of OL^- alone, indicating that calcium ions can advantageously promote the adsorption of OL^- on the surface of quartz. Mulliken population analysis along with electron density difference analysis showed electron transfer from calcium atoms to oxygen atoms on the hydroxylated surface and the oxygen atoms of OL^-, where Ca atoms serve as a bridge between the pre-hydroxylated quartz (101) surface and OL^- ions.

Author Contributions: Conceptualization, C.Z.; methodology, Z.X., S.C., and J.Z.; software, Z.X. and J.H.; validation, C.Z., D.C., P.C., and W.S.; formal analysis, Z.X., M.T., and Q.Z.; investigation, Z.X., M.T. and Q.Z.; resources, Y.H., D.C., P.C., and W.S.; data curation, C.Z. and Z.X.; writing—original draft preparation, Z.X. and J.H.; writing—review and editing, C.Z., D.C., P.C., and W.S.; visualization, Z.X. and J.H.; supervision, C.Z., D.C., P.C., and W.S.; project administration, C.Z. and Z.X.; funding acquisition, C.Z., Y.H., D.C., P.C., and W.S.

Funding: This research was funded by the Natural Science Foundation of China (Grant number 51704330), The Fund of State Key Laboratory of Mineral Processing (Grant number BGRIMM-KJSKL-2017-13), National key research and development plan (Grant number 2016YFE0101300), Open-End Fund for the Valuable and Precision Instruments of Central South University (Grant number CSUZC201805), National Key Scientific Research Project (Grant number 2018YFC1901601, 2018YFC1901602, and 2018YFC1901605), The Startup Fund of Central South University for Young Teachers (grant number 502044001), The graduate student independent exploration innovation project of Central South University (Grant number 2018zzts788), Natural Science Foundation of China (grant number 51374247), and The National 111 Project (Grant number B14034).

Acknowledgments: This work was performed in part using hardware and/or software provided by a Tianhe II supercomputer at the National Supercomputing Center in Guangzhou, and the High-Performance Computing Centers of Central South University and Nanjing University. The staff from the Supercomputing Center and High-Performance Computing Centers and the engineers from Beijing Paratera Technology Co., Ltd. provided effective support, and we offer our sincere gratitude.

Conflicts of Interest: The authors declare no conflict of interest.

References

1. Hao, H.; Li, L.; Yuan, Z.; Liu, J. Molecular arrangement of starch, Ca^{2+} and oleate ions in the siderite-hematite-quartz flotation system. *J. Mol. Liq.* **2018**, *254*, 349–356. [CrossRef]
2. Guo, W.; Zhu, Y.; Han, Y.; Wei, Y. Effects and activation mechanism of calcium ion on the flotation of quartz with fatty acid collector. *J. Northeast. Univ. Nat. Sci.* **2018**, *39*, 409–415.
3. Sahoo, H.; Rath, S.S.; Das, B.; Mishra, B.K. Flotation of quartz using ionic liquid collectors with different functional groups and varying chain lengths. *Miner. Eng.* **2016**, *95*, 107–112. [CrossRef]
4. Zhou, L. *The Foundation of Ore Petrology*; Metallurgical Industry Press: Beijing, China, 2007; pp. 145–148.
5. Vatalis, K.I.; Charalambides, G.; Ploutarch, B.N. Market of high purity quartz innovative applications. *Procedia Econ. Financ.* **2015**, *24*, 734–742. [CrossRef]
6. Wu, J.; Gao, X.; Chen, J.; Wang, C.; Zhang, S.; Dong, S. Review of high temperature piezoelectric materials, devices, and applications. *Acta Phys. Sin.* **2018**, *67*, 1–30.
7. Tao, W.; Haibing, L.; Songnan, Z.; Hongwei, Y.; Haijun, W.; Xiaodong, Y. Research of precision cleaning technology of large caliber optics. *Clean. World* **2018**, *34*, 43–49.
8. Saigusa, Y. Quartz-based piezoelectric materials. In *Advanced Piezoelectric Materials*, 2nd ed.; Woodhead Publishing: Sarston/Cambridge, UK, 2017; pp. 197–233.
9. Sandvik, K.L.; Larsen, E. Iron ore flotation with environmentally friendly reagents. *Miner. Metall. Process.* **2014**, *31*, 95–102. [CrossRef]
10. Ni, C.; Xie, G.; Jin, M.; Peng, Y.; Xia, W. The difference in flotation kinetics of various size fractions of bituminous coal between rougher and cleaner flotation processes. *Powder Technol.* **2016**, *292*, 210–216. [CrossRef]
11. Wang, L.; Liu, R.; Hu, Y.; Liu, J.; Sun, W. Adsorption behavior of mixed cationic/anionic surfactants and their depression mechanism on the flotation of quartz. *Powder Technol.* **2016**, *302*, 15–20. [CrossRef]
12. Zhu, Y.; Luo, B.; Sun, C.; Liu, J.; Sun, H.; Li, Y.; Han, Y. Density functional theory study of α-Bromolauric acid adsorption on the α-quartz (1 0 1) surface. *Miner. Eng.* **2016**, *92*, 72–77. [CrossRef]
13. Kou, J.; Guo, Y.; Sun, T.; Xu, S.; Xu, C. Adsorption mechanism of two different anionic collectors on quartz surface. *J. Cent. South Univ. Sci. Technol.* **2015**, *46*, 4005–4014.
14. Kou, J.; Xu, S.; Sun, T.; Sun, C.; Guo, Y.; Wang, C. A study of sodium oleate adsorption on Ca^{2+} activated quartz surface using quartz crystal microbalance with dissipation. *Int. J. Miner. Process.* **2016**, *154*, 24–34. [CrossRef]
15. Hu, Y.; He, J.; Zhang, C.; Zhang, C.; Sun, W.; Zhao, D.; Chen, P.; Han, H.; Gao, Z.; Liu, R.; et al. Insights into the activation mechanism of calcium ions on the sericite surface: A combined experimental and computational study. *Appl. Surf. Sci.* **2018**, *427*, 162–168. [CrossRef]
16. Wang, J.; Liu, Q.; Zeng, H. Understanding copper activation and xanthate adsorption on sphalerite by time-of-flight secondary ion mass spectrometry, X-ray photoelectron spectroscopy, and in situ scanning electrochemical microscopy. *J. Phys. Chem. C* **2013**, *117*, 20089–20097. [CrossRef]
17. Liu, W.; Zhang, S.; Wang, W.; Zhang, J.; Yan, W.; Deng, J.; Feng, Q.; Huang, Y. The effects of Ca(II) and Mg(II) ions on the flotation of spodumene using NaOL. *Miner. Eng.* **2015**, *79*, 40–46. [CrossRef]
18. Tian, M.; Gao, Z.; Sun, W.; Han, H.; Sun, L.; Hu, Y. Activation role of lead ions in benzohydroxamic acid flotation of oxide minerals: New perspective and new practice. *J. Colloid Interface Sci.* **2018**, *529*, 150–160. [CrossRef] [PubMed]
19. Li, F.; Zhong, H.; Wang, S.; Liu, G. The activation mechanism of Cu(II) to ilmenite and subsequent flotation response to α-hydroxyoctyl phosphinic acid. *J. Ind. Eng. Chem.* **2016**, *37*, 123–130. [CrossRef]
20. Luo, X.; Wang, Y.; Wen, S.; Ma, M.; Sun, C.; Yin, W.; Ma, Y. Effect of carbonate minerals on quartz flotation behavior under conditions of reverse anionic flotation of iron ores. *Int. J. Miner. Process.* **2016**, *152*, 1–6. [CrossRef]
21. Shi, Y.; Qiu, G.; Hu, Y.; Chen, C. Surface chemical reactions in oleate flotation of quartz. *Min. Metall. Eng.* **2001**, *21*, 43–45.
22. Gong, G.; Liu, J.; Han, Y.; Zhu, Y. An atomic scale investigation of the adsorption of sodium oleate on Ca^{2+} activated quartz surface. *Physicochem. Probl. Miner. Process.* **2019**, *55*, 426–436. [CrossRef]

23. Liu, A.; Fan, J.-c.; Fan, M.-q. Quantum chemical calculations and molecular dynamics simulations of amine collector adsorption on quartz (0 0 1) surface in the aqueous solution. *Int. J. Miner. Process.* **2015**, *134*, 1–10. [CrossRef]

24. Chen, J.; Lan, L.; Chen, Y. Computational simulation of adsorption and thermodynamic study of xanthate, dithiophosphate and dithiocarbamate on galena and pyrite surfaces. *Miner. Eng.* **2013**, *46–47*, 136–143. [CrossRef]

25. Liu, J.; Gong, G.; Han, Y.; Zhu, Y. New insights into the adsorption of oleate on cassiterite: A DFT study. *Minerals* **2017**, *7*, 236. [CrossRef]

26. Rath, S.S.; Sinha, N.; Sahoo, H.; Das, B.; Mishra, B.K. Molecular modeling studies of oleate adsorption on iron oxides. *Appl. Surf. Sci.* **2014**, *295*, 115–122. [CrossRef]

27. Zhao, C.; Chen, J.; Li, Y.; Huang, D.; Li, W. DFT study of interactions between calcium hydroxyl ions and pyrite, marcasite, pyrrhotite surfaces. *Appl. Surf. Sci.* **2015**, *355*, 577–581. [CrossRef]

28. Long, X.; Chen, Y.; Chen, J.; Xu, Z.; Liu, Q.; Du, Z. The effect of water molecules on the thiol collector interaction on the galena (PbS) and sphalerite (ZnS) surfaces: A DFT study. *Appl. Surf. Sci.* **2016**, *389*, 103–111. [CrossRef]

29. Bakó, I.; Hutter, J.; Pálinkás, G. Car–Parrinello molecular dynamics simulation of the hydrated calcium ion. *J. Chem. Phys.* **2002**, *117*, 9838–9843. [CrossRef]

30. Ikeda, T.; Boero, M.; Terakura, K. Hydration properties of magnesium and calcium ions from constrained first principles molecular dynamics. *J. Chem. Phys.* **2007**, *127*, 074503. [CrossRef]

31. Wang, J.; Xia, S.; Yu, L. Adsorption of Pb(II) on the kaolinite(001) surface in aqueous system: A DFT approach. *Appl. Surf. Sci.* **2015**, *339*, 28–35. [CrossRef]

32. Wang, X.; Liu, W.; Duan, H.; Wang, B.; Han, C.; Wei, D. The adsorption mechanism of calcium ion on quartz (101) surface: A DFT study. *Powder Technol.* **2018**, *329*, 158–166. [CrossRef]

33. Yin, Z.; Hu, Y.; Sun, W.; Zhang, C.; He, J.; Xu, Z.; Zou, J.; Guan, C.; Zhang, C.; Guan, Q.; et al. Adsorption mechanism of 4-Amino-5-mercapto-1,2,4-triazole as flotation reagent on chalcopyrite. *Langmuir* **2018**, *34*, 4071–4083. [CrossRef]

34. Farahat, M.; Hirajima, T.; Sasaki, K.; Doi, K. Adhesion of Escherichia coli onto quartz, hematite and corundum: Extended DLVO theory and flotation behavior. *Coll. Surf. B Biointerfaces* **2009**, *74*, 140–149. [CrossRef]

35. Luo, B.; Zhu, Y.; Sun, C.; Li, Y.; Han, Y. Flotation and adsorption of a new collector α-Bromodecanoic acid on quartz surface. *Miner. Eng.* **2015**, *77*, 86–92. [CrossRef]

36. De Leeuw, N.H.; Cooper, T.G. A computational study of the surface structure and reactivity of calcium fluoride. *J. Mat. Chem.* **2003**, *13*, 93–101. [CrossRef]

37. Pradip, P.; Rai, B. Molecular modeling and rational design of flotation reagents. *Int. J. Miner. Process.* **2003**, *72*, 95–110. [CrossRef]

38. Tian, M.; Zhang, C.; Han, H.; Liu, R.; Gao, Z.; Chen, P.; He, J.; Hu, Y.; Sun, W.; Yuan, D. Novel insights into adsorption mechanism of benzohydroxamic acid on lead (II)-activated cassiterite surface: An integrated experimental and computational study. *Miner. Eng.* **2018**, *122*, 327–338. [CrossRef]

39. Lima, R.M.F.; Brandao, P.R.G.; Peres, A.E.C. The infrared spectra of amine collectors used in the flotation of iron ores. *Miner. Eng.* **2005**, *18*, 267–273. [CrossRef]

40. Clark, S.J.; Segall, M.D.; Pickard, C.J.; Hasnip, P.J.; Probert, M.J.; Refson, K.; Payne, M.C. First principles methods using CASTEP. *Z. für Krist. -Cryst. Mater.* **2005**, *220*, 567–570. [CrossRef]

41. Levien, L.; Prewitt C, T.; Weidner D, J. Structure and elastic properties of quartz at pressure. *Am. Mineral.* **1980**, *65*, 920–930. Available online: http://rruff.geo.arizona.edu/AMS/minerals/Quartz (accessed on 22 February 2019).

42. Hu, Y.; Gao, Z.; Sun, W.; Liu, X. Anisotropic surface energies and adsorption behaviors of scheelite crystal. *Coll. Surf. A Physicochem. Eng. Asp.* **2012**, *415*, 439–448. [CrossRef]

43. Bandura, A.V.; Kubicki, J.D.; Sofo, J.O. Periodic density functional theory study of water adsorption on the α-Quartz (101) surface. *J. Phys. Chem. C* **2011**, *115*, 5756–5766. [CrossRef]

44. Sahoo, H.; Sinha, N.; Rath, S.S.; Das, B. Ionic liquids as novel quartz collectors: Insights from experiments and theory. *Chem. Eng. J.* **2015**, *273*, 46–54. [CrossRef]

45. National Center for Biotechnology Information. CID=23665730 [Online]; PubChem Compound Database. Available online: https://pubchem.ncbi.nlm.nih.gov/compound/23665730 (accessed on 22 February 2019).

46. Wang, G. Research on mechanism and application of several kinds of modified starch as hematite inhibitor. Master's Thesis, Central South University, Changsha, China, May 2013.
47. Tian, M.; Gao, Z.; Han, H.; Sun, W.; Hu, Y. Improved flotation separation of cassiterite from calcite using a mixture of lead (II) ion/benzohydroxamic acid as collector and carboxymethyl cellulose as depressant. *Miner. Eng.* **2017**, *113*, 68–70. [CrossRef]
48. Wang, D.; Hu, Y. *Solution Chemistry of Flotation*; Hunan Science and Technology Press: Beijing, China, 1988; pp. 235–238.
49. Lu, J.; Gao, H.; Jin, J.; Cen, D.; Ren, Z. Effect and mechanism of calcium ion on flotation of andalusite. *Chin. J. Nonferrous Met.* **2016**, *26*, 1311–1315.
50. Luo, N.; Wei, D.-z.; Shen, Y.-b.; Liu, W.-g.; Gao, S.-l. Effect of calcium ion on the separation of rhodochrosite and calcite. *J. Mat. Res. Technol.* **2018**, *7*, 96–101. [CrossRef]
51. Robert, P.; Per, S. Solution chemistry studies and flotation behaviour of apatite, calcite and fluorite minerals with sodium oleate collector. *Int. J. Miner. Process.* **1985**, *15*, 193–218.
52. Filippov, L.O.; Duverger, A.; Filippova, I.V.; Kasaini, H.; Thiry, J. Selective flotation of silicates and Ca-bearing minerals: The role of non-ionic reagent on cationic flotation. *Miner. Eng.* **2012**, *36–38*, 314–323. [CrossRef]
53. Liu, H.; Khoso, S.A.; Sun, W.; Zhu, Y.; Han, H.; Hu, Y.; Kang, J.; Meng, X.; Zhang, Q. A novel method for desulfurization and purification of fluorite concentrate using acid leaching and reverse flotation of sulfide. *J. Clean. Prod.* **2019**, *209*, 1006–1015. [CrossRef]
54. Xue, J.; Wang, S.; Zhong, H.; Li, C.; Wu, F. Influence of sodium oleate on manganese electrodeposition in sulfate solution. *Hydrometallurgy* **2016**, *160*, 115–122. [CrossRef]
55. Saikia, B.J.; Parthasarathy, G.; Sarmah, N.C. Fourier transform infrared spectroscopic estimation of crystallinity in SiO_2 based rocks. *Indian Acad. Sci.* **2008**, *31*, 775–779. [CrossRef]
56. Ren, L.; Qiu, H.; Zhang, Y.; Nguyen, A.V.; Zhang, M.; Wei, P.; Long, Q. Effects of alkyl ether amine and calcium ions on fine quartz flotation and its guidance for upgrading vanadium from stone coal. *Powder Technol.* **2018**, *338*, 180–189. [CrossRef]
57. John, P.P.; Kieron, B.; Wang, Y. Generalized gradient approximation for the exchange-correlation hole of a many-electron system. *Phys. Rev. B* **1996**, *54*, 16533–16539.
58. Perdew, J.P.; Chevary, J.A.; Vosko, S.H.; Jackson, K.A.; Pederson, M.R.; Singh, D.J.; Fiolhais, C. Atoms, molecules, solids, and surfaces: Applications of the generalized gradient approximation for exchange and correlation. *Phys. Rev. B* **1992**, *46*, 6671–6687. [CrossRef]
59. Hammer, B.H.; Hansen, L.B.; Norskov, J.K. Improved adsorption energetics within density-functional theory using revised Perdew-Burke-Ernzerhof functionals. *Phys. Rev. B* **1999**, *59*, 7413–7421. [CrossRef]
60. Perdew, J.P.; Ruzsinszky, A.; Csonka, G.I.; Vydrov, O.A.; Scuseria, G.E.; Constantin, L.A.; Zhou, X.; Burke, K. Restoring the Density-Gradient Expansion for Exchange in Solids and Surfaces. *Phys. Rev. Lett.* **2008**, *100*, 1–4. [CrossRef]
61. Perdew, J.P.; Burke, K.; Ernzerhof, M. Generalized gradient approximation made simple. *Phys. Rev. Lett.* **1996**, *77*, 3865–3868. [CrossRef]
62. Wu, Z.; Cohen, R.E. More accurate generalized gradient approximation for solids. *Phys. Rev. B* **2006**, *73*, 1–6. [CrossRef]
63. Wang, X.; Zhang, Q.; Li, X.; Ye, J.; Li, L. Structural and electronic properties of different terminations for quartz (001) surfaces as well as water molecule adsorption on it: A first-principles study. *Minerals* **2018**, *8*, 58. [CrossRef]

Article

Molecular Modeling of Interactions between N-(Carboxymethyl)-N-tetradecylglycine and Fluorapatite

Nan Nan, Yimin Zhu *, Yuexin Han and Jie Liu

College of Resources and Civil Engineering, Northeastern University, Shenyang 110819, China; nannan_neu@163.com (N.N.); Hanyuexin@mail.neu.edu.cn (Y.H.); liujie@mail.neu.edu.cn (J.L.)
* Correspondence: zhuyimin@mail.neu.edu.cn; Tel.: +86-24-8367-6828

Received: 28 March 2019; Accepted: 30 April 2019; Published: 6 May 2019

Abstract: In this study, a flotation collector N-(carboxymethyl)-N-tetradecylglycine (NCNT) was introduced for the purpose of energy-saving, and its adsorption ability on a fluorapatite (001) surface was investigated by density functional theory calculation. The results of frontier molecular orbital analysis of NCNT and adsorption energy between NCNT and fluorapatite (FAp) showed that NCNT possessed better activity and stronger interactions in the reagent–FAp system than oleic acid (OA). A simulation model revealed that the adsorption positions of NCNT on the fluorapatite surface are calcium atoms, at which NCNT chemisorbed on (001) fluorapatite surface via a bidentate geometry involving the formation of two Ca–O bonds. Flotation experiments verified that NCNT had a good recovery of 92.27% on FAp at pH 3.5, which was slightly lower than OA. Moreover, NCNT was used at 16 °C, which was much lower than the OA's service condition (25 °C).

Keywords: fluorapatite; density functional theory; frontier molecular orbital; flotation mechanism

1. Introduction

Fluorapatite (FAp, $Ca_{10}(PO_4)_6F$) is the most important of all apatites, including chlorapatite and hydroxyapatite, due to its large natural reserves. It is a major raw material of phosphate fertilizer and, as the global consumption of phosphate fertilizer reaches 90 million tons per year, FAp has high commercial value around the world. Hence, it is very important to exploit and concentrate FAp [1].

There are many methods for FAp concentration, with the most widely recognized being flotation [2]. This method is very effective for separating minerals from gangue, based on differences in hydrophobicity between mineral surfaces [3]. To meet production requirements, such differences in hydrophobicity can be magnified by the addition of a collector. Three types of collectors, including anionic, cationic, and amphoteric collectors, are commonly used in FAp flotation processes. Anionic collectors are a widely used phosphorite collector. Alkyl hydroxamic acid [4], vegetal oil [5,6], and new compounds such as Atrac [7], have been employed to separate apatite from dolomite, silicate, and magnetite, respectively. Mixed anionic collectors have been adopted to improve the phosphate flotation recovery [8]. A cationic collector is usually used in a reverse flotation process to separate phosphorite from silicate [9]. An amphoteric collector is relatively new compared to the other two collectors. Recently, dodecyl-N-carboxyethyl-N-hydroxyethyl-imidazoline [10], dodecyl-N-carboxyethyl-N-hyroxyethyl-imidazoline [11] and alpha-benzol amino benzyl phosphoric acid [12] have been synthesized, and used in phosphate mineral separations. Among the three types of collectors, anionic collectors have the advantage of being less expensive. However, sodium oleate, the most commonly used anionic collector, requires relatively high temperature in the flotation progress, therefore it is costly and unfriendly to the environment.

In the present study, to counter the energy cost problem, a new anionic collector, N-(carboxymethyl)-N-tetradecylglycine (NCNT) was introduced, which is low-cost and can be applied at room temperature. NCNT's interaction mechanism with FAp was investigated using molecular simulation methods.

Molecular simulation is a common practice for flotation studies at the atomic level and includes first principle calculations, the Monte Carlo method, and the molecular dynamic method. A considerable number of theoretical works on apatite have been published. The bulk structure properties of $Ca_5(PO_4)_3X$ (X = F, Cl or Br) have been investigated by the density functional theory (DFT) method with the generalized gradient approximation [13]. The structure and energy of FAp surfaces have been modeled by Mkhonto using the classical energy minimization technique [14]. The adsorption system of different adsorbates, including water [15], glycine [16,17], alkyl hydroxamate [18], diphosphonic acid [19], citric acid [20], and some small organic adsorbates [21], on apatite surfaces had been well investigated using the molecular dynamic simulation method and semi-empirical quantum mechanical calculations. However, to our present knowledge, the DFT method has not yet been introduced into the surface structure computing and adsorption system simulation of FAp [22,23].

In this study, the energies of FAp surfaces, optimization convergence tests on surface depth and vacuum were performed to obtain the optimal surface model. Nest, frontier molecular orbital and adsorption energies were computed to compare the interaction strengths of NCNT and OA on FAp surfaces. Then, the adsorption mechanism of NCNT on FAp (001) surface was explained from the partial density of states and electron density differences. Finally, flotation experiments were executed to investigate the NCNT's flotation ability.

2. Computational Details and Experimental Method

Calculations were carried out using the Cambridge serial total energy package [24,25] (CASTEP) and DMol3 modules [26,27]. The CASTEP module was employed to optimize geometry, calculate energies, and analyze the properties of adsorbates and mineral surfaces. The DMol3 module was used to calculate frontier orbital energies. Parameter consistency was maintained in geometry optimization and property analysis.

2.1. NCNT Characterization

The frontier molecular orbitals of adsorbates were calculated using the DMol3 modules, after their optimization by CASTEP. The calculation parameters of the collectors were consistent with the FAp bulk and surface optimization.

2.2. FAp Bulk Structure Optimization

Different exchange-correlation functionals of the generalized gradient approximation (GGA), cutoff energies for the plane wave basis set and Brillouin zone *k*-points [28,29] were tested. The lattice parameters of the computed bulk were compared to the experimental values from XRD spectra of the real mineral to verify the validity of the simulation parameters. Ultrasoft pseudopotentials were adopted and the calculated atomic orbitals were H—$1s^1$, C—$2s^22p^2$, N—$2s^22p^3$, O—$2s^22p^4$, F—$2s^22p^5$, P—$3s^23p^3$, Ca—$3s^23p^64s^2$ [30]. Convergence criteria were set as follows: when the Broyden–Fletcher–Goldfarb–Shanno (BFGS) algorithm was used, 2×10^{-5} eV for energy, 0.05 eV/Å for maximum force, 0.1 GPa for maximum stress, and 2×10^{-3} Å for maxmum displacement, the SCF tolerance was set at 2×10^{-6} eV/atom.

2.3. Surface Energy Calculation

Surface energy was the measurement of bond destruction that occurred during surface creation. A small surface energy of a certain crystal surface means that the surface has more stable thermodynamic stability. For the same surface, different slab terminations might impact the surface thermodynamic stability and adsorption activity with the collector. Therefore, it was necessary to investigate the surface energies of different slab terminations. The surface energy was calculated by Equation (1).

$$E_{surf} = [E_{slab} - (N_{slab}/N_{bulk})\,E_{bulk}]/2A \tag{1}$$

where E_{slab} and E_{bulk} refer to the total energies of the FAp surface slab and bulk unit cell, respectively; N_{slab} and N_{bulk} are the numbers of atoms contained in the slab and bulk unit cells, respectively; A is the unit area; and 2 signifies two surfaces along the z-axis in the FAp surface slab.

2.4. Adsorption Energy Calculation

The relative affinity of the interactions of the optimized FAp mineral surface and various adsorbates was quantified in terms of the total adsorption energy (E_{ads}), defined as Equation (2).

$$E_{ads} = E_{complex} - (E_{adsorbate} + E_{mineral}) \tag{2}$$

where $E_{complex}$ is defined as the energy of the optimized adsorption system of the collector and FAp, and $E_{adsorbate}$ and $E_{mineral}$ are the energy of the adsorbate and FAp surface, respectively [31].

2.5. Flotation Experiment

Flotation experiments were performed using an XFG 1150 flotation machine (Jilin Exploration Machinery Plant, Changchun, China). The procedure was as follows: the pure FAp mineral sample (2.00 g) was mixed with deionized water (30 mL) and then added in the flotation cell. The slurry was stirred for 3 min at a rotation rate of 1992 r/min. The pH regulator (0.1% HCl or NaOH solution) and collector (NCNT/OA) were added into the cell at 2 min intervals. After the flotation procedure, the concentrate and the tailing were dried separately for the recovery rate calculations.

3. Results

3.1. Structure and Property of NCNT

NCNT contained two acetic acid groups connected by an nitrogen (N) atom and a shorter hydrophobic carbon chain (C_{14}) than OA for the purpose of good adsorbability on FAp as well as better solubility. The structure of NCNT and OA are shown in Figure 1.

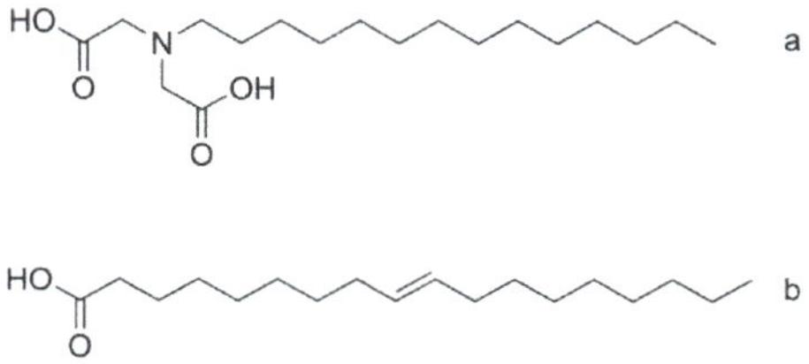

Figure 1. Molecular structure of NCNT (**a**) and OA (**b**).

The frontier molecular orbitals of NCNT were calculated to represent the reaction activity between NCNT and the FAp surface. Frontier molecular orbitals were presented by Fukui and constituted with the highest occupied molecular orbital (HOMO) and the lowest unoccupied molecular orbital (LUMO). The energy gap ($E_{HOMO-LUMO}$) between these two molecular orbitals was a criterion of compound activity, with a higher value indicating better compound activity [32]. The calculation results for the adsorbates are shown in Table 1.

Table 1. Energy gap ($E_{HOMO\text{-}LUMO}$) of the adsorbates and adsorption energies on FAp (001) surface.

Adsorbate	NCNT	NCNT^{2-}	OA	OA^{-}	H$_2$O	OH^{-}
E_{HOMO} (eV)	−5.357	2.786	−5.78	0.336	−6.454	4.78
E_{LUMO} (eV)	−1.178	3.718	−0.967	1.486	−0.275	9.757
$E_{HOMO\text{-}LUMO}$ (eV)	−4.179	−0.932	−4.813	−1.15	−6.179	−4.977

According to the values in Table 1, the $E_{HOMO\text{-}LUMO}$ increased as H$_2$O < OH^{-} < OA < NCNT < OA^{-} < NCNT^{2-}. According to frontier molecular orbital theory, the activity of the ionization state of NCNT (NCNT^{2-}) was remarkably stronger than its molecular state, indicating that the ionization state of NCNT was more likely to interact with FAp. This was mainly because lone pair electrons from a p orbital of an oxygen (O) atom in −OH and the π bond of C=O formed a p–π conjugation. This conjugation induced the electrons of the O atom in −OH to drift toward the π bond and make the dissociation of the H atom in OH^{-} easier. On the other hand, NCNT^{2-} more easily reacted with the FAp surface than with OA^{-}, water, and hydroxyl ions, indicating that the adsorption ability of NCNT^{2-} might be stronger than that of OA^{-}, and it could repel the hydration shell, and then absorb onto the FAp surface.

The HOMO of NCNT^{2-} and OA^{-} showed that a large proportion of the HOMO of NCNT^{2-} was provided by O atoms and a small portion was provided by N atoms (Figure 2). The HOMO of OA^{-} was afforded only by O atoms.

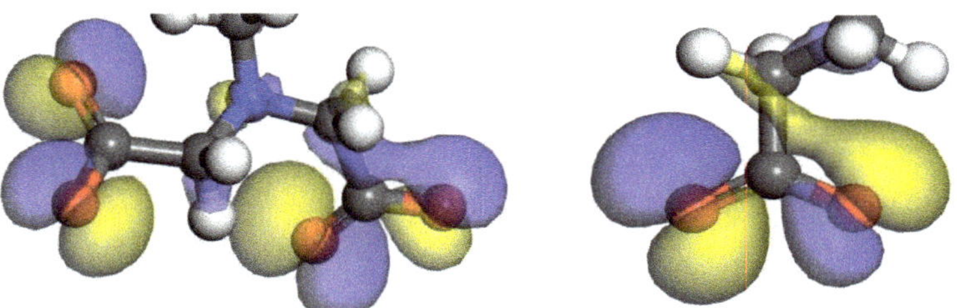

Figure 2. HOMO constitution of the ionic form: NCNT^{2-} (**left**) and OA^{-} (**right**). Carbon, nitrogen, oxygen and hydrogen atoms are in gray, dark blue, red and white, respectively.

3.2. Simulation Parameter Screening and Rationality Verification of FAp Bulk

To obtain a rational model, the simulation parameters were screened as follows [33]. The geometry optimizations of FAp bulk were executed under the condition of 340 eV cutoff energy and Brillouin zone k-point of 2 × 2 × 2 using different DFT functionals, including Perdew–Burke–Ernzerhof (PBE), revised Perdew–Burke–Ernzerhof (RPBE), Perdew–Wang's 1991 (PW91), Wu–Cohen (WC) revised and Perdew–Burke–Ernzerhof solids (PBESOL). The results were compared with experimental values [34] and are listed in Table 2.

Table 2. Comparison of lattice parameters of FAp bulk obtained from different exchange-correlation functionals with cutoff energy 340 eV and k-point 2 × 2 × 2 condition.

Data Resource	Functional	a/Å	b/Å	c/Å	Difference/%
Experimental		9.370	9.370	6.880	-
Simulation	GGA-PBE	9.305	9.305	6.858	0.593
	GGA-RPBE	9.450	9.459	6.941	0.862
	GGA-PW91	9.294	9.294	6.840	0.749
	GGA-WC	9.212	9.212	6.770	1.66
	GGA-PBESOL	9.199	9.199	6.785	1.71

The calculated lattice parameters agreed well with the experimental results when the GGA-PBE functional was adopted, and the difference between simulation and experimental results was 0.593%, which was acceptable in theoretical calculations. The total energies of cutoff energy and Brillouin zone k-point for convergence testing were calculated using this functional (Figure 3).

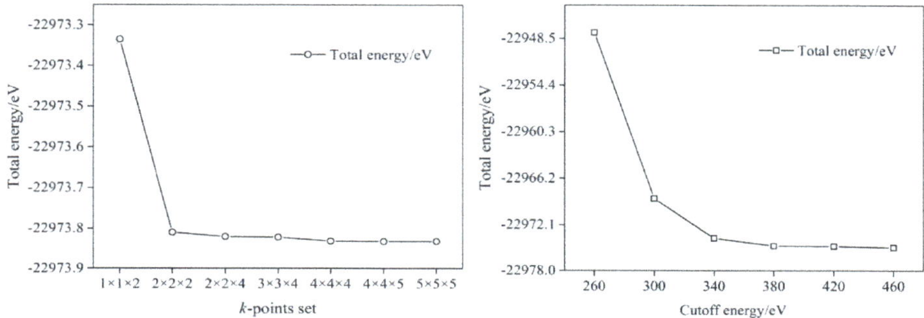

Figure 3. The k-points convergence test with cutoff energy of 340 eV (**left**) and cutoff energy convergence test with k-point $2 \times 2 \times 2$ (**right**).

With a cutoff energy of 340 eV and using the PBE functional, total energy decreased with increased cutoff energy. When the k-point was beyond $2 \times 2 \times 2$, the total energy converged very well. When the cutoff energy was >340 eV, the difference in total energies was <1 eV. Thus, the k-point $2 \times 2 \times 2$ and cutoff energy at 340 eV were suitable for calculation accuracy.

To ensure that the present simulation methods were credible, the simulated X-ray diffraction (XRD) was compared with experimental results, which showed that the simulated FAp XRD was in good agreement with the experimental spectra (Figure 4). This indicated that the simulated bulk was reliable and the parameter settings mentioned above rational and reasonable.

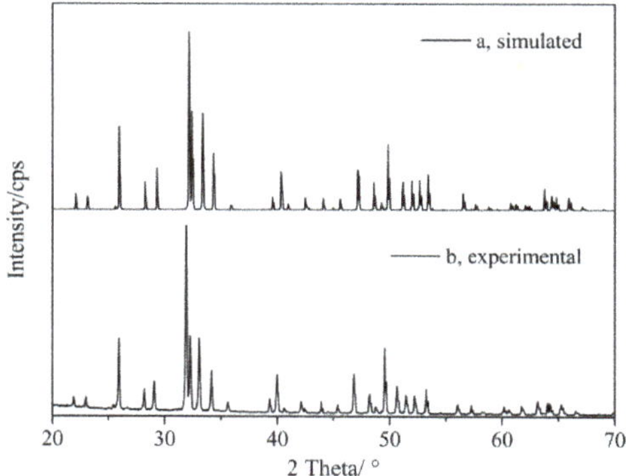

Figure 4. Comparison of (**a**) simulated and (**b**) experimental XRD spectra.

3.3. Structure Optimization of FAp Surface

According to the literature [35], the FAp (001) and (100) surfaces have been experimentally shown to be the dominant cleavage surfaces. As different slab terminations might generate different surface structures and thus affect surface energy, the surface energies of different slab terminations on FAp (001) and (100) surfaces were calculated. The slab termination with lowest surface energy was the most thermodynamically stable structure, which was then used as the surface model, with the slab sliced from geometry optimized FAp bulk. Vacuum thickness was set at 10 Å and the other surface simulation parameters employed were the same as for the bulk except that the *k*-point was set as $2 \times 2 \times 1$ (Figure 5).

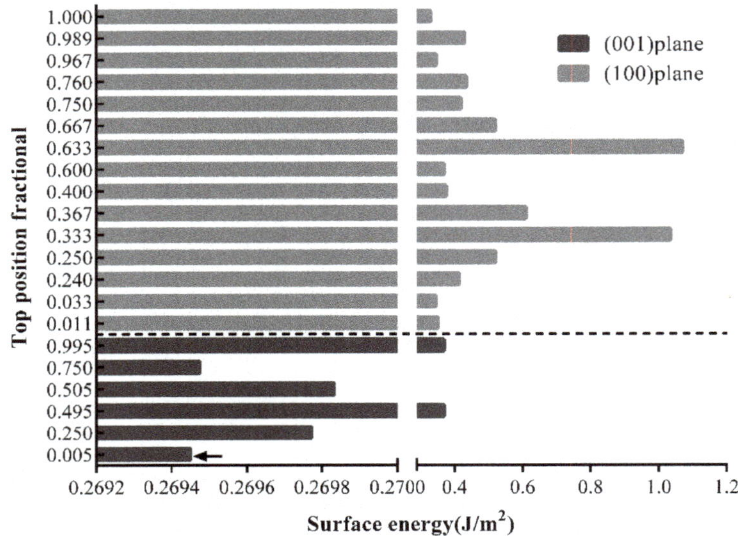

Figure 5. Surface energy of different top position fractionals on (100) and (001) surface.

In general, the surface energies of (001) were smaller than those of (100), indicating that (001) was the more thermodynamically stable surface. This conclusion confirmed the literature report that used the quantum mechanical method [14]. Moreover, when the top position was 0.005, (001) had the smallest surface energy (0.26945 J·m^{-2}). At this point, the slab termination was Ca–O–Ca.

Surface relaxation and reconstruction are common phenomena during the cleavage of solid materials. The top atoms on the surface lose their surrounding atoms, which leads to an unbalanced interaction between atoms. Thus, after cleavage, the top atoms relax to seek new balanced positions. In the inner surface, the relaxation phenomenon recedes, accompanied with increased distance from the top surface; therefore, surface energy will tend to be stable along with the increased slab depth. In the simulation calculation, the two surfaces along the *c*-axis (top and bottom) will act upon each other. This influence will reduce to negligible when increasing the vacuum thickness to a certain value. For these reasons, convergence tests of surface depth and vacuum thickness were performed to insure the reliability of the surface simulation model. Results of surface energy convergence tests are listed in Table 3.

Table 3. Convergence tests of slab depth and vacuum thickness.

Slab depth (Å)	6.858	10.287	13.716	17.145	20.574
Surface energy (J·m^{-2})	0.3107	0.5483	0.5769	0.5856	0.6022
Vacuum thickness (Å)	10	12	14	16	18
Surface energy (J·m^{-2})	0.3150	0.3167	0.3184	0.3186	0.3188

When the surface depth was larger than 10.287 Å, variation in surface energy was <0.03 J·m^{-2}. When the vacuum thickness was larger than 14 Å, surface energy changes tended to be small. Thus, the optimized parameters of the FAp (001) surface were 10.287 Å for surface depth and 18 Å for vacuum thickness in consideration of collector size. The optimized FAp (001) surface model is shown in Figure 6.

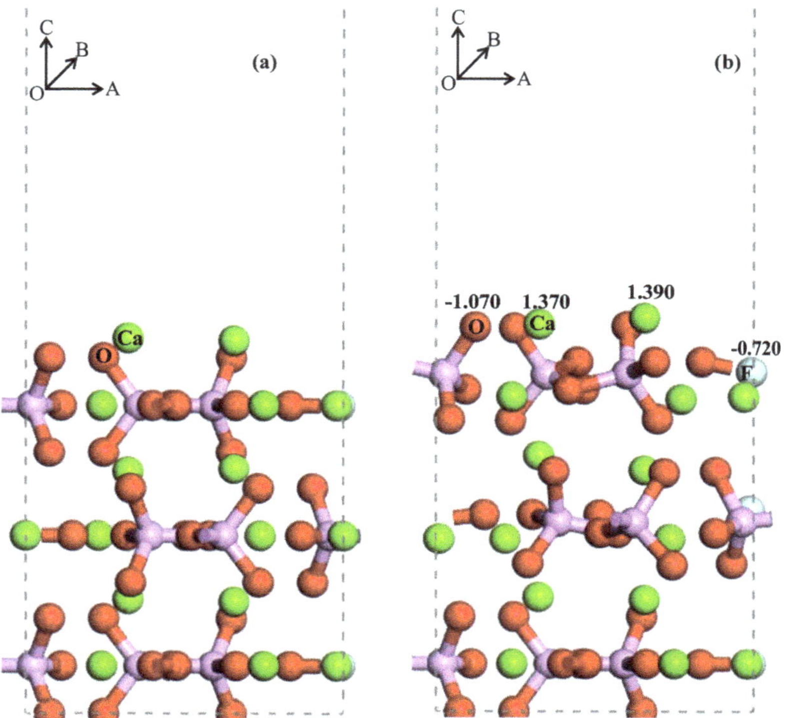

Figure 6. Surface structure of FAp (001) before (**a**) and after optimization (**b**). (Ca—green, phosphorus—purple, O—red, fluorine—light blue)

3.4. Adsorption Site and Adsorption Configurations

To obtain the adsorption site of the collector on the FAp (001) surface, mulliken charges were calculated (Figure 6). On the FAp (001) surface, calcium (Ca) atoms have positive charges, while O atoms and fluorine atoms have negative charges. According to the electronic theory of acid and alkali, a metal cation is a kind of Lewis acid, which is an electron pair acceptor, and an anion is a Lewis alkali, which is an electron pair donor. Based on this theory, Ca on the FAp surface was a Lewis acid and it could accept an electron from an anion collector (HOMO). Thus, Ca was the adsorption site on the FAp (001) surface.

After confirming the adsorption site, the adsorption configurations of $NCNT^{2-}$ and OA^- on FAp surfaces were considered with multiple possibilities and the most stable one was selected (Figure 7). In the resulting representation, the angle of C–N–C between the two acetates of $NCNT^{2-}$ was 111.289° due to the existence of the N atom. This angle provided $NCNT^{2-}$ with a specific steric configuration that allowed the $NCNT^{2-}$ to adsorb in a bidentate mode on the Ca–O–Ca terminated (001) surface, forming two Ca–O bonds with two surface Ca. This structure had the possibility of making the $NCNT^{2-}$–FAp interaction stronger than the OA^-–FAp interaction.

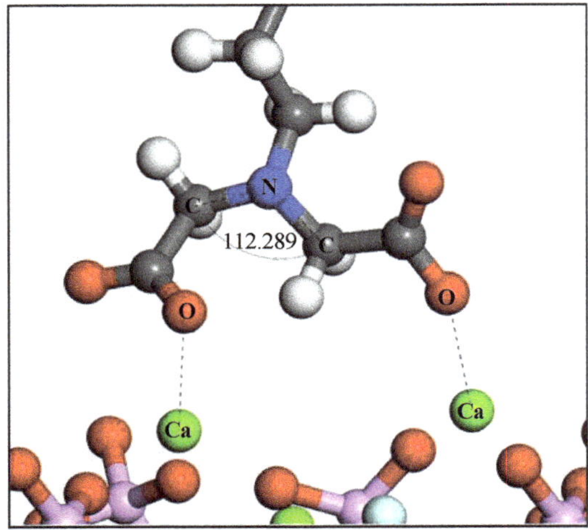

Figure 7. Adsorption configuration of NCNT^{2-} on FAp. (Ca—green, phosphorus—purple, O—red, H—white, fluorine—light blue, N—dark blue).

3.5. Adsorption Energy

Comparison of adsorption energies (E_{ads}) was the most credible evidence of interaction strengths in the computer simulation approach. A more negative E_{ads} indicated that interaction of the collector on the FAp surface was stronger. Therefore, adsorption energies of adsorbates on the FAp (001) surface were calculated, and it was observed that the adsorption energy of NCNT^{2-} was 8.936 J·m^{-2}, larger than that of OA$^-$, which indicated that the interaction strength of the NCNT–FAp system was a bit stronger than in the OA–FAp system (Table 4). These results showed that NCNT^{2-} expelled the hydration shell, such that it adsorbed on the FAp surface. On the other hand, the adsorption energy of NCNT^{2-} on the FAp (001) surface was smaller than that of H$_2$O and OH$^-$.

Table 4. Adsorption energy of the adsorbates on FAp (001) surface.

Adsorbate	NCNT^{2-}	OA$^-$	H$_2$O	OH$^-$
E_{ads} (J·m^{-2})	−210.297	−219.233	−192.71	−189.65

3.6. Adsorption Mechanism Study of NCNT^{2-}

The overall picture showed the partial density of the state of the O atom of NCNT^{2-} and the Ca atom bonding with the collector on the FAp surface (Figures 8 and 9). Before NCNT^{2-} was adsorbed on the FAp (001) surface, a valence band of O was constituted of teh p state, which was localized. The phenomenon that the valence band appeared near the Fermi surface indicated that the compound had good activity. After adsorption, the O p state shifted in the lower energy direction and the Ca conduction band was constituted of d and s states in the initial state. After bonding with the O, the conduction band of Ca disappeared. In short, after adsorption, the partial density of the state of O and Ca both moved in the lower energy direction. This phenomenon indicated that O and Ca become stable because of the O–Ca bond.

The calculated electron density difference depicted electron transmission in the bonding progress (Figure 10). The results showed that, after adsorption, Ca atoms on the surface gained electrons, while O atoms lost electrons. This phenomenon suggested that charges transfer from Ca atoms to O atoms. In addition, the electron depletion zone was around O atoms and the electron accumulation zone was around Ca atoms, which indicated that Ca and O bonded together by ionic bonding.

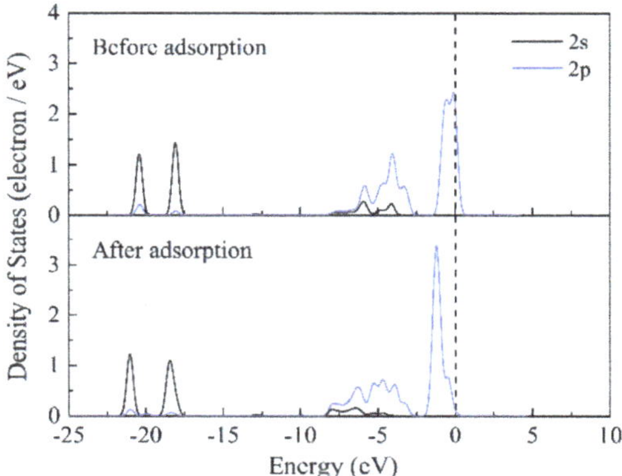

Figure 8. Partial density of states for O of FAp surface.

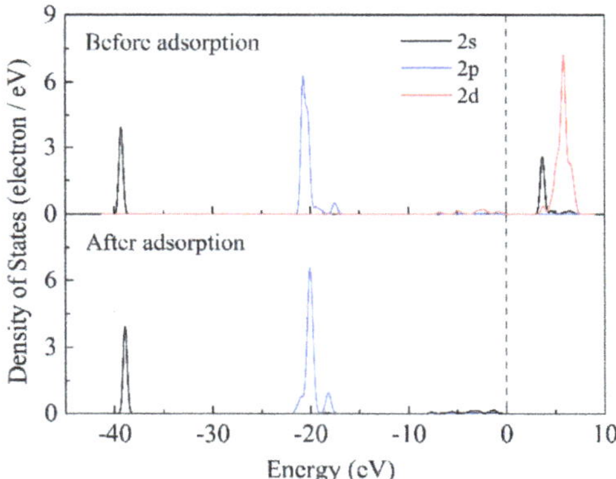

Figure 9. Partial density of states for Ca of NCNT^{2-}.

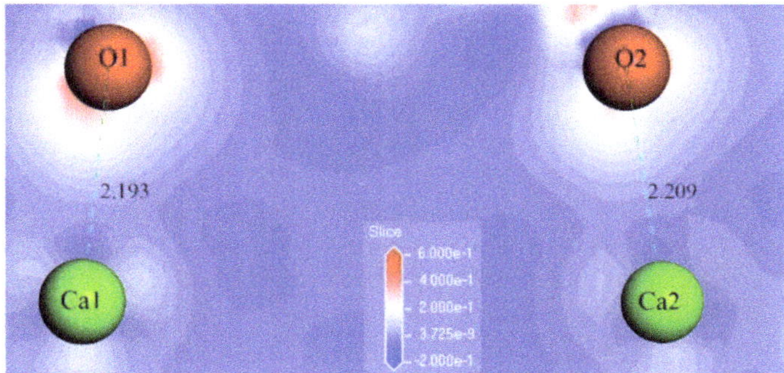

Figure 10. Electron density difference plot of adsorbed NCNT^{2-} on FAp (001) surface. Red contours correspond to electron density depletion, while blue contours represent electron density accumulation.

3.7. Flotation Results

Flotation results demonstrated that the best FAp recovery with the NCNT collector was 92.27%, which was slightly lower than the recovery with OA (94.52%, Figure 11). It was noteworthy that the optimal pH of NCNT was 3.5, which might be an inconvenience for processing equipment. When the temperature was dropped to 16 °C, FAp recovery with the NCNT collector still surpassed 90%, while the recovery with OA was reduced to 70.02%. This outcome indicated that, as expected, NCNT could be employed under a relatively low temperature, thus conserving energy.

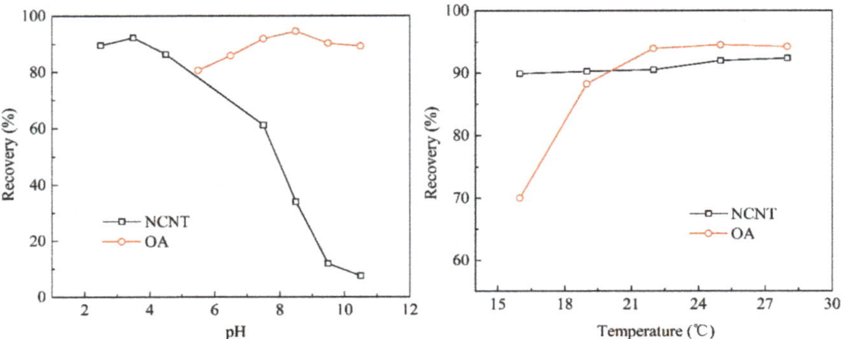

Figure 11. Effect of slurry Ph (**left**) and flotation temperature (**right**) on the recovery of FAp.

4. Discussion

The traditional collector sodium oleate requires heating because of its poor solubility [36]. The factors that affect the surfactant solubility include a shorter carbon chain length, multiple hydrophilic groups, increased branches and so on. Compared to oleic acid, the hydrophobic groups of NCNT were reduced from C18 to C14, and a carboxylic acid group was added to increase water solubility of the molecule, thereby achieving its low temperature tolerance. In order to quantify the solubility of the collector, we introduced the log$P$ parameter, which is commonly used in the chemical industry. Log$P$ is the logarithm of the ratio of the partition coefficient of a substance between n-octanol (oil) and water which is a measure of the relative distribution of the substance between the oil phase and water phases. A small log$P$ value means that the substance has higher hydrophilicity, so it is very soluble in water. According to the database Scifinder, the log$P$ of NCNT is 6.402, which is less than the 7.421 value of oleic acid. Therefore, as expected, the solubility of NCNT in water is significantly better than oleic acid.

Through simulation, the $E_{HOMO\text{-}LUMO}$ of NCNT^{2-} was larger than that of OA$^-$, and the interaction strength of the NCNT–FAp system was a bit stronger than that of the OA–FAp system. These results indicated that NCNT's molecular structure met the demands for a sufficient interaction strength with the apatite surface. However, a strong interaction strength might not directly lead to good flotation ability. Meanwhile, the collector's increased hydrophilicity caused concern regarding its flotation ability. Flotation experiments certified that NCNT had good recovery, 92.27%, on FAp, which was slightly lower than with OA. The temperature condition experiment revealed that, due to good hydrophilicity, NCNT could be employed at 16 °C, which was much lower than OA's service condition of 25 °C. Despite the advantages of NCNT, this new collector has its disadvantages. The first is poor foamability. In the flotation experiment, the foam generated by NCNT was not quite as much as with OA. It was projected here that the addition of a foaming agent might help NCNT attain a better flotation performance. The second was that the optimal pH condition for NCNT was 3.5. By mixing with other reagents, a highly efficient and low-energy mixed collector could be expected to be attained, which will be the focus of a future research project.

5. Conclusions

From the above analysis, the following conclusions were drawn:

1. The calculated FAp bulk lattice parameters agreed well with the experimental values when simulation parameters adopted were GGA-PBE functional, cutoff energy at 340 eV, and k-point 2 ×

2×2. The (001) surface of FAp was more stable than the (100) surface. Employing the parameters of surface depth (10.287 Å) and vacuum thickness (18 Å) obtained a rational structure for the FAp (001) surface.

2. According to the results of $E_{LUMO-HOMO}$ and adsorption energies, the new collector NCNT had better activity and stronger interactions in the reagent–FAp system than did OA.

3. Combining the results of the partial density of states, electronic density difference and the angle between two oxalic acids, the mechanism of NCNT adsorption on the FAp surface was described as the chemisorption of NCNT on the (001) surface via a bidentate geometry which formed two Ca–O bonds.

4. Flotation results confirmed the simulation results, which indicated that the NCNT collector could be used in FAp flotation at a lower temperature than OA.

In summary, NCNT was shown to be an energy-saving collector. If its disadvantages can be remedied by combination with other reagents, this new collector has potential for applications in the flotation industry.

Author Contributions: Conceptualization, Y.Z.; Data curation, N.N.; Formal analysis, N.N.; Funding acquisition, Y.Z. and J.L.; Investigation, N.N.; Project administration, Y.H.; Writing—original draft, N.N.; Writing—review & editing, N.N.

Funding: This research was funded by National Natural Science Foundation of China (grant number 51474055, 51774069 and 51674066).

References

1. Sharpley, A.; Rekolainen, S.; Tunney, H.; Carton, O. *Phosphorus Loss from Soil to Water*; CABI Publishing: Cambridge, UK, 1997.

2. Sis, H.; Chander, S. Reagents used in the flotation of phosphate ores: A critical review. *Miner. Eng.* **2003**, *16*, 577–585. [CrossRef]

3. Wills, B.A.; Napier-Munn, T. *Mineral Processing Technology*, 7th ed.; Elsevier Science & Technology Books: Amsterdam, The Netherlands, 2006.

4. Yu, J.; Ge, Y.Y.; Hou, J.T. Behavior and Mechanism of Collophane and Dolomite Separation Using Alkyl Hydroxamic Acid as a Flotation Collector. *Physicochem. Probl. Mineral Pro.* **2016**, *52*, 155–169.

5. Borges, A.A.; da Luz, J.A.M.; Ferreira, E.E. Characterization of the magnetic phase from carbonatitic phosphate ore. *Rem: Revista Escola de Minas* **2008**, *61*, 29–34.

6. Santos, E.P.; Dutra, A.J.B.; Oliveira, J.F. The effect of jojoba oil on the surface properties of calcite and apatite aiming at their selective flotation. *Int. J. Miner. Process.* **2015**, *143*, 34–38. [CrossRef]

7. Su, F.W.; Rao, K.H.; Forssberg, K.S.E.; Samskog, P.O. The influence of temperature on the kinetics of apatite flotation from magnetite fines. *Int. J. Miner. Process.* **1998**, *54*, 131–145. [CrossRef]

8. Cao, Q.B.; Cheng, J.H.; Wen, S.M.; Li, C.X.; Bai, S.J.; Liu, D. A mixed collector system for phosphate flotation. *Miner. Eng.* **2015**, *78*, 114–121. [CrossRef]

9. Filippov, L.O.; Duverger, A.; Filippova, I.V.; Kasaini, H.; Thiry, J. Selective flotation of silicates and Ca-bearing minerals: The role of non-ionic reagent on cationic flotation. *Miner. Eng.* **2012**, *36–38*, 314–323. [CrossRef]

10. Karlkvist, T.; Patra, A.; Rao, K.H.; Bordes, R.; Holmberg, K. Flotation selectivity of novel alkyl dicarboxylate reagents for apatite-calcite separation. *J. Colloid Interface Sci.* **2015**, *445*, 40–47. [CrossRef]

11. Ahmed, E.; Salah, E.M.; Mohamed, A.K.; Nagui, A.K.; Ayman, E.M. Dolomite-apatite separation by amphoteric collector in presence of bacteria. *J. Cent. South Univ.* **2013**, *20*, 1645–1652. [CrossRef]

12. Hu, Y.H.; Xu, Z.H. Interactions of amphoteric amino phosphoric acids with calcium-containing minerals and selective flotation. *Int. J. Miner. Process.* **2003**, *72*, 87–94. [CrossRef]

13. Li, C.-X.; Duan, Y.-H.; Hu, W.-C. Electronic structure, elastic anisotropy, thermal conductivity and optical properties of calcium apatite $Ca_5(PO_4)_3X$ (X = F, Cl or Br). *J. Alloy Compd.* **2015**, *619*, 66–77. [CrossRef]

14. Mkhonto, D.; de Leeuw, N.H. A computer modelling study of the effect of water on the surface structure and morphology of fluorapatite: Introducing a $Ca_{10}(PO_4)_6F_2$ potential model. *J. Mater. Chem.* **2002**, *12*, 2633–2642. [CrossRef]

15. Pareek, A.; Torrelles, X.; Angermund, K.; Rius, J.; Magdans, U.; Gies, H. Structure of interfacial water on fluorapatite (100) surface. *Langmuir* **2008**, *24*, 2459–2464. [CrossRef]
16. Pareek, A.; Torrelles, X.; Angermund, K.; Rius, J.; Magdans, U.; Gies, H. Competitive Adsorption of Glycine and Water on the Fluorapatite (100) Surface. *Langmuir* **2009**, *25*, 1453–1458. [CrossRef] [PubMed]
17. Schepers, T.; Brickmann, J.; Hochrein, O.; Zahn, D. Atomistic simulation study of calcium, phosphate and fluoride ion association to the teleopeptide-tails of collagen—Initial steps to biomineral formation. *Zeitschrift für anorganische und allgemeine Chemie* **2007**, *633*, 411–414. [CrossRef]
18. Pradip; Rai, B.; Rao, T.K.; Krishnamurthy, S.; Vetrivel, R.; Mielczarski, J.; Cases, J.M. Molecular modeling of interactions of alkyl hydroxamates with calcium minerals. *J. Colloid Interface Sci.* **2002**, *256*, 106–113. [CrossRef]
19. Pradip; Rai, B.; Rao, T.K.; Krishnamurthy, S.; Vetrivel, R.; Mielczarski, J.; Cases, J.M. Molecular modeling of interactions of diphosphonic acid based surfactants with calcium minerals. *Langmuir* **2002**, *18*, 932–940. [CrossRef]
20. Filgueiras, M.R.T.; Mkhonto, D.; de Leeuw, N.H. Computer simulations of the adsorption of citric acid at hydroxyapatite surfaces. *J. Cryst. Growth* **2006**, *294*, 60–68. [CrossRef]
21. Mkhonto, D.; Ngoepe, P.E.; Cooper, T.G.; de Leeuw, N.H. A computer modelling study of the interaction of organic adsorbates with fluorapatite surfaces. *Phys. Chem. Miner.* **2006**, *33*, 314–331. [CrossRef]
22. Kohn, W.; Sham, L.J. Self-Consistent Equations Including Exchange and Correlation Effects. *Phys. Rev.* **1965**, *140*, A1133–A1138. [CrossRef]
23. Xie, X.-D.; Lu, D. *Energy Band Theory of Solids*; Fudan University Press: Shanghai, China, 1998.
24. Clark, S.J.; Segall, M.D.; Pickard, C.J.; Hasnip, P.J.; Probert, M.J.; Refson, K.; Payne, M.C. First principles methods using CASTEP. *Zeitschrift für Kristallographie-Crystalline Materials* **2005**, *220*, 567–570. [CrossRef]
25. Segall, M.D.; Lindan, P.J.D.; Probert, M.J.; Pickard, C.J.; Hasnip, P.J.; Clark, S.J.; Payne, M.C. First-principles simulation: Ideas, illustrations and the CASTEP code. *J. Phys. Condens. Mat.* **2002**, *14*, 2717–2744. [CrossRef]
26. Delley, B. From molecules to solids with the DMol3 approach. *J. Chem. Phys.* **2000**, *113*, 7756–7764. [CrossRef]
27. Delley, B. An all-electron numerical method for solving the local density functional for polyatomic molecules. *J. Chem. Phys.* **1990**, *92*, 508–517. [CrossRef]
28. Monkhorst, H.J.; Pack, J.D. Special points for Brillouin-zone integrations. *Phys. Rev. B* **1976**, *13*, 5188–5192. [CrossRef]
29. Pack, J.D.; Monkhorst, H.J. Special points for Brillouin-zone integrations-a reply. *Phys. Rev. B* **1977**, *16*, 1748–1749. [CrossRef]
30. Vanderbilt, D. Soft self-consistent pseudopotentials in a generalized eigenvalue formalism. *Phys. Rev. B Condens. Matter.* **1990**, *41*, 7892–7895. [CrossRef] [PubMed]
31. Zhu, Y.M.; Luo, B.B.; Sun, C.Y.; Liu, J.; Sun, H.T.; Li, Y.J.; Han, Y.X. Density functional theory study of alpha-Bromolauric acid adsorption on the alpha-quartz (101) surface. *Miner. Eng.* **2016**, *92*, 72–77. [CrossRef]
32. Sauer, J.; Sustmann, R. Mechanistic Aspects of Diels-Alder Reactions: A Critical Survey. *Angew. Chem. Int. Ed.* **1980**, *19*, 779–807. [CrossRef]
33. Perdew, J.P.; Chevary, J.A.; Vosko, S.H.; Jackson, K.A.; Pederson, M.R.; Singh, D.J.; Fiolhais, C. Atoms, molecules, solids, and surfaces: Applications of the generalized gradient approximation for exchange and correlation. *Phys. Rev. B* **1992**, *46*, 6671–6687. [CrossRef]
34. Hendricks, S.B.; Jefferson, M.E.; Mosley, V.M. The Crystal Structures of some Natural and Synthetic Apatite-Like Substances. *Zeitschrift für Kristallographie-Crystalline Materials* **1932**, *81*, 352–369. [CrossRef]
35. Deer, W.A.; Howie, R.A.; Zussman, J. *An Introduction to the Rock-Forming Minerals*; Longman Group: Harlow, UK, 1992.
36. Zhu, H.; Qin, W.; Chen, C.; Liu, R. Interactions Between Sodium Oleate and Polyoxyethylene Ether and the Application in the Low-Temperature Flotation of Scheelite at 283 K. *J. Surfactants Deterg.* **2016**, *19*, 1289–1295. [CrossRef]

Article

New Insights into the Configurations of Lead(II)-Benzohydroxamic Acid Coordination Compounds in Aqueous Solution: A Combined Experimental and Computational Study

Jianyong He [1,†] , Haisheng Han [1,†] , Chenyang Zhang [1,2,*] , Yuehua Hu [1] , Dandan Yuan [3] , Mengjie Tian [1] , Daixiong Chen [2,*] and Wei Sun [1,*]

[1] Key Laboratory of Hunan Province for Clean and Efficient Utilization of Strategic Calcium-Containing Mineral Resources, School of Minerals Processing and Bioengineering, Central South University, Changsha 410083, Hunan, China; hjy2016@csu.edu.cn (J.H.); hanhai5086@csu.edu.cn (H.H.); hyh@csu.edu.cn (Y.H.); 155606018@csu.edu.cn (M.T.)

[2] Key Laboratory of Hunan Province for Comprehensive Utilization of Complex Copper-Lead Zinc Associated Metal Resources, Hunan Research Institute for Nonferrous Metals, Changsha 410100, Hunan, China

[3] Institute of Theoretical and Computational Chemistry, Nanjing University, Nanjing 210023, Jiangsu, China; yuandandan05@gmail.com

* Correspondence: zhangchenyang@csu.edu.cn (C.Z.); cdxcsu@gmail.com (D.C.); sunmenghu@csu.edu.cn (W.S.); Tel.: +86-731-8883-0482 (C.Z. & W.S.); +86-731-8523-9033 (D.C.)

† These authors have equally contributed to this work.

Received: 22 July 2018; Accepted: 19 August 2018; Published: 25 August 2018

Abstract: Novel collector lead(II)-benzohydroxamic acid (Pb(II)–BHA) complexes in aqueous solution were characterized by using experimental approaches, including Ultraviolet-visible (UV-Vis) spectroscopy and electrospray ionization-mass spectrometry (ESI-MS), as well as first-principle density functional theory (DFT) calculations with consideration for solvation effects. The Job plot delineated that a single coordinated $Pb(BHA)^+$ should be formed first, and that the higher coordination number complexes can be formed subsequently. Moreover, the Pb(II)–BHA species can aggregate with each other to form complicated structures, such as $Pb(BHA)_2$ or highly complicated complexes. ESI-MS results validated the existence of $Pb\text{-}(BHA)_{n=1,2}$ under different solution pH values. Further, the first-principles calculations suggested that $Pb(BHA)^+$ should be the most stable structure, and the Pb atom in $Pb(BHA)^+$ will act as an active site to attack nucleophiles. These findings are meaningful to further illustrate the adsorption mechanism of Pb(II)–BHA complexes, and are helpful for developing new reagents in mineral processing.

Keywords: Pb(II)–BHA; lead chemistry; metal–organic collectors; DFT calculation; surface activation

1. Introduction

Metal–organic coordination complexes have been widely used in the materials, chemistry. Recently, their promising applications in mining have attracted research attention [1–8]. For instance, lead(II)-benzohydroxamic acid (Pb(II)–BHA) complexes are effective collectors in the beneficiation of oxide minerals, including tungsten minerals, cassiterite, and rutile [9,10]. Given the excellent selectivity and good collecting ability of Pb(II)–BHA, the scheelite flotation process could be simplified remarkably without the addition of sodium silicate, overcoming the shortage of heating in the routine flotation of scheelite [9–11]. The beneficiation of scheelite is currently one of the most challenging problems worldwide in the field of mineral processing. Conventionally, separating scheelite from calcium bearing minerals, such as fluorite and calcite, by using anionic collectors (especially for fatty acid) is difficult due to their highly similar properties in calcium-bearing surfaces [12,13]. Fortunately, our group first introduced the Pb complexes of benzohydroxamic acid (Pb(II)–BHA complexes) to effectively separate scheelite from calcium-bearing gangues by properly regulating the Pb/BHA ratio and pH [9,11]. Furthermore, as displayed in Figure 1a, the novel flotation scheme (Scheme i in Figure 1b), with Pb(II)–BHA complexes used as the collector, has shown better performance than the traditional activation flotation scheme (Scheme ii in Figure 1b), which first added Pb(II) as the activator and subsequently added BHA as the collector [10,14–23]. The effective microstructures of Pb(II)–BHA complexes in the solution, however, as well as their interaction mechanisms with oxide minerals remain unclear. Considering the using of Pb(II) ions may result in environmental/public health issues, a further understanding of this activating mechanism can be helpful for exploring the alternative reagents.

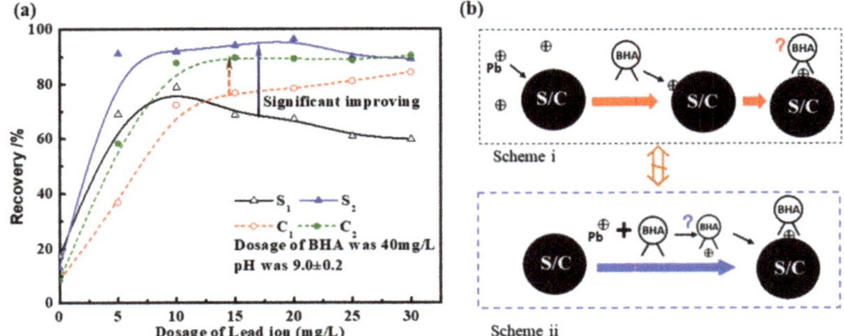

Figure 1. (a) The flotation recovery of scheelite (S) [9] and cassiterite (C) [24] as a function of dosage of Pb(II) ion ($_{1/2}$ in (a) represents the flotation scheme in i and ii in (b)); (b) two flotation schemes.

The hydroxamic acid group (–CO–NH–OH) is the functional group of the BHA [25–27]. The functional group has different properties that remain poorly characterized; in fact, a reliable assignment of the correct structure is challenging because the several possible conformations strongly depend on concentration, temperature, and the nature of the solvent [28]. The hydroxamate collectors, such as benzohydroxamic acid (BHA), naphthenic hydroxamate, and amide hydroxamate, have been useful as highly selective flotation collectors in recent years. The role to function as collectors in mineral flotation has been documented by Lynch et al. [29]. The chelate between hydroxamic acids and the metal ion, however, has remained poorly investigated. Currently, first-principle calculations based on density functional theory (DFT) and some advanced experimental technologies are used to obtain more information on the molecular structure of BHA [30]. Wander et al.'s benchmarking calculations indicate that the DFT calculation can achieve near chemical accuracy of hydrolysis constants for metal ions in most cases [31]. Nuclear magnetic resonance and DFT calculations performed by Garcia et al. show that the adopted BHA conformation of BHA aqueous solution is a closed Z (cis) configuration in aqueous solution [32]. Both Z (cis) and E (trans) conformations in Scheme 1 regularly co-exist in solvent. The concentration and environmental factors determine the ratio of Z type to E type conformations to some extent, and potential barriers are present among different conformations [33,34].

Scheme 1. BHA conformers.

BHA can chelate with metal ions, such as copper(II) cadmium(II), cobalt(II), nickel(II), manganese(II), lead(II), zinc(II), aluminum(III), iron(III), and bismuth(III), thus taking on diverse chelate structures [26,35]. Very few heavy metal ion complexes of BHA chelates have been extensively investigated for their special uses, such as the bismuth(III) complex that has activity against *Helicobacter pylori*; however, reports about the novel Pb(II)–BHA complexes are also few. Understanding the microstructures of Pb(II)–BHA complexes is essential to improve the technique working in flotation practice and fundamental field of lead chemistry [36–39].

This current study aims to investigate configurations of Pb(II)–BHA coordination complexes in aqueous solution. Accordingly, UV-Vis spectroscopy and electrospray ionization mass spectrometry (ESI-MS) were performed to characterize the configuration of Pb(II)–BHA complexes. Furthermore, first-principle DFT calculations were performed to understand the constituents and properties of Pb(II)–BHA complexes at the molecular level. The frontier molecular orbital [40] and natural atomic orbitals (NAOs) [41] were used to describe the reactivity of the studied Pb(II)–BHA complex. This work has shed new light on effective microstructures of Pb(II)–BHA coordination complexes for mineral flotation.

2. Methodology

2.1. Experimental Details

2.1.1. Materials

Analytical grade BHA (>98%) was purchased from Tokyo Chemical Industry Co., Ltd. in Japan. Analytical grade lead nitrate (>99%) was used. pH regulators were prepared with the stock solution of sodium hydroxide (>96%) and hydrochloric acid (36~38%). All purity is in mass percent. The 18.2 MΩ pure water produced by Arium Mini Plus (Sartorius Weighing Technology, Goettingen, Germany) was used in this work.

2.1.2. UV-Vis Tests

The UV-Vis spectrum of the Pb(II)-hydroxide system has been investigated elsewhere [42]. Here we will not assign the characteristic peaks of the Pb(II)–BHA complex. The Job plot [43], proposed by Job, provides qualitative and quantitative insight into the stoichiometry with the underlying association of ligand- and solvent-dependent reaction rates. The Job plot was used to track the full reaction path. Moreover, we changed the guest and host solution to comprehensively understand the stoichiometry of the product. In this work, the concentration of the Pb(II)–BHA complex, as determined by integration of the intensity of specific wavenumber, was plotted against the mole fraction X_A (the guest solution is BHA solution) and X_B (the guest solution is Pb(II) ion solution). A Shimadzu UV2600/2700 Ultraviolet spectrophotometer was used to obtain the UV-Vis spectra at a fixed concentration of Pb(II) ion at 0.1 mM using the equimolar continuous change method and molar ratio method.

2.1.3. ESI-MS Tests

ESI-MS patterns were collected in positive ion mode with a Bruker Q-TOF Qualification Standard Kit, using solutions of the 0.1 mM/L mixture of lead nitrate and BHA. ESI-MS patterns were used to obtain the proof of coordinated compounds of Pb(II)–BHA complexes at the molar ratio of 1:1 (*v:v*, at the natural pH 4.4) and 1:2 (*v:v*, at pH 13.0). The natural pH was selected for exclusion of guest species so that we can get a relatively simple Pb–BHA structure in aqueous solution. A high pH of 13.0 was selected in according to a report that high pH can result in more complicated solution species [10]. The two case studies, regarded as representatives of both acid and alkaline solutions, are supposed to provide evidence of Pb–BHA complexes.

2.2. Computational Methods

All calculations were performed with the Gaussian 09 (version D.01) quantum chemistry package, based on the B3LYP method: A three-parameter hybrid functional by replacing a certain amount of the PW91 generalized gradient approximation (GGA) correlation functional with the LYP GGA correlation functional [44–46]. The implicit solvation effects were considered using the polarized continuum models (PCM) in the calculation [47]. On the other hand, the water molecules in the first hydration of Pb(II)–BHA complex were also considered with the explicit solvation model. The aug-cc-pVDZ basis set was employed as all-electron basis set in all types of calculation on the light atoms H, C, N, and O, except for the Pb atom in Pb(II)–BHA complex systems. The aug-cc-pVDZ-PP basis set with a relativistic pseudopotential was used for the Pb atom. The basis set, obtained from EMSL Basis Set Exchange web site, has already been verified as producing acceptable thermodynamic information of hydrated Pb(II) ions [30,48,49]. The Los Alamos effective core potential double-ξ (LanL2Dz) was used for the primary geometry optimization of modeled benzohydroxamic acid and its Pb(II) complex in ionic form. LanL2Dz replaced 78 core electrons with relativistic effective core potential (RECP) [50]; therefore, only two valence electrons of Pb(II) ions were described [51]. To refine the structure and calculate the molecular orbitals, we further used the larger aug-cc-pVDZ basis set for the light atoms, such as hydrogen, oxygen, nitrogen, and carbon, and we used the aug-cc-pVDZ–PP with RECP of the inner 60 electrons for Pb(II) ions to calculate the frequency [52]. The default convergence parameters (with maximum force within 4.5×10^{-4}, force RMS within 1.8×10^{-3}) in the Gaussian 09 software were retained to optimize the structure. All calculations were successfully converged, without virtual frequencies in the vibration analysis.

There are three protonation sites in the BHA molecule. The carbonyl oxygen site is found to be the preferred site for protonation by Arora et al. [53] in aqueous solution. In the building of BHA anion models [54], the proton at the carbonyl oxygen was removed according to the preferred deprotonation site reported by Begoña et al. [32]. The Pb(II) ion was set as the metal center ion which would be the coordination center of the bidentate BHA anions. The envisaged conformations were ligated by two, three and four BHA anions. Thermodynamic values for Gibbs free energy [55] were obtained using the PCM with the context:

$$\Delta G_r = \sum G_{prod} - \sum G_{react} \tag{1}$$

where G_{prod} and G_{react} are the free energy or the products and the reactants included in the reaction, respectively.

The Gauss View was used as a visualization tool in this paper. In addition, all calculations including the mapped molecular orbitals in this work were performed at the theory level of PCM-B3LYP/aug-cc-pVDZ on light atoms (C, H, O, and N) and PCM -B3LYP/aug-cc-pVDZ-PP on Pb atoms [56]. Molecular orbital contours for the highest occupied molecular orbital (HOMO) and the lowest unoccupied molecular orbital (LUMO) of the cluster model were computed at the same theoretical level. The contributions of the Pb atom to the frontier molecular orbitals were calculated with a multifunctional wave function analyzer Multiwfn [57] based on the NAO method.

3. Results and Discussion

3.1. Experimental

3.1.1. UV-Vis Results

Because of the superposition of UV-Vis absorbance peaks of the products and the reactants, carefully processing the collected UV-Vis data is essential. Initially, the first and the last points were fixed at zero absorbance because the complexity of the product might be different when one component is in excess, and the further fitting procedures would exclude the two points [58]. The starting second point and the last second point were connected to form a background. After deducing the background absorbance, the Job plots were plotted, as shown in Figure 2b,d. Here, Job plots were fitted using the method of initial tangents using experimental data close to the beginning points (the first and last points were not included due to the formation of hydrated Pb complexes and the solvation of BHA) [59]. As illustrated in Figure 2, X_A and X_B were the mole fractions of Pb(II) ions and BHA according to dosing method 1 and 2, respectively. The wavelength at 230 nm represents the significant changes in the solution components. The same trends can also be obtained by using other wavelengths between 230 nm and 240 nm (Figure 3).

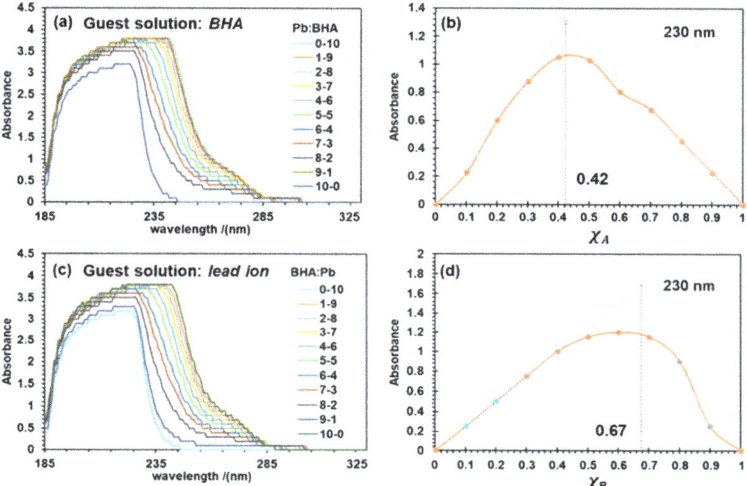

Figure 2. (**a**,**c**) are the UV-Vis absorbance spectra spanning the wavelength from 185 nm to 325 nm for the different dosing strategies; (**b**,**d**) are the corresponding Job plots at 230 nm of (**a**,**c**), respectively. X_A and X_B are the mole fractions of Pb(II) ions and BHA according to dosing strategies 1 and 2, respectively. All stock solutions are prepared at 1 mM/L concentration.

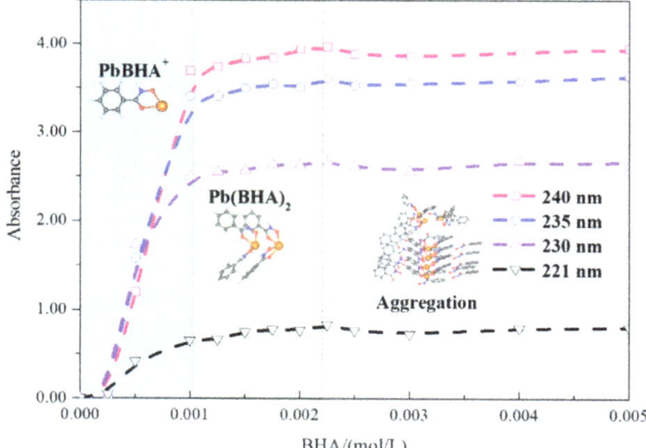

Figure 3. UV-Vis absorbance changes of 1 mM/L Pb(II) ion solution with respect to the continuous addition of BHA (at wavelengths of 221, 230, 235, and 240 nm), based on the continuous concentration change method. Spheres for atoms Pb, C, H, O, and N in the inset are colored in orange, grey, white, red, and blue, respectively.

Figure 2a,c shows that the increasing dosage of Pb(II) ions and BHA has strengthened and extended the absorbance peaks. Job plots have been obtained from the newly mixed Pb(II)–BHA mixture, with the collected characteristic absorbance peaks of UV-Vis spectra in aqueous solution having both dosing methods. The absorbance peak at 230 nm has been plotted with respect to the range of the molar ratios of Pb(II):BHA from 1:9 to 9:1. The obtained Job plots show that the stoichiometry of the complexes for the first one with BHA as a guest solution is $X_A = 0.42$ (Figure 2b), which supports a stoichiometry of Pb(II):BHA between 1:1 to 1:2. Meanwhile, the reversing dosing method with the Pb(II) ion solution as the guest solution obtained $X_B = 0.67$ (Figure 2d), clearly indicating that a stoichiometry of 1:2 corresponds to the structure of $Pb(BHA)_2$. These results support the formation of $Pb(BHA)_2$ complex when excessive BHA solution is added.

At the natural pH value of 4.4, the main components of Pb^{2+} solution are the Pb^{2+} cations [42]; as for the BHA solution, the mean components are BHA^- anions [60]. Hence, both Pb^{2+} and BHA^- should be the main reactants involved in the coordination reaction. In addition, the stability constants obtained in our previous work [10] for the Pb–BHA complexes are 9.14 ± 0.05 $((Pb)(BHA)^+)$ and 12.63 ± 0.01 $((Pb)(BHA)_2)$. Apparently, these experimental results support the existence of $Pb(BHA)^+$ and $Pb(BHA)_2$. Considering the order that we prepared the mixture for the UV-Vis tests can influence the reaction paths, the accepted potential reaction mechanism are described as follows:

$$BHA^- + Pb^{2+} \rightarrow Pb(BHA)^+ \xrightarrow{BHA^-} Pb(BHA)_2 \tag{2}$$

$$2BHA^- + Pb^{2+} \rightarrow Pb(BHA)_2 \tag{3}$$

where Equation (2) can be explained as the stepwise formation of the single coordinated $Pb(BHA)^+$ complex, when a small quantity of BHA^- solution is used as a guest specie, and Equation (3) is ideal for interpreting the formation of $Pb(BHA)_2$, when Pb(II) solution is used as a guest specie. The water molecules and other ions are excluded to better identify the highest coordination number of BHA with Pb(II) as central metal ion.

Figure 3 shows the continued UV-Vis test results at the characteristic peaks of 221, 230, 235, and 240 nm that can explain the mechanism of forming high coordination complexes. These drawn curves show a consistent trend. For a clear illustration, we have divided these curves into three stages. At the beginning stage, the absorbance increases rapidly. At the end of the rising stage at 1 mM/L of BHA, a 1:1 (*v:v*) Pb(II)–BHA complex is formed, which is consistent with both the obtained result of the $Pb(BHA)^+$ complex in the Job plots and the computational section. Afterward, the curves show a slow rising stage region with a small slope from 1 mM/L to 2.5 mM/L. The fluctuation of these collected data suggests that the component in the solution should be intricate. Interestingly, these trends end with the same proportion of Pb:BHA = 1:2.5, at which the $Pb(BHA)_2$ complex can exist as indicated by the Job plots. When increasing the BHA concentration, these curves adopt a horizontal line-like region of the similar absorbance intensity to pure BHA solution.

The three stages in the curve can be seen as the stepwise formation of high coordination complexes, as interpreted by Equation (2). Note that the higher coordination components with more than 2 BHA as ligands are not sufficiently supported by the Job plots. Better experimental evidence has been collected from the electrospray ionization-mass spectrometry (ESI-MS) pattern.

3.1.2. ESI-MS Results

To find out possible Pb(II)–BHA complexes formed in the solution, the collected ESI-MS data of reaction mixtures of lead nitrate and BHA with water as the solvent at the positive mode are shown in Figure 4. Figure 4a shows that when the solution pH is approximately 13.0 in the mixture, the main product is the single coordination complex $Pb(BHA)^+$ with m/z values of 343. Figure 4b shows that Pb(II)–BHA complex at the molar ratio of 1:2 can produce m/z values of 362 and 497 responding to the $(Pb(H_2O)BHA)^+$ and $(Pb(OH)BHA_2)^+$ in aqueous solution, respectively. The later $(Pb(OH)BHA_2)^+$ configuration is quite abnormal, in which lead ion possesses a positive tetravalence. The reason for this is still unclear. However, this does not affect the following discussions and conclusions. The inset in Figure 4b shows the predicated spectra of $(Pb(OH)BHA_2)^+$ is in good agreement with the experimental one, further suggesting the existence of the double coordinated complexes. Proof of the single coordinated Pb(II) complex and double coordinated complex with BHA as ligands is provided, indicating that the $Pb(BHA)^+$, $Pb(BHA)_2$, and the possible high coordination complexes are closely related to the pH value. This finding is consistent with the result obtained from the continuous concentration change method. Meanwhile, these findings of the hydrated Pb(II) of $(PbOH(H_2O)_6)^+$ and the Pb(II)–BHA complex, as shown in Figure 4b, suggested that the hydration shell of Pb(II) may be destroyed due to its coordination reaction with BHA.

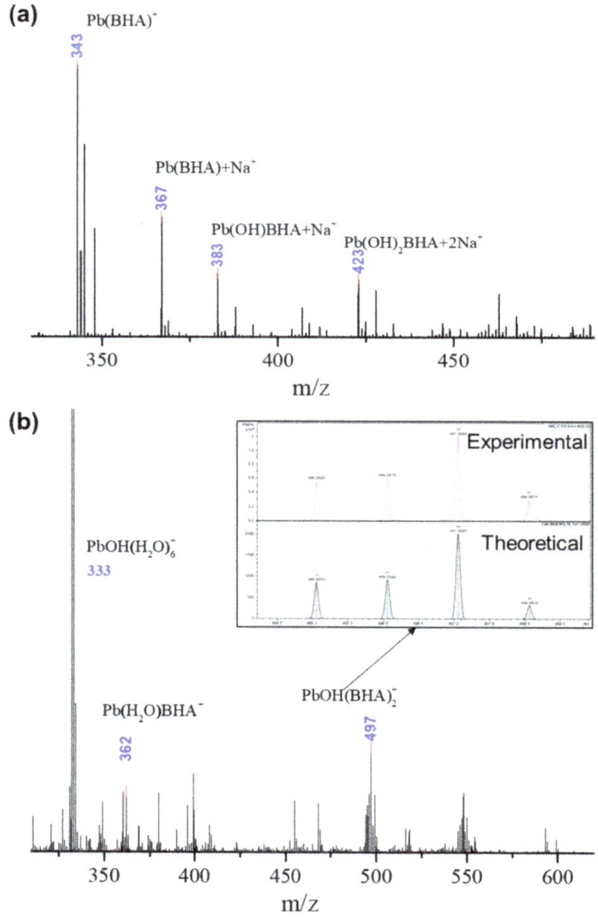

Figure 4. ESI-MS results of the mixture of (**a**) Pb:BHA molar ratio of 1:1 at a pH value of 13.0 and (**b**) Pb:BHA molar ratio of 1:2 at the natural pH of 4.4. The inset in (**b**) is the comparative results of the theoretical spectra and the experimental spectra. (Solution concentrations are 1 mM/L).

The ESI-MS results have validated the existence of the Pb(II) complex with one and two BHA as ligands. The experimental approaches are not sufficient, however, to provide exact structures or properties at the microscopic scale; thus, the structures of Pb(II)–BHA complexes, have been further explored by using first-principle DFT calculations at high accuracy, including the solvation effects.

3.2. Theoretical Prediction

3.2.1. Prediction of Stable BHA Isomers and Pb(II)–BHA Complexes

All optimized BHA and BHA$^-$ (the anion of the BHA molecule with the dissociation of the proton) structures and their complexes with the Pb(II) ion are shown in the first two rows in Figure 5a at a B3LYP/aug-cc-pVDZ theoretical level. These optimized structures are divided into four categories called Ei, Zi, Ea, and Za. From the calculation results as shown in Figure 5b, it is clear that the corresponding Za type is most stable configuration for BHA, BHA$^-$, Pb(BHA)$^{2+}$, and Pb(BHA)$^+$. The interesting finding is that BHA prefers to coordinate with Pb(II) to form a "Pb–O–C–N–O" five-membered ring, not a "Pb–O–C–N" four-membered ring, which is consistent with the previous report [26].

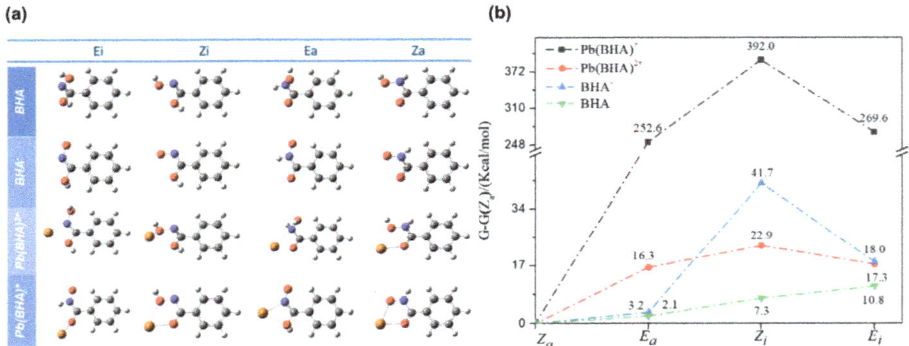

Figure 5. (**a**) Optimized structures of BHA, BHA$^-$, Pb(BHA)$^{2+}$, and Pb(BHA)$^+$. (**b**) Comparison of Gibbs free energy of BHA, BHA$^-$, Pb(BHA)$^{2+}$, and Pb(BHA)$^+$ isomers with the Gibbs free energy of Za type as zero point. Spheres for Pb, C, H, O, and N atoms are colored in orange, gray, white, red, and blue, respectively.

Because Za-type BHA is the most stable structure, it has been adopted in the following calculations and discussions. To better understand the interaction of Pb–BHA with water molecules, the influence of water molecules on the studied system has been further investigated using the explicit water molecules.

3.2.2. Single Pb–BHA in Aqueous Solution

The hydration of lead-BHA complex with different coordination water molecules has been investigated to consider the real aqueous environment. Figure 6a shows that water molecules were added one by one to coordinate with the Pb(BHA)$^+$ complex. This procedure can be described by Equations (4) and (5) below:

$$Pb(H_2O)_m^{2+} + BHA^- \leftrightarrow Pb(H_2O)_m(BHA)^+ \tag{4}$$

$$Pb(BHA)^+(H_2O)_m + H_2O \leftrightarrow Pb(BHA)^+H_2O_{(m+1)} \tag{5}$$

where m(0→4). Here, the hydrated lead(II) cation is a six folded complex in the first hydration shell, according to the previous report [51].

As shown in Figure 6a, the electronical density distribution of LUMO (bottom (a)) shifted from the lead ion to the BHA molecule, indicating the active site of the lead-BHA complex can be influenced by the coordinated water molecules. The increasing number of water molecules from 0 to 3 results in small differences in the electronical density distribution of LUMO and HOMO, which suggests the active site of Pb(BHA)$^+$ is stable. A fairly dispersed distribution of LUMO and HOMO on the whole complex has been observed, after adding more than 3 water molecules, which should be due to the influence of the surrounding water molecules.

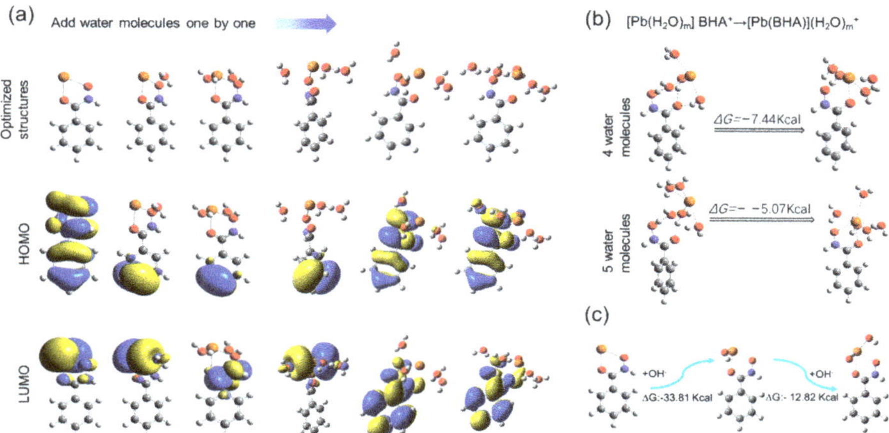

Figure 6. The optimized structure (top (**a**)) of hydrated Pb(BHA)$^+$ with the addition of water molecules one by one and the corresponding highest occupied molecular orbital (HOMO) (middle (**a**)) and lowest unoccupied molecular orbital (LUMO) (bottom (**a**)). (**b**) Gibbs free energy difference between the hydrated Pb(BHA)$^+$ and hydrated Pb^{2+} +BHA complex with four and five water molecules. (**c**) The hydroxylation of Pb(BHA)$^+$. The contours have been computed at the PCM-B3LYP level of theory with a threshold of 0.001 au, and spheres for Pb, C, H, O, and N atoms are colored in orange, gray, white, red, and blue, respectively.

Figure 6b shows the optimized structures of a BHA reacted with a hydrated lead(II) ion (left of Figure 6b, denoted as (Pb(H$_2$O)$_m$)BHA)$^+$ and the hydration structure of Pb(BHA)$^+$ (right of Figure 6b, denoted as (Pb–BHA)(H$_2$O)$_m$)$^+$ with four (m = 4) and five (m = 5) water molecules, respectively. The Gibbs free energy differences between ((Pb(H$_2$O)$_m$)BHA)$^+$ and ((Pb–BHA)(H$_2$O)$_m$)$^+$ are −7.44 kcal/mol and −5.07 kcal/mol with respect to four (m = 4) and five (m = 5) water molecules, respectively. These calculated results show that ((Pb–BHA)(H$_2$O)$_m$)$^+$ is more favorable in thermodynamics.

As is tabulated in Table 1, all the Gibbs free energy differences of the coordination reactions of (Pb–BHA)$^+$(H$_2$O)$_m$ (m > 0) with another water molecule shown are positive, although all the Gibbs free energy changes of the coordination reactions of Pb^{2+} with water molecule are negative. This indicates that Pb^{2+} prefers to coordinate with more than five water molecules, but (Pb–BHA)$^-$ will not like to coordinate with more than two water molecules (ΔG > 10 kcal/mol for the third water molecule), and in the hydration configurations of ((Pb–BHA)(H$_2$O)$_m$)$^+$, (Pb–BHA)$^-$ and (Pb–BHA)$^-$H$_2$O should be the dominate species. These findings are consistent with the ESI-MS results.

Table 1. The changes in the reaction Gibbs free energy (ΔG) and the difference of Pb–O mean distance in the hydration system with coordination number (CN) m of water molecules varied from 1 to 5.

CN of Water Molecules (m)	ΔG/(Kcal/mol)		Pb–O Mean Distance (Å)	
	$(Pb–BHA)^+(H_2O)_m$	$Pb(H_2O)_m^{2+}$	$(Pb–BHA)^+(H_2O)_m$	$Pb(H_2O)_m^{2+}$
1	1.41	-50.48	2.53	2.34
2	1.38	-36.75	2.64	2.38
3	7.40	-27.34	2.69	2.42
4	4.64	-17.65	2.77	2.49
5	8.35	-14.56	2.83	2.55

On the other hand, with the addition of the coordinated water molecules, the average bond length of Pb–O in both hydration systems of $(Pb–BHA)^+(H_2O)_m$ and $Pb(H_2O)_m^{2+}$ increased. The bond length of Pb–O in $(Pb–BHA)^+H_2O$ is 2.53 Å, which is similar to the mean bond length of Pb–O in $Pb(H_2O)_5$ (2.55 Å). In structure chemistry, the basic idea is that the longer bond length, the less stable the corresponding complex. Thus, the changes of mean bond length of Pb–O in $(Pb–BHA)^+(H_2O)_m$ and $Pb(H_2O)_m^{2+}$ also confirmed that $(Pb–BHA)^+(H_2O)_m$ are less stable than $(Pb–BHA)^+$ and $Pb(H_2O)_m^{2+}$.

In addition, the scheelite flotation [16] are usually performed at the alkaline pH, hydroxyl ion (OH^-) are a key component in the pulp. Thus, the coordination reactions of $Pb(BHA)^+$ with OH^- were also investigated in this work. Herein, the coordination of the $Pb(BHA)^+$ with OH^- has been simulated in a stepwise order, as shown in Figure 6c. The reaction site has been chosen according to the LUMO of $Pb(BHA)^+$. The optimized structures of $Pb(BHA)(OH)$ and $(Pb(BHA)(OH)_2)^-$ and the reaction Gibbs free energies changes (ΔG) are also given in Figure 6c. The hydroxyl binding with $Pb(BHA)^+$ and $PbOH(BHA)$ respond with binding energies of -33.81 Kcal/mol and -12.82 Kcal/mol, respectively. These results suggest that the coordination reaction of hydroxyl with the $Pb(BHA)^+$ are thermodynamically favorable. At a natural pH of 4.4, both Pb^{2+} and BHA^- should be the major reactants involved in the coordination reaction. At a real flotation plant, however, a higher pH value close to 9 is used. At such pH or a higher pH, both $Pb(BHA)(OH)$ and $(Pb(BHA)(OH)_2)^-$ could exist in the flotation solution. These results are in good accordance with experimental m/z value of 383 and 423 in ESI-MS patterns, which refer to $(Pb(BHA)(OH)+Na)^+$ and the $(Pb(BHA)(OH)_2 + 2Na)^+$ species in Figure 4, respectively.

Consistently positive ΔG of the coordination reactions between $(Pb–BHA)^+(H_2O)m$ (m > 1) and another water molecule should be ascribed to the result of the stronger coordination of BHA^- with Pb^{2+} (ΔG is -356.55 kcal/mol as listed in Table 1). The coordination involving charge transfer from the ligand BHA^- to the metal ion Pb^{2+} makes the properties and electronic structure of Pb–BHA less active than the independent lead(II) ion. The results agree well with the transformation in Figure 6b, where the bonded BHA is strong enough to protect the Pb^{2+} from the influence of surrounding water molecules.

Based on these results and discussions (the BHA^- can coordinate with the Pb(II) smoothly), the following work has been conducted under the implicit model rather than explicit water molecules.

3.2.3. High Coordination Complexes

In thermodynamics, the isomer with the lowest Gibbs free energy (G) could be the most stable and most efficient isomer; Za type structures should be the optimal configurations and the dominant components according to the Gibbs free energies shown in Figure 5b, which is consistent with the results obtained by Begoña et al. [32]. The Za type structure as the ligand has been used for the subsequent calculations. The initial Pb(II)–BHA coordination complexes are modeled with the Pb(II) ion as the central metal ion, based on the stepwise mechanism in Equation (2). To obtain the possible high coordination compounds, we increased the coordination number of BHA ligand to 4.

As shown in Figure 7, all Pb(II)–BHA coordination complexes possess hemidirected geometry (in such configuration, BHA ligands occupy merely half of the space surrounding the Pb(II) atom) [7,51,61]. As shown in Table 2, the reaction between the BHA with the Pb(II) (Figure 7a) corresponds to a total reaction free energy change of −356.55 Kcal/mol, indicating the Pb–BHA$^-$ is a fairly favorable specie in thermodynamics. For the second BHA$^-$ ligand (Figure 7b), the total reaction free energy change is −27.20 Kcal/mol, excluding the reaction free energy of Pb(BHA)$^+$. The reaction energy (−356.55 Kcal/mol) of BHA$^-$ with Pb^{2+} is much higher than that of water molecules with Pb (−50.48 Kcal/mol), indicating the coordination reaction of BHA$^-$ with Pb^{2+} will be more efficient than that of H$_2$O with Pb^{2+}. Thereafter, the coordination reaction of the third BHA$^-$ produces a positive change in Gibbs free energy of 0.03 Kcal/mol, implying that the binding of the third water molecule with the former Pb(BHA)$_2$ is not favorable in thermodynamics. Moreover, as is shown in Figure 7c, when CN reaches 4, the intramolecular aggregation occurred due to a hydrogen bonding interaction (The highlighted N–H...O bond in Figure 7c) between the added BHA$^-$ and another adjoining BHA$^-$, with the hydrogen bond (N–H...O) length of 1.84 Å [26,62] and the N–H...O angle of 153.9°. A higher CN than 3 should not be stable, but a Pb(II)–BHA complex with three or more BHA ligands can appear due to the intermolecular interactions, including the Van der Waals' force, H-bonding interaction.

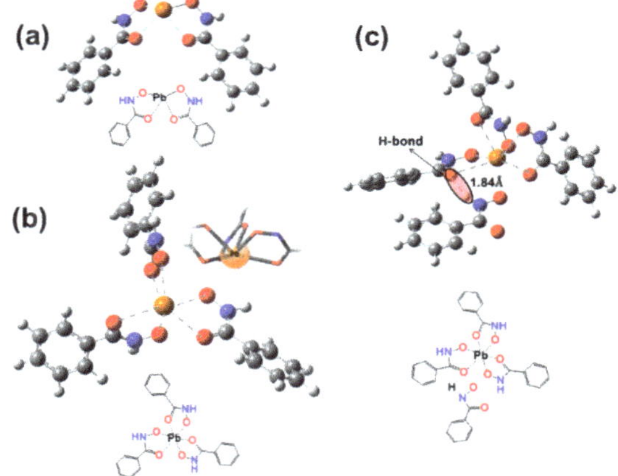

Figure 7. (a–c) Optimized structures of Pb(II)–BHA with coordination numbers (CN) of Figures 2a, 3b and 4c. Spheres for Pb, C, H, O, and N atoms are colored in orange, gray, white, red, and blue, respectively.

Table 2. The changes in Gibbs free energy (ΔG) and Pb–O mean distance in Pb(II)–BHA complexes with CN of BHA ligands ranging from 1 to 3.

CN	Reaction	ΔG(Kcal/mol)	Pb–O (Å)	ε_{gap} (eV)
1	Pb^{2+} + BHA$^-$	−356.55	2.26	4.17
2	Pb(BHA)$^+$ + BHA$^-$	−27.20	2.40	4.50
3	Pb(BHA)$_2$ + BHA$^-$	+0.03	2.58	4.17

3.2.4. Frontier Molecular Orbital Analysis

In Figure 8 the ligands of Pb(BHA)$_2$ (b) and (Pb(BHA)$_3$)$^-$ (c) occupied just half of the space surrounding the Pb(II) ions. Pb(BHA)$_2$ and (Pb(BHA)$_3$)$^-$ adopted the hemidirected Pb(II)–BHA structures due to the lone pair electron contributed by BHA ligands [51,63]. Additionally, the structure of BHA can form some dimer structures and even interact with themselves. The adjoining BHA may result in aggregation due to intermolecular interaction [32].

As illustrated in Figure 8a, the LUMO of Pb(BHA)$^+$ is mainly located at the Pb atom indicating that the Pb site of the Pb(BHA)$^+$ should be active in electrophilic reaction. The LUMO of Pb(BHA)$_2$ in Figure 7b and the (Pb(BHA)$_3$)$^-$ in Figure 8c, however, spreads over the whole molecule, suggesting that the molecule is not active in electrophilic reaction. Because the Pb(II)–BHA is used as selective collector of oxide mineral, where the hydrated oxide mineral surface is an electron rich system, the Pb(BHA)$^+$ can be more actively absorb onto the surface than other species in the Pb(II)–BHA solution. Additionally, as is shown in Table 3, the Pb atom contributes 91.93% to the LUMO, a rate much higher than that in other Pb(II)–BHA complexes, based on the NAO [64] method. These results have further verified that the Pb atom in Pb(BHA)$^+$ can be the active site to adsorb onto the oxide minerals surface. This is consistent with the reported results [9,10,24,65].

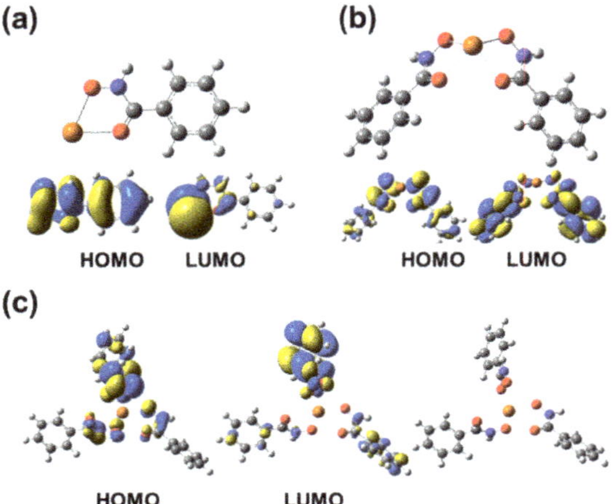

Figure 8. (a–c) The frontier molecular orbitals (HOMO and LUMO) of Pb(II)–BHA complexes with 1 a, 2b, and 3c BHA ligands. The orbitals are generated at the B3LYP/aug-cc-pVDZ theory level with a threshold of 0.001 au, and the solvation effect is included with PCM model. Spheres for Pb, C, H, O, and N atoms are colored in orange, gray, white, red, and blue, respectively.

Table 3. The contribution of the Pb atom to frontier molecular orbitals (HOMO+1, HOMO, LUMO, and LUMO−1) based on the NAO [64] method.

Pb–BHA	Orbital Composition Assigned to Pb atom/%			
	HOMO−1	HOMO	LUMO	LUMO+1
Pb(BHA)$^+$	0.03	3.55	91.93	77.54
Pb(BHA)$_2$	1.14	1.28	1.14	0.02
Pb(BHA)$_3^-$	1.01	0.75	0.13	0.00

4. Conclusions

In the present study, experimentally, Job plots considering two dosing strategies indicate that the Pb coordination compounds in solution may adopt the stoichiometry of Pb(BHA)$^+$ and Pb(BHA)$_2$. UV-Vis results obtained by the equimolar continuous change method show that the Pb(BHA)$^+$ and the high coordination number compounds should be formed stepwise in aqueous solution. Thereafter, the electrospray ionization-mass spectrometry (ESI-MS) results provide strong proofs that the Pb(BHA)$^+$ and (PbOH(BHA)$_2$)$^+$ should be the stable species at the alkaline pH in solution; and the Pb(BHA)$^+$ should be the major component in aqueous solution at the natural pH. Furthermore, first-principles density functional theory (DFT) calculations show that: the Za type BHA should be the major structure; the Pb(BHA)$^+$ can be stable in aqueous solution; and the formation of higher coordination number complexes are not favorable. Finally, the frontier molecular orbitals analyses show that the Pb atom of Pb(BHA)$^+$ is the largest contributor to the LUMO. This suggests that the Pb atom in the structure should be the active site for accepting nucleophile and can also be the active site to adsorb onto the surface of oxide minerals. Both experimental results and theoretical results are in good accordance. These findings are meaningful to further illustrate the adsorption mechanism of Pb(II)–BHA complexes in mineral processing.

Author Contributions: C.Z. and W.S. conceived and designed the calculations; D.C. designed the experiments; C.Z. and J.H. performed the DFT calculations and analyzed computational results; H.H. and J.H. performed the experiments and analyzed experimental results; Y.H. analyzed the experimental data and modified the paper; J.H. and H.H. investigated literatures; D.Y. and M.T. constructed the computational model and drew the figures; C.Z. and J.H. wrote the paper.

Funding: The work was supported by Natural Science Foundation of China (No. 51704330); the Found of State Key Laboratory of Mineral Processing (No. BGRIMM-KJSKL-2017-13); Natural Science Foundation of China (No. 51374247); the Startup Fund of Central South University for Young Teachers (502044001); The Innovation Program for Postgraduate Students of Central South University (No. 2018zzts802); the National 111 Project (No. B14034); Collaborative Innovation Center for Clean and Efficient Utilization of Strategic Metal Mineral Resources; Innovation Driven Plan of Central South University (No. 2015CX005); Key Laboratory of Hunan Province for Clean and Efficient Utilization of Strategic Calcium-containing Mineral Resources (No. 2018TP1002); and the National Science and Technology Support Project of China.

Acknowledgments: This work was carried out in part using hardware and/or software provided by a Tianhe II supercomputer at the National Supercomputing Center in Guangzhou, and the High-Performance Computing Centers of Central South University and Nanjing University. ESI-MS measurements were conducted at the Modern Analysis and Testing Center of CSU.

Conflicts of Interest: The authors declare no conflicts of interest.

References

1. Hou, G.L.; Chen, B.; Transue, W.J.; Yang, Z.; Grutzmacher, H.; Driess, M.; Cummins, C.C.; Borden, W.T.; Wang, X.B. Spectroscopic characterization, computational investigation, and comparisons of ecx- (e = as, p, and n; x = s and o) anions. *J. Am. Chem. Soc.* **2017**, *139*, 8922–8930. [CrossRef] [PubMed]

2. Kuta, J.; Clark, A.E. Trends in aqueous hydration across the 4f period assessed by reliable computational methods. *Inorg. Chem.* **2010**, *49*, 7808–7817. [CrossRef] [PubMed]

3. Aguilo, E.; Moro, A.J.; Gavara, R.; Alfonso, I.; Perez, Y.; Zaccaria, F.; Guerra, C.F.; Malfois, M.; Baucells, C.; Ferrer, M.; et al. Reversible self-assembly of water-soluble gold(i) complexes. *Inorg. Chem.* **2017**, *57*, 1017–1028. [CrossRef] [PubMed]

4. Akturk, E.S.; Scappaticci, S.J.; Seals, R.N.; Marshak, M.P. Bulky beta-diketones enabling new lewis acidic ligand platforms. *Inorg. Chem.* **2017**, *56*, 11466–11469. [CrossRef] [PubMed]

5. Bhatta, S.R.; Mondal, B.; Vijaykumar, G.; Thakur, A. Ict-isomerization-induced turn-on fluorescence probe with a large emission shift for mercury ion: Application in combinational molecular logic. *Inorg. Chem.* **2017**, *56*, 11577–11590. [CrossRef] [PubMed]

6. Choi, K.M.; Kim, D.; Rungtaweevoranit, B.; Trickett, C.A.; Barmanbek, J.T.D.; Alshammari, A.S.; Yang, P.; Yaghi, O.M. Plasmon-enhanced photocatalytic CO_2 conversion within metal–organic frameworks under visible light. *J. Am. Chem. Soc.* **2017**, *139*, 356–362. [CrossRef] [PubMed]

7. Jalilehvand, F.; Sisombath, N.S.; Schell, A.C.; Facey, G.A. Lead(II) complex formation with L-Cysteine in aqueous solution. *Inorg. Chem.* **2015**, *54*, 2160–2170. [CrossRef] [PubMed]

8. Li, J.; Yu, X.; Xu, M.; Liu, W.; Sandraz, E.; Lan, H.; Wang, J.; Cohen, S.M. Metal–organic frameworks as micromotors with tunable engines and brakes. *J. Am. Chem. Soc.* **2017**, *139*, 611–614. [CrossRef] [PubMed]

9. Han, H.; Hu, Y.; Sun, W.; Li, X.; Chen, K.; Zhu, Y.; Nguyen, A.V.; Tian, M.; Wang, L.; Yue, T.; et al. Novel catalysis mechanisms of benzohydroxamic acid adsorption by lead ions and changes in the surface of scheelite particles. *Miner. Eng.* **2018**, *119*, 11–22. [CrossRef]

10. Tian, M.; Hu, Y.; Sun, W.; Liu, R. Study on the mechanism and application of a novel collector-complexes in cassiterite flotation. *Colloid Surf. A Physicochem. Eng. Aspects* **2017**, *522*, 635–641. [CrossRef]

11. Han, H.; Hu, Y.; Sun, W.; Li, X.; Cao, C.; Liu, R.; Yue, T.; Meng, X.; Guo, Y.; Wang, J.; et al. Fatty acid flotation versus bha flotation of tungsten minerals and their performance in flotation practice. *Int. J. Miner. Process.* **2017**, *159*, 22–29. [CrossRef]

12. Gao, Z.; Li, C.; Sun, W.; Hu, Y. Anisotropic surface properties of calcite: A consideration of surface broken bonds. *Colloid Surf. A: Physicochem. Eng. Aspects* **2017**, *520*, 53–61. [CrossRef]

13. Li, C.; Gao, Z. Effect of grinding media on the surface property and flotation behavior of scheelite particles. *Powder Technol.* **2017**, *322*, 386–392. [CrossRef]

14. Sun, L.; Hu, Y.; Sun, W. Effect and mechanism of octanol in cassiterite flotation using benzohydroxamic acid as collector. *Trans. Nonferrous Met. Soc. China* **2016**, *26*, 3253–3257. [CrossRef]

15. Gao, Z.Y.; Sun, W.; Hu, Y.H. New insights into the dodecylamine adsorption on scheelite and calcite: An adsorption model. *Miner. Eng.* **2015**, *79*, 54–61. [CrossRef]

16. Han, H.-S.; Liu, W.-L.; Hu, Y.-H.; Sun, W.; Li, X.-D. A novel flotation scheme: Selective flotation of tungsten minerals from calcium minerals using Pb–BHA complexes in shizhuyuan. *Rare Met.* **2017**, *36*, 533–540. [CrossRef]

17. Hu, Y.; Qiu, G.; Miller, J.D. Hydrodynamic interactions between particles in aggregation and flotation. *Int. J. Miner. Process.* **2003**, *70*, 157–170. [CrossRef]

18. Zhao, G.; Wang, S.; Zhong, H. Study on the activation of scheelite and wolframite by lead nitrate. *Minerals* **2015**, *5*, 247–258. [CrossRef]

19. Filippova, I.V.; Filippov, L.O.; Duverger, A.; Severov, V.V. Synergetic effect of a mixture of anionic and nonionic reagents: Ca mineral contrast separation by flotation at neutral ph. *Miner. Eng.* **2014**, *66*, 135–144. [CrossRef]

20. Deng, L.; Zhao, G.; Zhong, H.; Wang, S.; Liu, G. Investigation on the selectivity of *N*-((hydroxyamino)-alkyl) alkylamide surfactants for scheelite/calcite flotation separation. *J. Ind. Eng. Chem.* **2016**, *33*, 131–141. [CrossRef]

21. Chen, Z.; Ren, Z.; Gao, H.; Lu, J.; Jin, J.; Min, F. The effects of calcium ions on the flotation of sillimanite using dodecylammonium chloride. *Minerals* **2017**, *7*, 28. [CrossRef]

22. Materna, K.L.; Crabtree, R.H.; Brudvig, G.W. Anchoring groups for photocatalytic water oxidation on metal oxide surfaces. *Chem. Soc. Rev.* **2017**, *46*, 6099–6110. [CrossRef] [PubMed]

23. Gao, Y.; Gao, Z.; Sun, W.; Yin, Z.; Wang, J.; Hu, Y. Adsorption of a novel reagent scheme on scheelite and calcite causing an effective flotation separation. *J. Colloid Interf. Sci.* **2018**, *512*, 39–46. [CrossRef] [PubMed]

24. Tian, M.; Gao, Z.; Han, H.; Sun, W.; Hu, Y. Improved flotation separation of cassiterite from calcite using a mixture of lead(II) ion/benzohydroxamic acid as collector and carboxymethyl cellulose as depressant. *Miner. Eng.* **2017**, *113*, 68–70. [CrossRef]

25. Schraml, J. Derivatives of hydroxamic acids. *Appl. Organomet. Chem.* **2000**, *14*, 604–610. [CrossRef]

26. Codd, R. Traversing the coordination chemistry and chemical biology of hydroxamic acids. *Coord. Chem. Rev.* **2008**, *252*, 1387–1408. [CrossRef]

27. Flipo, M.; Charton, J.; Hocine, A.; Dassonneville, S.; Deprez, B.; Deprez-Poulain, R. Hydroxamates: Relationships between structure and plasma stability. *J. Med. Chem.* **2009**, *52*, 6790–6802. [CrossRef] [PubMed]

28. Al-Saadi, A.A. Conformational analysis and vibrational assignments of benzohydroxamic acid and benzohydrazide. *J. Mol. Struct.* **2012**, *1023*, 115–122. [CrossRef]

29. Lynch, A.J.; Watt, J.S.; Fich, J.A.; Harbort, G.J. History of flotation technology. In *Forth Flotation: A Century of Innovation*; SME: Littleton, CO, USA, 2007; pp. 65–91.

30. Lei, X.L.; Pan, B.C. The geometries and proton transfer of hydrated divalent lead ion clusters $[Pb(H_2O)n]^{2+}$ (n = 1–17). *J. Theor. Comput. Chem.* **2012**, *11*, 1149–1164. [CrossRef]

31. Wander, M.C.F.; Rustad, J.R.; Casey, W.H. Influence of explicit hydration waters in calculating the hydrolysis constants for geochemically relevant metals. *J. Phys. Chem. A* **2010**, *114*, 1917–1925. [CrossRef] [PubMed]

32. García, B.; Ibeas, S.; Leal, J.M.; Secco, F.; Venturini, M.; Senent, M.L.; Niño, A.; Muñoz, C. Conformations, protonation sites, and metal complexation of benzohydroxamic acid. A theoretical and experimental study. *Inorg. Chem.* **2005**, *44*, 2908–2919. [CrossRef] [PubMed]

33. Adiguzel, E.; Yilmaz, F.; Emirik, M.; Ozil, M. Synthesis and characterization of two new hydroxamic acids derivatives and their metal complexes. An investigation on the keto/enol, E/Z and hydroxamate/hydroximate forms. *J. Mol. Struct.* **2017**, *1127*, 403–412. [CrossRef]

34. Caudle, M.T.; Crumbliss, A.L. Dissociation kinetics of (*N*-methylacetohydroxamato) iron(III) complexes: A model for probing electronic and structural effects in the dissociation of siderophore complexes. *Inorg. Chem.* **1994**, *33*, 4077–4085. [CrossRef]

35. O'Brien, E.C.; Farkas, E.; Gil, M.J.; Fitzgerald, D.; Castineras, A.; Nolan, K.B. Metal complexes of salicylhydroxamic acid (H_2Sha), anthranilic hydroxamic acid and benzohydroxamic acid. Crystal and molecular structure of $[Cu(phen)_2(Cl)]Cl \times H_2Sha$, a model for a peroxidase-inhibitor complex. *J. Inorg. Biochem.* **2000**, *79*, 47–51. [CrossRef]

36. Bodwin, J.J.; Cutland, A.D.; Malkani, R.G.; Pecoraro, V.L. Cheminform abstract: The development of chiral metallacrowns into anion recognition agents and porous materials. *Coord. Chem. Rev.* **2001**, *216*, 489–512. [CrossRef]

37. Pereira, C.F.; Howarth, A.J.; Vermeulen, N.A.; Almeida Paz, F.A.; Tomé, J.P.C.; Hupp, J.T.; Farha, O.K. Towards hydroxamic acid linked zirconium metal–organic frameworks. *Mater. Chem. Front.* **2017**, *1*, 1194–1199. [CrossRef]

38. He, J.; Han, H.; Zhang, C.; Xu, Z.; Yuan, D.; Chen, P.; Sun, W.; Hu, Y. Novel insights into the surface microstructures of lead(ii) benzohydroxamic on oxide mineral. *Appl. Surf. Sci.* **2018**, *458*, 405–412. [CrossRef]

39. Feng, Q.; Zhao, W.; Wen, S.; Cao, Q. Activation mechanism of lead ions in cassiterite flotation with salicylhydroxamic acid as collector. *Sep. Purif. Technol.* **2017**, *178*, 193–199. [CrossRef]

40. Fukui, K.; Yonezawa, T.; Nagata, C. Theory of substitution in conjugated molecules. *Bull. Chem. Soc. Jpn.* **1954**, *27*, 423–427. [CrossRef]

41. Glendening, E.D.; Landis, C.R.; Weinhold, F. Natural bond orbital methods. *Wiley Interdiscip. Rev. Comput. Mol. Sci.* **2012**, *2*, 1–42. [CrossRef]

42. Perera, W.N.; Hefter, G.A.; Sipos, P.M. An investigation of the lead(II)−hydroxide system. *Inorg. Chem.* **2001**, *40*, 3974–3978. [CrossRef] [PubMed]

43. Renny, J.S.; Tomasevich, L.L.; Tallmadge, E.H.; Collum, D.B. Method of continuous variations: Applications of job plots to the study of molecular associations in organometallic chemistry. *Angew. Chem. Int. Ed.* **2013**, *52*, 11998–12013. [CrossRef] [PubMed]

44. Frisch, M.J.; Trucks, G.W.; Schlegel, H.B.; Frisch, M.J.; Trucks, G.W.; Schlegel, H.B.; Scuseria, G.E.; Robb, M.A.; Cheeseman, J.R.; Scalmani, G.; et al. *Gaussian 09, Revision A. 02*; Gaussian Inc.: Wallingford, CT, USA, 2009.

45. Becke, A.D. Becke's three parameter hybrid method using the lyp correlation functional. *J. Chem. Phys.* **1993**, *98*, 5648–5652. [CrossRef]

46. Yu, H.S.; Li, S.L.; Truhlar, D.G. Perspective: Kohn-sham density functional theory descending a staircase. *J. Chem. Phys.* **2016**, *145*, 130901. [CrossRef] [PubMed]

47. Ho, J.; Ertem, M.Z. Calculating free energy changes in continuum solvation models. *J. Phys. Chem. B* **2016**, *120*, 1319–1329. [CrossRef] [PubMed]

48. Feller, D. The role of databases in support of computational chemistry calculations. *J. Comput. Chem.* **1996**, *17*, 1571–1586. [CrossRef]

49. Malekghassemi, M. An Exploration of Molecular Mechanics and Quantum Chemical Methods. Available online: http://citeseerx.ist.psu.edu/viewdoc/download?doi=10.1.1.567.4830&rep=rep1&type=pdf (accessed on 18 August 2018).

50. Lee, Y.S.; Ermler, W.C.; Pitzer, K.S. Ab initio effective core potentials including relativistic effects. V. SCF calculations with ω–ω coupling including results for Au^{2+}, Tlh, Pbs, and PbSe. *J. Chem. Phys.* **1993**, *73*, 159–165. [CrossRef]

51. Wander, M.C.F.; Clark, A.E. Hydration properties of aqueous Pb(II) ion. *Inorg. Chem.* **2008**, *47*, 8233–8241. [CrossRef] [PubMed]

52. Ochterski, J.W. Vibrational Analysis in Gaussian. Available online: https://www.lindsayengineering.com/img/resources/vib.pdf (accessed on 18 August 2018).

53. Arora, R.; Issar, U.; Kakkar, R. Theoretical study of the molecular structure and intramolecular proton transfer in benzohydroxamic acid. *Comput. Theor. Chem.* **2017**, *1105*, 18–26. [CrossRef]

54. Steinberg, G.M.; Swidler, R. The benzohydroxamate anion. *J. Org. Chem.* **1965**, *30*, 2362–2365. [CrossRef]

55. Ochterski, J.W. Thermochemistry in Gaussian. Available online: http://www.lct.jussieu.fr/manuels/Gaussian03/g_whitepap/thermo/thermo.pdf (accessed on 18 August 2018).

56. Dennington, R.; Keith, T.; Millam, J. Gaussview, Version 5. Available online: https://gaussview.soft112.com (accessed on 18 August 2018).

57. Lu, T.; Chen, F. Multiwfn: A multifunctional wave function analyzer. *J. Comput. Chem.* **2012**, *33*, 580–592. [CrossRef] [PubMed]
58. Rekharsky, M.; Inoue, Y.; Tobey, S.; Metzger, A.; Anslyn, E. Ion-pairing molecular recognition in water: Aggregation at low concentrations that is entropy-driven. *J. Am. Chem. Soc.* **2002**, *124*, 14959–14967. [CrossRef] [PubMed]
59. Bruneau, E.; Lavabre, D.; Levy, G.; Micheau, J.C. Quantitative analysis of continuous-variation plots with a comparison of several methods: Spectrophotometric study of organic and inorganic 1:1 stoichiometry complexes. *J. Chem. Educ.* **1992**, *69*, 833. [CrossRef]
60. Wise, W.M.; Brandt, W.W. Spectrophotometric determination of vandium(v) with benzohydroxamic acid and 1-hexanol. *Anal. Chem.* **1955**, *27*, 1392–1395. [CrossRef]
61. Leon-Pimentel, C.I.; Amaro-Estrada, J.I.; Saint-Martin, H.; Ramirez-Solis, A. Born-Oppenheimer molecular dynamics studies of Pb(II) micro hydrated gas phase clusters. *J. Chem. Phys.* **2017**, *146*, 084307. [CrossRef] [PubMed]
62. Azizi, A.; Ebrahimi, A. The $X^-\cdots$benzohydrazide complexes: The interplay between anion-π and H-bond interactions. *Struct. Chem.* **2017**, *28*, 687–695. [CrossRef]
63. Moncomble, A.; Cornard, J.P.; Meyer, M. A quantum chemistry evaluation of the stereochemical activity of the lone pair in Pb^{II} complexes with sequestering ligands. *J. Mol. Model.* **2017**, *23*, 24. [CrossRef] [PubMed]
64. Lu, T.; Chen, F. Calculation of molecular orbital composition. *Acta Chim. Sin.* **2011**, *69*, 2393–2406.
65. Wei, Z.; Hu, Y.; Han, H.; Sun, W.; Wang, R.; Wang, J. Selective flotation of scheelite from calcite using Al-Na$_2$SiO$_3$ polymer as depressant and Pb–BHA complexes as collector. *Miner. Eng.* **2018**, *120*, 29–34. [CrossRef]

Article

Turbulence Models for Single Phase Flow Simulation of Cyclonic Flotation Columns

Shiqi Meng [1], Xiaoheng Li [1], Xiaokang Yan [1,*], Lijun Wang [2], Haijun Zhang [3] and Yijun Cao [3,*]

[1] School of Chemical Engineering and Technology, China University of Mining and Technology, Xuzhou 221116, China
[2] School of Power Engineering, China University of Mining and Technology, Xuzhou 221116, China
[3] Chinese National Engineering Research Center of Coal Preparation and Purification, Xuzhou 221116, China
* Correspondence: xk-yan@cumt.edu.cn (X.Y.); yijuncao@126.com (Y.C.)

Received: 10 June 2019; Accepted: 25 July 2019; Published: 30 July 2019

Abstract: Cyclonic fields are important for cyclonic static microbubble flotation columns (FCSMCs), one of the most important developments in column flotation technology, particularly for separation of fine particles, where the internal flow field has enormous influence on flotation performance. PIV (particle image velocimetry) and CFD (computational fluid dynamics) are the most effective methods to study flow fields. However, data is insufficient for FCSMC flow fields and similar cyclonic equipment, with turbulence model simulations producing different views to measured data. This paper employs an endoscope and PIV to measure axial and cross sections for single-phase swirling flow fields in FCSMCs. We then compare various turbulence model simulations (Reynolds stress model (RSM), standard k-ε, realizable k-ε, and RNG (renormalization group) k-ε) to the measured data. The RSM (Reynolds stress model) predicts cyclonic flow field best in flotation columns with 16.22% average relative velocity deviation. Although the realizable k-ε model has less than 30% relative deviation in radial and tangential directions, axial deviations reach 78.11%. Standard k-ε and RNG k-ε models exhibited approximately 40% and 30% radial and tangential deviation, respectively, and cannot be used even for trend predictions for axial velocity. k-ε models are based on isotropic assumptions with semi-empirical formulas summarized from experiments, whereas RSM fundamentally considers laminar flow and Reynolds stress, and hence is more suitable for anisotropic performance. This study will contribute to flotation column and other cyclonic flow field equipment research.

Keywords: CFD; cyclonic flow field; flotation column; endoscopic laser PIV; turbulence models

---

1. Introduction

Cyclonic structures are employed in cyclonic static microbubble flotation columns (FCSMCs), which have sprung up since it was patented in 1999 [1] and been successfully applied to refining stages, to facilitate centrifugal separation [2], bubble dispersion [3], and the reduction of sorting granularity [4]. Figure 1 shows a typical FCSMC structure, incorporating an inverted cone in its middle part. Air is introduced and crushed into microbubbles by bubble generators in the pipe flow. The mixture of slurry and bubbles enters the flotation column from two symmetrical inlets on the swirl inverted cone, forming a swirl flow field. Hydrophobic mineral particles attach to the bubbles and are carried up to be concentrated. Under the centrifugal force in the swirling flow field, the high-density and coarse particles move outward and are discharged from the tail port. The remaining minerals enter the mid port and are separated again in the following pipe flow unit. This complex process involve lift, drag, and other interaction since gas and particles move in liquid. Under liquid forces and interface forces, collisions, adhesions, and detachments between particles and bubbles happen, as shown in Figure 2. Generally, this process happened in a turbulence flow environment [5] and are affected by the turbulent flow field. Therefore, the turbulence in cyclonic flow fields is an important research topic for this type of flotation column and other similar equipment.

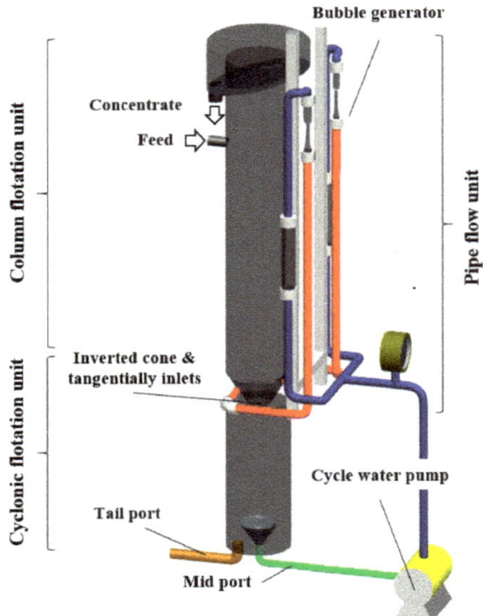

Figure 1. Typical cyclonic static microbubble flotation column structure.

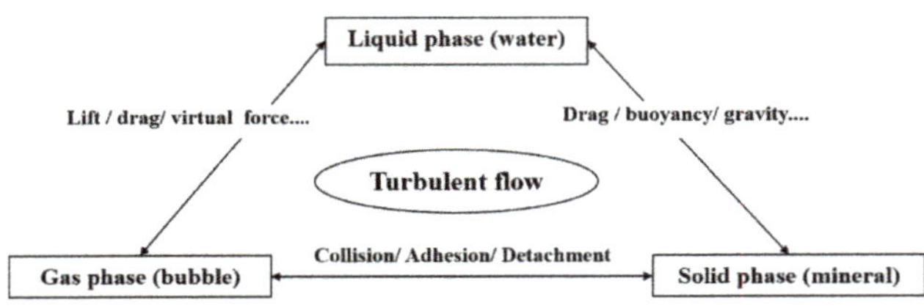

Figure 2. Forces and interactions inside an operating flotation column.

Fine mineral particles and water are thoroughly mixed in a stirred tank to form a homogeneous slurry, and then introduced into the flotation column. Flotation feed particles are generally <0.074 mm in diameter, and flow well in the slurry. This work will provide an important reference for the study of multiphase flow in a flotation column and will have a positive impact on the industrial application of flotation columns.

Flow fields have been studied for many years with continuous developments in experimental fluid dynamics (EFD) and computational fluid dynamics (CFD). Particle image velocimetry (PIV) is an advanced EFD-based velocity measurement method developed in the late 1970s. PIV can record velocity distribution information for a large number of spatial points in the same transient state, providing flow field structure and characteristics [6].

Although PIV has been widely used for cyclonic flow field measurement, data remains inadequate to fully reflect detailed flow features. For example, Marins et al. [7] used PIV to measure axial and tangential velocity of a hydrocyclone without an air core to obtain the radial velocity. Cui et al. [8] used PIV to measure axial and radial velocity in the axial section of a hydrocyclone with an air core, but cross-section measurement was not possible due to overflow. Tangential velocity is particularly difficult to measure for flotation columns since the column top is open and the swirling flow interferes with the light path. Yan et al. [9] and Wang et al. [10] used PIV to measure single-phase cyclonic flow fields for flotation columns, but only obtained axial velocity by measuring one quasi axial section of the flotation column under a fixed volume flow rate.

Computational fluid dynamics provides effective methods to numerically study flow fields and has been successfully used to calculate various cyclonic equipment parameters. However, numerical simulations tend to simplify boundary conditions and material properties, and discretization differences can lead to different results. Therefore, numerical simulation requires experimental data to verify its reliability. Reasonable numerical simulation has a guiding effect on experimental and theoretical analysis, which can make up for the lack of experimental work. The turbulence model selection has significant influence on CFD simulation accuracy for single-phase cyclonic flow. Due to the large resources occupied by DNS (direct numerical simulation) and LES (large eddy simulation), the Reynolds stress model (RSM) and k-ε turbulence models are more commonly used for column simulation. Khan et al. [11] proposed a three-dimensional (3D) model of a bubble column simulated using k-ε and RSM models. They performed LDA (laser doppler anemometry measurements) in the column to verify the simulation, and confirmed that k-ε and RSM models could effectively predict bubble column internal flow fields. However, several issues remain for application of these turbulence models to cyclonic flow simulations. Wang et al. [10] compared single-phase PIV experiment results with CFD predictions, and concluded that the RSM model was more suitable for single-phase cyclonic field simulation in a FCSMC. Zhang et al. [12] and Yan et al. [13] used the standard k-ε turbulence model to simulate single-phase internal flow fields within flotation columns and analyze their benefits. Swain et al. [14] used RSM and standard k-ε models to simulate gas–liquid two-phase flow in a hydrocyclone, and concluded that both models could predict the hydrocyclone flow pattern. Azadi et al. [15] used RNG k-ε and Reynolds stress models to calculate cyclonic flow fields, and showed that RSM predictions produced smaller deviations but required higher computational power than the RNG k-ε model.

Two main problems have prevented flow field application to cyclonic flotation columns: Measured data is not available for the tangential section and turbulent numerical models remain uncertain.

Therefore, the current study first improved the traditional PIV experimental method for flotation columns, and then developed an experimental platform for measuring the swirling field axial and cross sections inside a flotation column, providing detailed flow field information including axial, radial and tangential velocity. We subsequently performed 3D and unsteady single-phase simulations for cyclonic flotation columns, and investigated different turbulence model suitability to accurately model single-phase flow fields inside the flotation column. This study provides a new reference for numerical studies and PIV tests on flotation columns and similar equipment flow fields.

2. PIV Experiment

2.1. Experiment Apparatus

We constructed a cyclonic flotation column as sown in Figure 1 for PIV tests, with height = 1400 mm and inner diameter = 190 mm. The flotation column was constructed from glass to ensure transparency. The experimental medium was deionized water at room temperature. A water-filled cube box surrounded the column flotation unit to eliminate cylindrical surface refraction. Water from the pump was divided into two symmetrical lines. After passing through a bubble generator with the air inlet closed, water tangentially flowed into the inverted cone, forming a cyclonic field. Since the single-phase experiment did not include particles, we neglected feeding and discharging effects.

2.2. Experiment Method

We measured flotation column axial and cross sections to obtain axial, radial, and tangential velocity. The flotation column position was fixed during the experiment, and measurements were performed by adjusting the relative positions of the laser light sheet and CCD camera. The scanning surface formed by the laser light sheet should be kept perpendicular to the imaging direction of the

CCD camera. The tracer particles used for the PIV test were hollow glass spheres having a diameter of 10 μm, and its density was 1.05 g/cm^3, which shows good water following performance. The software used for the PIV test was Dynamic Studio, which is a mature image system software platform for many fluid mechanics experiments, such as PIV, LIF, combustion, and atomization analysis. In the PIV test, Dynamic Studio processed image information into vectors, velocity, and other hydrodynamic parameters of the flow field.

Figure 3 shows how the laser light sheet, CCD camera, and test sections were placed to measure the flotation column axial section. Several bubbles clustered near the column axis, which would interfere with the laser light sheet, reduce the flow field and luminance at the axis and in the other half of the flotation column, impacting measurement accuracy. This approach is similar to that employed by Wang et al. [10]. Therefore, following some initial experiments, we adopted the quasi-axial section, deviating 10 mm from the central axial section as the measurement plane to ensure more accurate experimental data, as shown in the dash-solid section in Figure 3 (260–340 mm above the inverted cone). The left side was illuminated by laser with sufficient intensity. Four circulating flow types were selected for experiments, considering industrial applications. Table 1 shows optimal experimental parameters derived from several trial experiments and adjustments, and we averaged 200 photograph pairs. Experimental data at 300 mm height was extracted for comparison with CFD simulations.

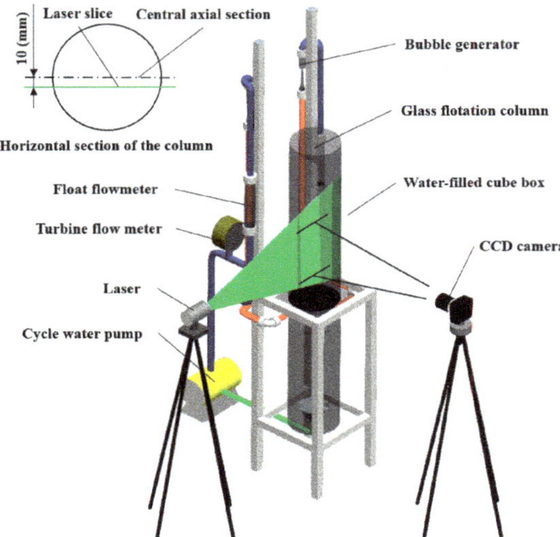

Figure 3. Particle image velocimetry (PIV) measurement of a quasi-axial section.

Table 1. PIV experimental parameters for cyclonic static microbubble flotation column vertical sections.

Measuring Surface	Recycling Flow (m^3/h)	Measurement Parameters	Laser Interval (μs)	Trigger Frequency
Vertical section,	0.25		1200	
260–340 mm above	0.50	Axial velocity,	800	9.0 Hz
the cyclonic	1.00	radial velocity	600	
flotation unit	1.50		300	

To obtain tangential velocity, we installed the CCD camera above the column, providing downward looking images to measure the horizontal section. However, the water swirls when the flotation column is in operation, and the free surface between air and water inside the column flotation zone, forms an inverted cone. Laser-illuminated tracer particles are refracted when passing through this free surface, causing deviations in the CCD images. To solve this problem, we employed an endoscope for the PIV test, allowing the lens to be inserted into the water to test the flow field, and hence eliminating most optical interference. Velarde [16] and Helmi [17] used an endoscopic laser PIV technique for dense gas–solid fluidized beds in 2016 and 2018, respectively, which inspired the current proposed PIV measurement system for cyclonic flow fields in the flotation column.

Figure 4 shows how an endoscope was installed onto the CCD camera and extended below the free surface. The measured horizontal section was 300 mm above the inverted cone. Table 2 shows optimal experimental parameters derived from many preliminary experiments, with 200 photograph pairs captured for averaging. Similar to flotation column axial section measurements discussed above, we selected data along a straight line in the radial direction, and the same four circulating flows were employed, as shown in Table 2.

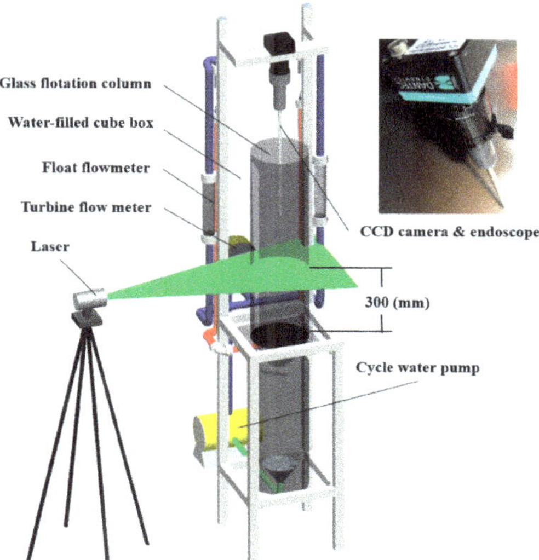

Figure 4. Horizontal section PIV measurement.

Table 2. Cyclonic static microbubble flotation column (FCSMC) horizontal section PIV measurement parameters.

Measuring Surface	Recycling Flow (m³/h)	Measurement Parameter	Laser Interval (μs)	Trigger Frequency
Column cross-section, 300 mm above the column flotation unit	0.25 0.50 1.00 1.50	Tangential velocity	1000 600 400 200	9.0 Hz

3. Numerical Simulation

3.1. Turbulence Models

Turbulence models greatly impact results from single-phase numerical simulations (Table 3). Turbulence model expression establishment and application differ, producing inconsistent predictions for many flow field types. We compared the RSM, standard k-ε model, realizable k-ε model, and RNG k-ε models. As one of the most widely used turbulence models, the standard k-ε model is based on transport equations for turbulent kinetic energy, k, and its dissipation rate, ε [18], which are defined as

$$\frac{\partial(\rho k)}{\partial t} + \frac{\partial(\rho k u_i)}{\partial x_i} = \frac{\partial}{\partial x_j}\left[\left(\mu + \frac{\mu_t}{\sigma_k}\right)\frac{\partial k}{\partial x_j}\right] + G_k + G_b - \rho\varepsilon - Y_M + S_k, \tag{1}$$

$$\frac{\partial(\rho\varepsilon)}{\partial t} + \frac{\partial(\rho\varepsilon u_i)}{\partial x_i} = \frac{\partial}{\partial x_j}\left[\left(\mu + \frac{\mu_t}{\sigma_\varepsilon}\right)\frac{\partial\varepsilon}{\partial x_j}\right] + \frac{C_{1\varepsilon}\varepsilon}{k}(G_k + C_{3\varepsilon}G_b) - C_{2\varepsilon}\rho\frac{\varepsilon^2}{k} + S_\varepsilon. \tag{2}$$

Table 3. Symbol and interpretation.

Symbol	Interpretation	Symbol	Interpretation
k	Turbulence kinetic energy	φ	Press strain
ε	Turbulence dissipation rate	D	Turbulence diffusion
ρ	Density	p	Pressure
u	Velocity	μ	Viscosity
C	Constant	t	Time
P	Turbulence production term	δ	Kronecker delta function
$i,1$	x component	$j,2$	y component
$k,3$	z component (in the RSM)	S_k,S_ε	User-defined source terms.
G_k	The generation of turbulence kinetic energy due to the mean velocity gradients, calculated as described in Modeling Turbulent Production in the k-ε Models.		
G_b	The generation of turbulence kinetic energy due to buoyancy, calculated as described in Effects of Buoyancy on Turbulence in the k-ε Models.		
Y_M	The contribution of the fluctuating dilatation in compressible turbulence to the overall dissipation rate, calculated as described in Effects of Compressibility on Turbulence in the k-εModels.		
$\sigma_k,\sigma_\varepsilon$	The turbulent Prandt1 numbers for k and ε.		

The RNG k-ε model was proposed by Yakhot and Orzga [19] in 1986. It is similar in form to the standard k-ε model, but includes some refinements. The RNG model has an additional term in its equation that improves the accuracy for rapidly strained flows. In addition, the effect of swirl on turbulence is included in the model, enhancing accuracy for swirling flows. What is more, it provides an analytical formula for turbulent Prandtl numbers, while the standard k-ε model uses user-specified, constant values. Therefore, these features make the RNG k-ε model more accurate and reliable for a wider class of flows than the standard k-ε model. The equations are given as:

$$\frac{\partial(\rho k)}{\partial t} + \frac{\partial(\rho k u_i)}{\partial x_i} = \frac{\partial}{\partial x_j}\left[a_k\mu_{eff}\frac{\partial k}{\partial x_j}\right] + G_k + G_b - \rho\varepsilon - Y_M + S_k, \tag{3}$$

$$\frac{\partial(\rho\varepsilon)}{\partial t} + \frac{\partial(\rho\varepsilon u_i)}{\partial x_i} = \frac{\partial}{\partial x_j}\left(a_\varepsilon\mu_{eff}\frac{\partial\varepsilon}{\partial x_j}\right) + \frac{C_{1\varepsilon}\varepsilon}{k}(G_k + C_{3\varepsilon}G_b) - C_{2\varepsilon}\rho\frac{\varepsilon^2}{k} - R_\varepsilon + S_\varepsilon. \tag{4}$$

The realizable k-ε model was proposed by Moin et al. [20] who believe that the standard k-ε model may cause stress errors when the time-averaged strain rate is strong. The model [21] implements an alternative formulation for turbulent viscosity [22] based on a modified transport equation for dissipation derived from an exact equation for mean-square vorticity fluctuation transportation:

$$\frac{\partial(\rho k)}{\partial t} + \frac{\partial(\rho k u_j)}{\partial x_j} = \frac{\partial}{\partial x_j}\left[\left(\mu + \frac{\mu_t}{\sigma_k}\right)\frac{\partial k}{\partial x_j}\right] + G_k + G_b - \rho\varepsilon - Y_M + S_k, \tag{5}$$

$$\frac{\partial(\rho\varepsilon)}{\partial t} + \frac{\partial(\rho\varepsilon u_j)}{\partial x_j} = \frac{\partial}{\partial x_j}\left[\left(\mu + \frac{\mu_t}{\sigma_\varepsilon}\right)\frac{\partial\varepsilon}{\partial x_j}\right] + \rho C_1 E\varepsilon - \rho C_2\frac{\varepsilon^2}{k + \sqrt{v\varepsilon}} + C_{1\varepsilon}\frac{\varepsilon}{k}C_{3\varepsilon}G_b + S_\varepsilon. \tag{6}$$

RSM closes the Reynolds-averaged Naiver–Stokes equations by abandoning the isotropic eddy-viscosity hypothesis and solving transport equations for Reynolds stresses, together with an equation for the dissipation rate [23–25]. The control equations are as follows:

Continuity equations:

$$\frac{\partial\rho}{\partial t} + \frac{\partial}{\partial}(\rho u_i) = 0. \tag{7}$$

Momentum equation:

$$\frac{\partial u_i}{\partial t} + \frac{\partial}{\partial x_j}\left(u_i u_j\right) = -\frac{1}{\rho}\frac{\partial p}{\partial x_i} + \frac{1}{\rho}\frac{\partial}{\partial x_j}\left[\mu\left(\frac{\partial u_i}{\partial x_j} + \frac{\partial u_j}{\partial x_i}\right)\right] + \frac{\partial}{\partial x_j}\left(-\overline{u_i' u_j'}\right),$$ (8)

where

$$u_i = \overline{u_i} + u_i'.$$ (9)

Turbulence production term P_{ij} is defined as

$$P_{ij} = \rho\left(\overline{u_i' u_k'}\frac{\partial u_j}{\partial x_k} + \overline{u_j u_k'}\frac{\partial u_j}{\partial x_k}\right).$$ (10)

Turbulence dissipation rate ε_{ij} is given as

$$\varepsilon_{ij} = 2\mu\overline{\frac{\partial u_i'}{\partial x_k}\frac{\partial u_j'}{\partial x_k}}.$$ (11)

Press strain φ_{ij} is defined as

$$\varphi_{ij} = p\overline{\left(\frac{\partial u_i'}{\partial x_j} + \frac{\partial u_j'}{\partial x_i}\right)}.$$ (12)

Turbulence diffusion $D_{T,ij}$ is given as

$$D_{T,ij} = -\frac{\partial}{\partial x_k}\left(\rho\overline{u_i' u_j' u_k'} + \overline{pu_i'}\delta_{jk} - \mu\frac{\partial}{\partial x_k}\overline{u_i' u_j'}\right).$$ (13)

3.2. Geometry and Mesh

The 3D FCSMC model was established for the experimental device, as shown in Figure 5, and geometric model processing was consistent with Wang et al. [10] and Yan et al. [13]. A structured mesh was constructed aside from the inverted cone, which was unstructured mesh (see Figure 5), and grid independence was verified. Figure 6 shows axial velocity along the horizontal direction at 300 mm column height for different mesh schemes. We selected a model incorporating 379,658 grids.

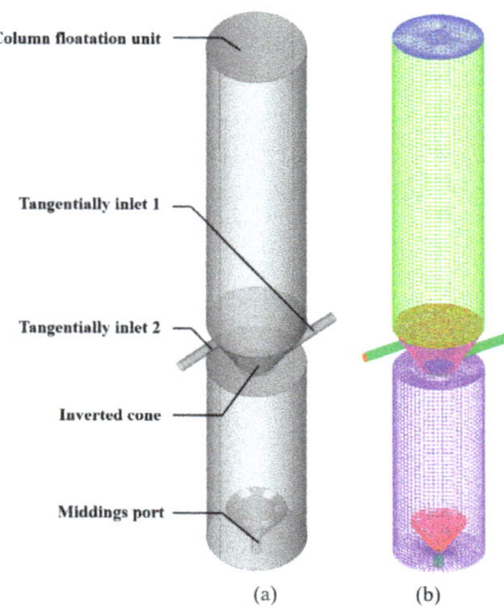

Figure 5. Flotation column 3D model (**a**) and mesh (**b**).

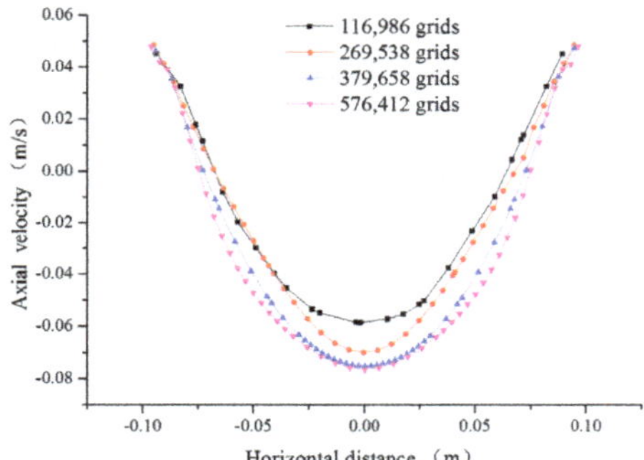

Figure 6. Grid independence: The curve of axial velocity along the horizontal direction on a section 300 mm above the inverted cone at 1.50 m³/h flux.

3.3. Boundary Conditions and Solution

In order to match the numerical simulation with the PIV experiment, we set the boundary conditions according to the experiment. Two tangential inlets (Figure 5, Inlet 1 and 2) in the cyclonic unit are defined as velocity inlets, with fluid speeds of 0.11, 0.22, 0.44, and 0.66 m/s for 0.25, 0.50, 1.00, and 1.50 m³/h flow rates, respectively. Since the fluid was single phase, we assumed that all feed was discharged from the middling port, with the amount of feed equal to the amount of discharge. Therefore, the middling port outlet was defined as a negative velocity inlet with 0.22, −0.44, −0.88, and −1.32 m/s, respectively. The acceleration of gravity is 9.81 m/s, which is vertically downward along the axis. Table 4 shows all boundary conditions and calculation values.

Table 4. Circulation flow boundary conditions.

Boundary	Type	Circulating Volume Rate (m³/h)			
		0.25	0.50	1.00	1.50
Inlet 1	Inlet velocity (m/s)	0.11	0.22	0.44	0.66
Inlet 2	Inlet velocity (m/s)	0.11	0.22	0.44	0.66
Concentrate outlet	Symmetry	-	-	-	-
Middling port outlet	Inlet velocity (m/s)	−0.22	−0.44	−0.88	−1.32

Commercial CFD code ANSYS Fluent was employed to carry out the simulation. We used the SIMPLE algorithm for pressure–velocity coupling, the least squares cell based on gradient, the PRESTO! scheme for pressure discretization, and the second order upwind scheme for momentum discretization, turbulent kinetic energy, turbulent dissipation rate, and Reynolds stresses. The calculation used an unsteady solver, hence once the parameters and residuals of the monitored point reached statistical stability, the results of the previous 20 s duration were averaged and used for subsequent analysis.

3.4. CFD and PIV Comparison

3.4.1. Velocity Vectors

Figure 7a shows the vertical PIV experimental vectors located at the plane 260–340 mm above the inverted cone and parallel to the 10 mm axial section. Simulation and PIV results were compared at 1.00 m³/s to verify prediction accuracy. PIV showed gradually increasing water vector magnitude as distance from the column wall increased, followed by a sharp reduction, with water flowing approximately horizontally toward the axis. Figure 7b shows that RSM model vectors were consistent with the PIV results, whereas other models (Figure 7c–e) produced divergent directions close to the axis, being diagonally downwards.

Figure 8 compares PIV (Figure 8a) and model (Figure 8b–e) results for the horizontal plane 300 mm above the inverted cone. All the models are consistent with PIV in that the water rotates counterclockwise around the center.

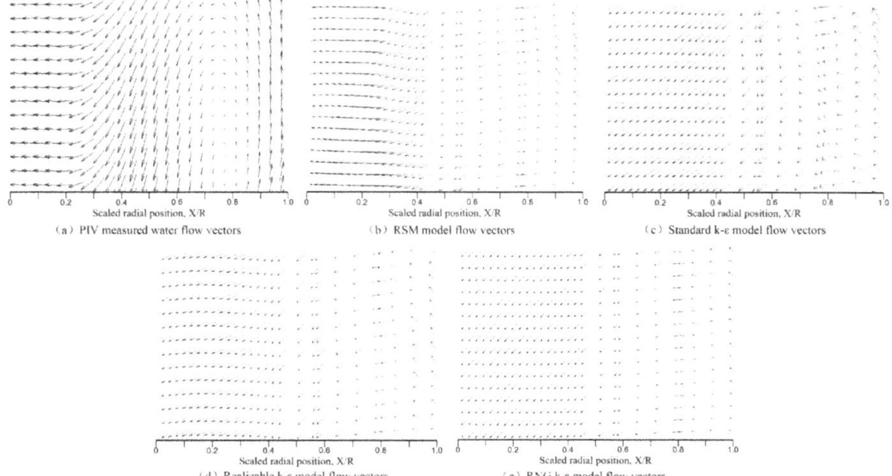

Figure 7. Experimental and simulation water flow vectors in quasi-axis sections. (**a**) PIV measured water flow vectors; (**b**) RSM model flow vectors; (**c**) Standard k-ε model flow vectors; (**d**) Realizable k-ε model flow vectors; (**e**) RNG k-ε model flow vectors.

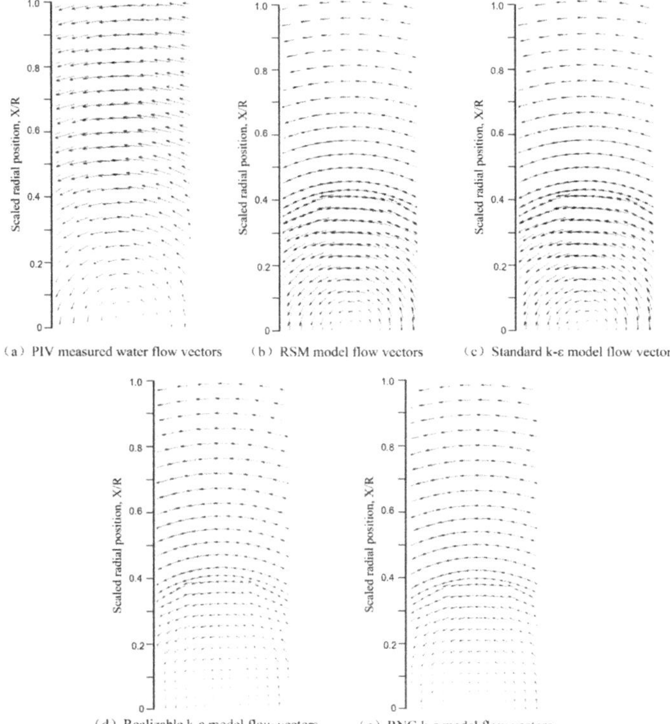

Figure 8. Experiment model water flow vectors in horizontal sections. (**a**) PIV measured water flow vectors; (**b**) RSM model flow vectors; (**c**) Standard k-ε model flow vectors; (**d**) Realizable k-ε model flow vectors; (**e**) RNG k-ε model flow vectors.

3.4.2. Quantitative Analysis

Figures 9 and 10 compare the simulated and measured velocity profiles along the horizontal direction for a cyclic flux of 0.25, 0.50, 1.00, and 1.50 m^3/h, respectively, at z = 300 mm, 10 mm off-axis. As for Figure 11, the velocity distribution of the simulation and measurement in the horizontal direction from the axis to the wall at z = 300 mm is compared. The RSM and realizable k-ε turbulent model axial velocity distribution trends exhibit the in "$\sqrt{}$" shape, consistent with PIV results. However, the standard k-ε and RNG k-ε model trends diverge widely.

Figures 10 and 11 show radial and tangential velocity results, respectively. Distribution trends along the horizontal direction are consistent with PIV results for all turbulence models.

Thus, RSM and realizable k-ε models exhibit good prediction performance for velocity distributions in all directions, whereas the standard k-ε model and the RNG k-ε model can only predict velocity distribution trends in radial and tangential directions.

Tables 5–7 summarize absolute deviations between simulation and PIV results (excluding data within 3 mm of the zero velocity point) for axial, radial, and tangential velocities, respectively. Since the standard k-ε model and the RNG k-ε model cannot predict velocity distribution trends in the axial direction, only radial and tangential errors were analyzed. The RSM model achieved the smallest error for all directions.

Table 5. Simulation model axial velocity absolute deviation from PIV measurement.

Flow Rate (m^3/h)	Turbulence Model			
	RSM (%)	Standard k-ε (%)	Realizable k-ε (%)	RNG k-ε (%)
0.25	19.89	87.12	106.15	102.86
0.50	18.76	71.87	68.42	80.11
1.00	14.39	90.42	67.22	86.85
1.50	16.39	83.58	70.65	88.26
Mean absolute deviation	17.36	83.25	78.11	89.52

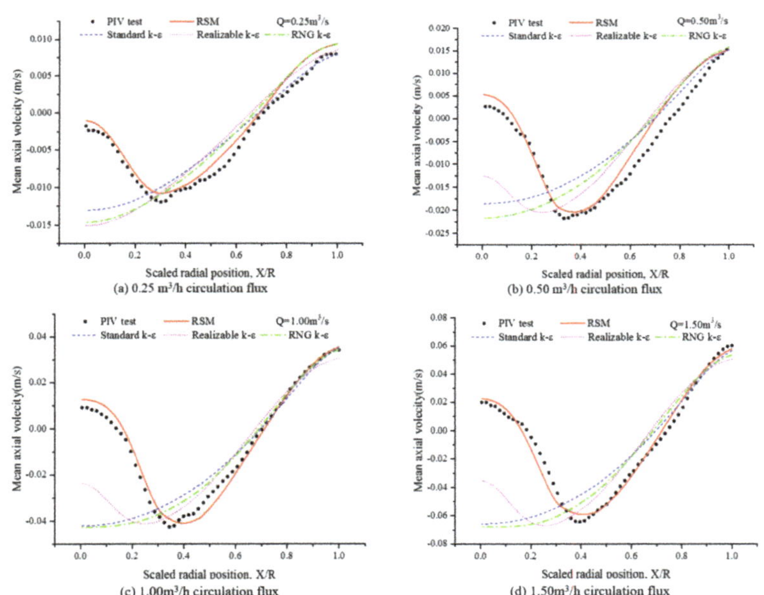

Figure 9. Axial velocity distribution calculated by turbulence models and PIV measurement along the horizontal direction, 300 mm high, and 10 mm from the axial section.

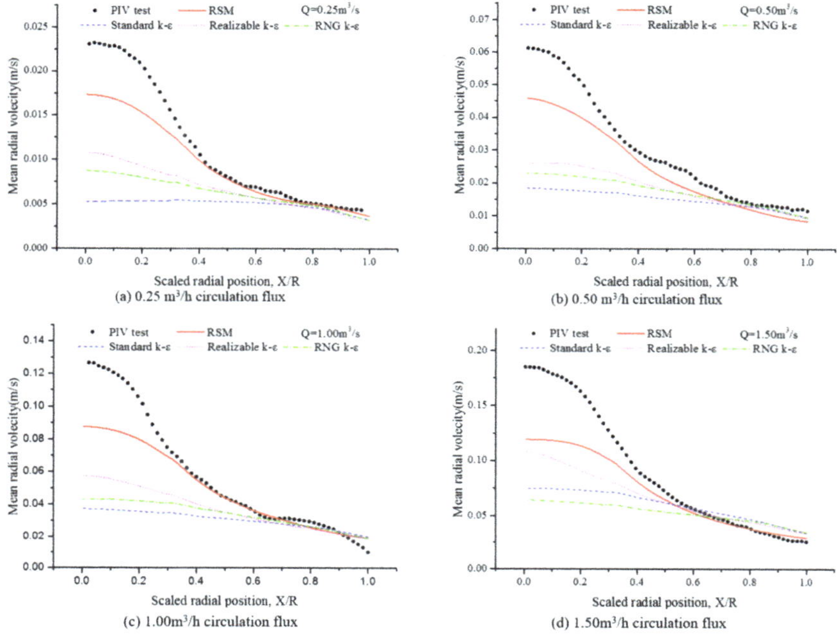

Figure 10. Radial velocity distribution calculated by turbulence models and PIV measurement along the horizontal direction, 300 mm high, and 10 mm from the axial section.

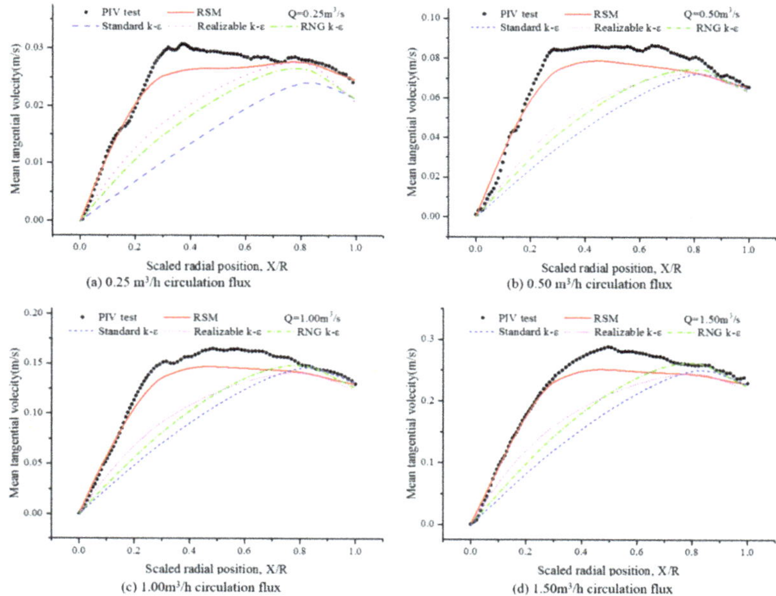

Figure 11. Tangential velocity distribution calculated by turbulence models and PIV measurement along the horizontal direction, 300 mm high.

Table 6. Simulation model radial velocity absolute deviation from PIV measurement.

Flow Rate (m³/h)	Turbulence Model			
	RSM (%)	Standard k-ε (%)	Realizable k-ε (%)	RNG k-ε (%)
0.25	11.54	41.33	30.82	31.89
0.50	18.23	38.42	30.50	31.29
1.00	13.55	37.17	29.24	31.56
1.50	15.86	32.06	25.20	36.54
Mean absolute deviation	14.79	37.25	28.94	32.82

Table 7. Simulation model tangential velocity absolute deviation from PIV measurement.

Flow Rate (m³/h)	Turbulence Model			
	RSM (%)	Standard k-ε (%)	Realizable k-ε (%)	RNG k-ε (%)
0.25	16.87	46.84	27.39	32.86
0.50	13.87	37.74	27.88	30.79
1.00	10.39	36.49	24.61	28.14
1.50	12.77	34.60	24.60	26.61
Mean absolute deviation	13.48	38.92	26.12	29.60

The k-ε models are based on isotropic assumptions with semi-empirical formulas summarized from experimental results. However, the high-intensity swirling field has a large turbulent velocity gradient and fluctuating turbulent velocity, and hence is not even approximately isotropic. Therefore, k-ε models are less effective, and the swirling flow field inside the flotation column cannot be correctly modelled. The RSM model is the most accurate model among the four models considered. This model abandons the isotropic assumption, considering laminar flow and Reynolds stress, and hence is suitable for anisotropic cases.

4. Conclusions

We performed numerical simulations and PIV measurements for a single-phase swirling field in a lab-scale FCSMC at four circulation fluxes (0.25, 0.50, 1.00, and 1.50 m³/h). Axial and radial velocity distributions were measured and modelled on the flotation column quasi-axial section. Experimental errors due to refraction at the free surface were avoided by adding an endoscopic lens to the CCD camera, inserting the speculum below the free surface, and allowing tangential velocity distribution to be measured on the indicated section. The RSM, standard k-ε model, realizable k-ε model, and RNG k-ε model were used for CFD simulations for FCSMC single-phase swirling fields. Comparing the PIV experimental and simulation results and of the PIV, we made the following conclusions:

(1) Standard k-ε model and RNG k-ε model. Radial and tangential velocity distribution trends simulated by these models were consistent with PIV measurements, but axial velocity along the horizontal direction deviated widely from PIV results. Thus, axial errors were not considered. Mean absolute radial and tangential velocity deviations were 37.25% and 38.92%, for the standard k-ε model and slightly better for the RNG k-ε model (32.82% and 29.60%), respectively.

(2) RSM and realizable k-ε model. Radial, tangential, and axial velocity distribution trends simulated by these models were all consistent with PIV measurements. Mean absolute axial, radial, and tangential velocity deviations were 17.36%, 14.79%, and 13.48% for the RSM model, and 78.11%, 28.94%, and 26.12% for the realizable k-ε model, respectively. Therefore, the RSM model exhibited superior performance, particularly in the axial direction.

(3) The k-ε models [26] are based on isotropic assumptions with semi-empirical formulas summarized from experimental results. However, the internal flow field of the flotation column is a high-intensity vortex field with a large turbulent velocity gradient and fluctuating turbulent velocity, and hence is not approximately isotropic. The RSM model [27] abandons the isotropic assumption, considering laminar flow and Reynolds stress, and hence is suitable for anisotropic cases, especially the

swirling flow field inside the flotation column. In fact, the RANS (Reynolds Averaged Navier-Stokes) model [28] does not distinguish between the size and direction of the vortex, and it is limited by the model conditions. Thus, it is difficult to get perfect results for complex problems such as an inverse pressure gradient numerical simulation. In addition, the RANS model relies on boundary conditions, so the calculations are very computationally dependent. DNS and even LES are good choices without considering computational costs and widespread adoption.

Author Contributions: Conceptualization, X.Y., Y.C.; methodology, S.M., L.W.; software, S.M., X.L.; validation, X.L.; investigation, H.Z.; data curation, S.M., X.Y.; writing-original draft preparation, S.M.; writing-review and editing, S.M., X.Y.; visualization, S.M., X.L.; and funding, Y.C., H.Z., L.W.

Funding: This research was funded by the National Natural Science Foundation of China (U1704252), the Fundamental Research Funds for the Central Universities (2017XKZD02) and the National Natural Science Foundation of China (51722405).

Conflicts of Interest: The authors declare no conflict of interest.

References

1. Jiongtian, L. Cyclonic-Static Micro-Bubble Floatation Apparatus and Method. U.S. Patent 6073775, 19 January 1999.
2. Tian, J.; Ni, L.; Song, T.; Olson, J.; Zhao, J. An overview of operating parameters and conditions in hydrocyclones for enhanced separations. *Sep. Purif. Technol.* **2018**, *206*, 268–285. [CrossRef]
3. Li, X.; Xu, H.; Liu, J.; Zhang, J.; Li, J.; Gui, Z. Cyclonic state micro-bubble flotation column in oil-in-water emulsion separation. *Sep. Purif. Technol.* **2016**, *165*, 101–106. [CrossRef]
4. Zhang, H.; Liu, J.; Wang, Y.; Cao, Y.; Ma, Z.; Li, X. Cyclonic-static micro-bubble flotation column. *Miner. Eng.* **2013**, *45*, 1–3. [CrossRef]
5. Guichao, W.; Linhan, G.; Subhasish, M.; Evans, G.M.; Joshi, J.B.; Songying, C. A review of CFD modelling studies on the flotation process. *Miner. Eng.* **2018**, *127*, 153–177.
6. Buchhave, P. Particle image velocimetry—Status and trends. *Exp. Therm. Fluid Sci.* **1992**, *5*, 586–604. [CrossRef]
7. Marins, L.P.M.; Duarte, D.G.; Loureiro, J.B.R.; Moraes, C.A.C.; Freire, A.P.S. LDA and PIV characterization of the flow in a hydrocyclone without an air-core. *J. Pet. Sci. Eng.* **2010**, *70*, 168–176. [CrossRef]
8. Cui, B.Y.; Wei, D.Z.; Gao, S.L.; Liu, W.G.; Feng, Y.Q. Numerical and experimental studies of flow field in hydrocyclone with air core. *Trans. Nonferr. Met. Soc. China* **2014**, *24*, 2642–2649. [CrossRef]
9. Yan, X.; Meng, S.; Wang, A.; Wang, L.; Cao, Y. Hydrodynamics and separation regimes in a cyclonic-static microbubble flotation column. *Asia Pac. J. Chem. Eng.* **2018**, *13*, e2185. [CrossRef]
10. Wang, A.; Yan, X.; Wang, L.; Cao, Y.; Liu, J. Effect of cone angles on single-phase flow of a laboratory cyclonic-static micro-bubble flotation column: PIV measurement and CFD simulations. *Sep. Purif. Technol.* **2015**, *149*, 308–314. [CrossRef]
11. Khan, Z.; Bhusare, V.H.; Joshi, J.B. Comparison of turbulence models for bubble column reactors. *Chem. Eng. Sci.* **2017**, *164*, 34–52. [CrossRef]
12. Zhang, M.; Li, T.; Wang, G. A CFD study of the flow characteristics in a packed flotation column: Implications for flotation recovery improvement. *Int. J. Miner. Process.* **2017**, *159*, 60–68. [CrossRef]
13. Yan, X.; Liu, J.; Cao, Y.; Wang, L. A single-phase turbulent flow numerical simulation of a cyclonic-static micro bubble flotation column. *Int. J. Min. Sci. Technol.* **2012**, *22*, 95–100. [CrossRef]
14. Swain, S.; Mohanty, S. A 3-dimensional Eulerian–Eulerian CFD simulation of a hydrocyclone. *Appl. Math. Model.* **2013**, *37*, 2921–2932. [CrossRef]
15. Azadi, M.; Azadi, M.; Mohebbi, A. A CFD study of the effect of cyclone size on its performance parameters. *J. Hazard. Mater.* **2010**, *182*, 835–841. [CrossRef] [PubMed]
16. Velarde, I.C.; Gallucci, F.; Annaland, M.V.S. Development of an endoscopic-laser PIV/DIA technique for high-temperature gas–solid fluidized beds. *Chem. Eng. Sci.* **2016**, *143*, 351–363. [CrossRef]
17. Helmi, I.C.V.A.; Gallucci, F.; van Sint Annaland, M. Hydrodynamics of dense gas-solid fluidized beds with immersed vertical membranes using an endoscopic-laser PIV/DIA technique. *Chem. Eng. Sci.* **2018**, *182*, 146–161. [CrossRef]

18. Spalding, B.D. *Lectures in Mathematical Models of Turbulence*; Academic Press: Cambridge, MA, USA, 1972.
19. Victor Yakhot, S.A.O. Renormalization group analysis of turbulence: Basic theory. *J. Sci. Comput.* **1986**, *1*, 3–11. [CrossRef]
20. Peinke Moin, J.K.; Gerrit, O. *Progress in Large Eddy Simulation of Turbulent Flows*; AIAA Paper 97-15761; American Institute of Aeronautics and Astronautics: Reston, VA, USA, 1997.
21. Shih, T.H.; Liou, W.W.; Shabbir, A.; Yang, Z.; Zhu, J. *A New Kappa-Epsilon Eddy Viscosity Model for high Reynolds Number Turbulent Flows-Model Development and Validation*; Linthicum Heights Md Nasa Center for Aerospace Information |c: Linthicum Heights, MD, USA, 1995.
22. Orszag, S.A.; Yakhot, V.; Flannery, W.S.; Boysan, F.; Choudhury, D.; Maruzewski, J.; Patel, B. *Renormalization Group Modeling and Turbulence Simulations*; Elsevier: Amsterdam, The Netherlands, 1993.
23. Launder, B.E.; Reece, G.J.; Rodi, W. Progress in the development of a Reynolds-stress turbulence closure. *J. Fluid Mech.* **1975**, *68*, 537–566. [CrossRef]
24. Launder, B.E. Second-moment closure: Present … and future? *Int. J. Heat Fluid Flow* **1989**, *10*, 282–300. [CrossRef]
25. Gibson, M.M.; Launder, B.E. Ground effects on pressure fluctuations in the atmospheric boundary layer. *J. Fluid Mech.* **2006**, *86*, 491–511. [CrossRef]
26. Mendoza-Escamilla, V.X.; Alonzo-García, A.; Mollinedo, H.R.; González-Neria, I.; Yáñez-Varela, J.A.; Martinez-Delgadillo, S.A. Assessment of k-ε models using tetrahedral grids to describe the turbulent flow field of a PBT impeller and validation through the PIV technique. *Chin. J. Chem. Eng.* **2018**, *26*, 942–956.
27. Khan, Z.; Joshi, J.B. Comparison of k–ε, RSM and LES models for the prediction of flow pattern in jet loop reactor. *Chem. Eng. Sci.* **2015**, *127*, 323–333. [CrossRef]
28. Xiao, H.; Cinnella, P. Quantification of Model Uncertainty in RANS Simulations: A Review. *Prog. Aerosp. Sci.* **2019**, *108*, 1–31. [CrossRef]

Article

Study of K-Feldspar and Lime Hydrothermal Reaction: Phase and Mechanism with Reaction Temperature and Increasing Ca/Si Ratio

Shanke Liu [1,2,*], **Cheng Han** [1,2] **and Jianming Liu** [1,2]

[1] Key Laboratory of Mineral Resources, Institute of Geology and Geophysics, Chinese Academy of Sciences, Beijing 100029, China; daxu2224@hotmail.com (C.H.); jmliu@mail.iggcas.ac.cn (J.L.)

[2] Institutions of Earth Science, Chinese Academy of Sciences, Beijing 100029, China

* Correspondence: liushanke@mail.iggcas.ac.cn; Tel.: +86-10-8299-8571

Received: 15 December 2018; Accepted: 9 January 2019; Published: 14 January 2019

Abstract: To elucidate the physicochemical properties of the artificial silicate composite material, K-feldspar and lime were reacted in mild hydrothermal conditions (different reaction temperatures and various K-feldspar/lime ratios). Formed phases were investigated using various techniques, such as X-ray powder diffraction, the Rietveld method, scanning electron microscopy (SEM), and inductively coupled plasma-optical emission spectrometry. The analysis revealed that tobermorite, grossular (hydrogarnet), alpha-dicalcium silicate hydrate (α-C_2SH), amorphous calcium silicate hydrate, potassium carbonate, bütschliite, calcite, and calcium hydroxide formed with various conditions. Both the temperature and the Ca/Si molar ratio in the starting material greatly affected the formation of phases, especially the generation of tobermorite and α-C_2SH. The substitution of H_4O_4 $\leftrightarrow$ SiO_4 proceeded with the increase of the Ca/Si molar ratio rather than the reaction temperature and the reaction time. More hydrogen was incorporated in hydrogarnet through the substitution of H_4O_4 $\leftrightarrow$ SiO_4 with the increase of the Ca/Si molar ratio in the starting material. Due to the properties of tobermorite as a cation exchanger and its potential applications in hazardous waste disposal, experimental parameters should be optimized to obtain better performance of the artificial silicate composite material from K-feldspar and lime hydrothermal reaction. The dissolution mechanism of K-feldspar was also discussed.

Keywords: K-feldspar; tobermorite; hydrogarnet; hydrothermal reaction; phase analysis

1. Introduction

Silicates are extremely important materials, both naturally and artificially, for the development of science and technology. Although there is no known way of accurately ascertaining the abundance of every mineral within the earth's crust, based on a rough estimate, silicates comprise most of the earth's crust, as shown in Figure 1 [1]. In addition, silicates are the main component of ceramics, Portland cement, and glass. Silicate-based materials as amendments such as zeolites, clays, and cementitious materials show increasing interest in the stabilization/solidification of heavy-metal contaminated soils or environments [2–4].

Recently, an artificial silicate composite material, namely a nanosubmicron mineral-based soil conditioner (NMSC) prepared by the hydrothermal reaction of K-feldspar and lime, was reported [5–7]. Field and in-house tests indicated that the artificial silicate composite material showed excellent performance and served multiple functions that were closely related to its physicochemical properties and mineral components [8,9]. Of them, both carbonates and 11 Å tobermorite contributed to pH improvement and the inhibition of cadmium (Cd) uptake in rice. Tobermorite is one of the calcium silicate hydrate (CSH) compounds, and the main interest is related to its close relationship with the CSH phases formed during the hydration of Portland cement [10]. In addition, tobermorite may act as cation exchangers to absorb unhealthy elements and has potential applications in waste disposal [11–13]. Tobermorite is also a slow-release reservoir for some nutrients, such as [K$^+$] and [NH$_4$$^+$] [14]. As a primary product of the hydrothermal reaction, the weight percentage of tobermorite can reach 35%–40% by controlling the reaction conditions.

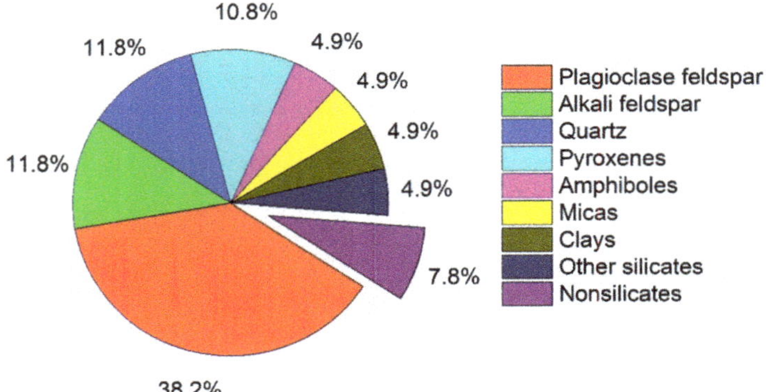

Figure 1. Estimated volume percentages for the most common minerals in the earth's crust. The data are from the reference [1], and are normalized to 100%.

In a CaO–Al$_2$O$_3$–SiO$_2$–H$_2$O (C–A–S–H) hydrothermal system, the abundance of formed minerals is closely related to the reaction conditions, such as the ratio of starting reactants [15–17]. Therefore, a study of mineral components and its abundance is a key point for the artificial silicate composite material under the different reaction conditions. We carried out a primary test on the hydrothermal reaction of K-feldspar and lime by a single-factor analysis [18]. However, at that time, the authors did not realize that the silicate composite material NMSC would show such excellent performance and serve multiple functions, as mentioned above [8,9]. Therefore, the main aim of the previous test was to improve the percentage of K-feldspar dissolution via optimizing experimental parameters. We further investigated the mechanism of the hydrothermal reaction between K-feldspar and different alkalis [including Ca(OH)$_2$] and identified minerals formed through an elaborative interface experiment [19]. Several researchers also studied the preparation of potassium fertilizer via the hydrothermal alternation of K-feldspar ore at 200 °C [20,21]. However, none of the abovementioned studies investigated the hydrothermal reaction of K-feldspar under alkaline conditions with reaction conditions, such as various reaction temperatures, various reactant ratios (the ratio of K-feldspar and lime in the starting material), and various reaction times. For a better understanding of the chemistry and mineralogy of the artificial silicate composite material, the investigation of K-feldspar and lime hydrothermal reaction must consider various conditions rather than a limited one.

Three groups of the hydrothermal experiments were carried out for investigating the K-feldspar and lime system with various conditions, i.e., the reaction temperature, the reactant ratio, and the reaction time. For each group of the hydrothermal experiment, only one factor was changed and others remained constant. The reaction temperatures of 130 °C to 250 °C were chosen for studying the effects of the reaction temperature on the hydrothermal reaction between K-feldspar and lime. In general, the operating temperature of a common autoclave in China is approximately 190 °C in order to ensure safety during the production. Therefore, based on a view from the commercialization point, the reaction temperature of 190 °C was chosen for studying the hydrothermal reaction with the reaction time (from 0 h to 36 h) and the reactant ratio (the ratio of K-feldspar and lime was 3:7 to 7:3, and 10 g total solid mixture was fixed). The artificial silicate composite material under various hydrothermal conditions was systematically investigated by combining various techniques, such as X-ray powder diffraction (XRPD), the Rietveld method, scanning electron microscopy (SEM), energy dispersive X-ray spectroscopy

(EDS), and inductively coupled plasma–optical emission spectrometry (ICP–OES). The material was then further analyzed from a viewpoint of chemistry and mineralogy. We published the report of the hydrothermal reaction with the reaction time in another journal [22]. Here, we only present the results of the hydrothermal reaction with the reaction temperature and the reactant ratio.

2. Materials and Methods

Materials and methods were briefly described in the publication of Liu et al. [18] and are stated in detail here for better understanding.

2.1. Materials

K-feldspar was obtained from a quarry near Geshi Town, Ningyang County, Shandong Province, China, and its chemical formula is $(K_{0.96}Na_{0.04})(Al_{1.00}Si_{3.00})O_{8.00}$. Lime (CaO, AR) was obtained from Sinopharm Chemical Reagent Co., Ltd. (Beijing, China).

2.2. Hydrothermal Method

K-feldspar blocks were crushed using a jaw crusher and then pulverized by a Vibratory Disc Mill RS 200 (Retsch, Hann, Germany). Lime blocks were manually crushed and ground in an agate mortar. Both materials were sieved to a particle size of less than 74 μm by a 200-mesh screen.

2.2.1. Reaction Temperature

A mixture of K-feldspar powder (5.5 g), lime powder (4.5 g), and deionized water (30 mL) was homogenously blended for five minutes by a magnetic agitator at the speed of 500 rpm, and then was heated in an autoclave (100 mL volume) for 20 h at 130, 160, 190, 220, and 250 °C. After the hydrothermal reaction ended, the reactors were naturally cooled to room temperature. For comparison, the same operation was carried out for a mixture of 5.5 g K-feldspar powder, 4.5 g lime powder, and 30 mL deionized water in an autoclave at room temperature (25 °C). Then, the production materials were removed from the autoclaves and dried at 105 °C for 12 h. Finally, the samples were manually crushed and ground in an agate mortar to prepare them for assessment. These corresponding production materials at 25, 130, 160, 190, 220, and 250 °C were labeled as L–1, L–2, L–3, L–4, L–5, and L–6, respectively.

2.2.2. Reactant Ratio

A 10 g solid powder mixture of K-feldspar and lime was cooked along with 30 mL deionized water in the autoclaves at 190 °C for 13.6 h. The mixture was obtained by homogeneously blending K-feldspar powder and lime powder in dry conditions, and K-feldspar/lime mass ratio in the mixture was 7/3, 6.5/3.5, 6/4, 5.5/4.5, 5/5, 4.5/5.5, 4/6, 3.5/6.5, and 3/7, respectively. After the hydrothermal reaction was stopped, the reactors were naturally cooled to room temperature. Then, the productions were removed from those autoclaves and were dried at 105 °C for 12 h. Lastly, they were crushed and ground for evaluation. The hydrothermal productions for various K-feldspar/lime mass ratios of 7/3, 6.5/3.5, 6/4, 5.5/4.5, 5/5, 4.5/5.5, 4/6, 3.5/6.5, and 3/7 were labeled as M–1, M–2, M–3, M–4, M–5, M–6, M–7, M–8, and M–9, respectively (Table S1).

2.3. Leaching Experiment

To analyze the chemical composition, a 1-g aliquot of each sample (fifteen samples, from L–1 to L–6 and from M–1 to M–9) was added to 100 ml of deionized water, and another 1-g aliquot was added to 100 ml of 0.5 mol·L^{-1} HCl. Both samples were placed in a constant-temperature oscillator for one hour and filtered through filter paper. The filtered solutions and residues were conserved for further analysis.

2.4. Analytical Methods

2.4.1. SEM and EDS

The morphology of the powdered materials was studied using a Leo 1450 VP SEM instrument (Carl Zeiss AG, Jena, Germany) under high vacuum on the uncoated specimen, operated at 20 kV and with a beam current of 1 nA. EDS using an energy dispersive X-ray spectrometer (mounted on the scanning electron microscope, Kevex superdry EDS, Valencia, CA, USA) was performed to measure an average composition of spherical and bar particles of Sample L–6. The H atom was not provided by EDS because of the limited resolution.

2.4.2. XRPD

To distinguish the phases formed during the hydrothermal reactions, XRPD patterns of all of the powdered samples (fifteen samples, L–1 to L–6 and M–1 to M–9) were obtained using a Dmax2400 X-ray diffractometer (Rigaku, Tokyo, Japan) at 40 kV and 80 mA in the 2 theta range from 3° to 65° at a step of 0.02° with a counting time of 0.3 s per step and a total scan time of 15 min 30 s. Quantitative phase analysis (QPA) of the XRPD results was conducted using GSAS [23] and EXPGUI [24] software with the Rietveld method [25]. The instrumental parameter file was obtained from the XRPD pattern of LaB$_6$ standard (SRM 660b, National Institute of Standards and Technology, USA).

2.4.3. ICP-OES

The element concentrations in the filtrated solutions were determined using ICP–OES with IRIS advantage inductively coupled argon plasma optical emission spectrometers (Thermo Fisher Scientific Inc., Waltham, MA, USA), and its limit of detection (LOD) was 3σ (σ is the standard deviation of blank determination), and the relative standard deviation (RSD) was approximately 0.5%–2.0%.

3. Results and Discussion

In an early study [18], the authors performed the hydrothermal reaction using the same process as that mentioned in the experimental section. Samples of L–1 to L–6 and samples of M–1 to M–9 in this work corresponded to from LSF12–1 to LSF12–6 and from LSF11–1 to LSF11–9, respectively, in a previous work [18]. In that report, we aimed to optimize the experimental parameters by the single-factor analysis in order to improve the percentage of K-feldspar dissolution. Therefore, we have not presented the results reported in that paper yet. Here, we only show new analysis and supply additional data to better understand the physicochemical properties of K-feldspar and lime hydrothermal reactions.

3.1. Phase

3.1.1. Crystal

With the reaction temperature and the reactant ratio (the Ca/Si molar ratio), the hydrothermal treatment of K-feldspar with lime exhibited complex product phases, including remaining K-feldspar, calcium hydroxide, calcite, grossular, tobermorite, alpha-dicalcium silicate hydrate (α-C$_2$SH), potassium carbonate, bütschliite, and amorphous calcium silicate hydrate (ACSH) (Figure 2, Table 1 and Table S2). In fact, the products were finally formed through a two-step process (Figure S1). First, calcium (alumino) silicate hydrates and potassium or calcium hydroxide were produced in the pressure-tight autoclaves during heating. Then, carbonates were formed through the reaction between their corresponding hydroxides and carbon dioxide during the drying of the sample at 105 °C in air.

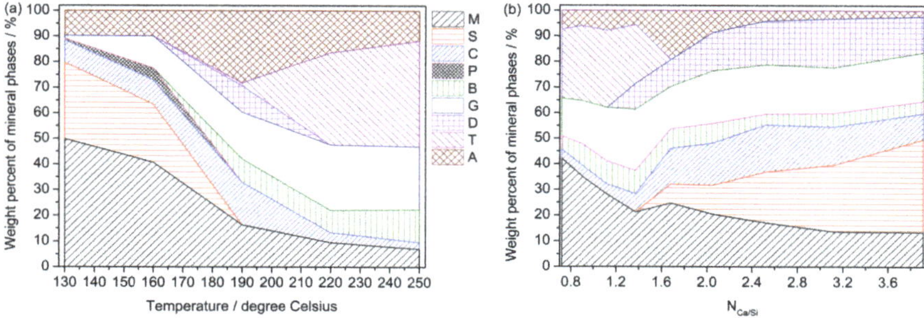

Figure 2. Results of quantitative phase analysis by the Rietveld method: (**a**) with reaction temperature; (**b**) with Ca/Si molar ratio (see the detail of the reactant ratio in Table S1). Abbreviations: M—K-feldspar [$KAlSi_3O_8$]; S—Calcium hydroxide [slaked lime (portlandite), $Ca(OH)_2$]; C—Calcite [$CaCO_3$]; P—Potassium carbonate [K_2CO_3]; B—Bütschliite [K_2CaCO_3]; G—Grossular [$Ca_3Al_2(SiO_4)_{1.53}(OH)_{5.88}$]; D—alpha-dicalcium silicate hydrate [α-C_2SH, $Ca_2(SiO_3OH)(OH)$]; T—Tobermorite [$Ca_{2.25}(Si_3O_{7.5}(OH)_{1.5})(H_2O)$]; A—Amorphous phase.

Table 1. An overview of mineral phases with various reaction conditions.

Single Factor	$N_{Ca/Si}$	Temperature	Time [b]
Experimental condition	Temp = 190 °C Time = 13.6 h $0.72 \leq N_{Ca/Si} \leq 3.91$	$130\,°C \leq Temp \leq 250\,°C$ Time = 20 h $N_{Ca/Si} = 1.37$	Temp = 190 °C $0\,h \leq Time \leq 36\,h$ $N_{Ca/Si} = 1.37$
M [a]	all	all	all
S	$N_{Ca/Si} > 1.37$	$130\,°C \leq Temp < 190\,°C$	Time < 16 h
C	all	all	all
P	/	$130\,°C \leq Temp < 190\,°C$	4 h < Time < 16 h
B	all	Temp > 160 °C	Time > 8 h
G	all	all	Time > 4 h
D	$N_{Ca/Si} > 1.12$	$16\,°C < Temp < 220\,°C$	8 h < Time < 32 h
T	$N_{Ca/Si} < 1.68$	Temp > 160 °C	Time > 16 h
A	all	all	all

Note: [a] Abbreviations: M—K-feldspar [$KAlSi_3O_8$]; S—Calcium hydroxide [slaked lime (portlandite), $Ca(OH)_2$]; C—Calcite [$CaCO_3$]; P—Potassium carbonate [K_2CO_3]; B—Bütschliite [K_2CaCO_3]; G—Grossular [$Ca_3Al_2(SiO_4)_{1.53}(OH)_{5.88}$]; D—alpha-dicalcium silicate hydrate [α-C_2SH, $Ca_2(SiO_3OH)(OH)$]; T—Tobermorite [$Ca_{2.25}(Si_3O_{7.5}(OH)_{1.5})(H_2O)$]; A—Amorphous phase; [b] the data of phase with reaction time are cited from the previous work [22].

With the increase of the reaction temperature, the amount of K-feldspar decreased in terms of the quantitative phase analysis data (Figure 2 and Table S2). However, K-feldspar were still discernible when the temperature was increased to 250 °C, which suggests that the K-feldspar structure is difficult to destroy under these hydrothermal conditions. All calcium hydroxide apparently reacted when the temperature reached 190 °C (Table S2). With the increasing of the Ca/Si ratio, calcium hydroxide appeared when the Ca/Si ratio reached 1.68, indicating that calcium hydroxide was apparently leftover. More and more calcium hydroxide was formed. The QPA results demonstrated that the amount of leftover K-feldspar decreased with the increasing Ca/Si molar ratio (Figure 2 and Table S2), and this was in accord with its decreasing ratio in the starting material. However, there still left about 13% K-feldspar when the Ca/Si ratio reached 3.91 (the weight ratio between K-feldspar and lime is 3:7), and approximately 60% K-feldspar reacted with lime in terms of an estimation. K-feldspar did not react with lime at room temperature (25 °C) because only calcium hydroxide and calcite were formed for L-1 while the sample was dried at 105 °C in air (Table S2). In fact, all carbonates were formed while the sample was dried at 105 °C in air. Calcite appeared in all the produced materials (Table 1). With the exception of potassium carbonate, the double carbonate bütschliite began to appear.

Bütschliite can be obtained in the systems of $CaCO_3$—K_2CO_3–H_2O and $CaCO_3$/$Ca(OH)_2$–KOH–H_2O at room temperature [26,27]. This was proven by the fact that bütschliite, calcium hydroxide, and calcite were identified as the phases contained in the solid residue. The solid residue was obtained by vaporing a mixed solution filtrated from a mixture of a 1-g aliquot of each sample (nine samples, M–1 to M–9) and 100 ml of deionized water (Figure S2). Therefore, potassium hydroxide, which was not detected by XRPD, reacted with calcium hydroxide and carbon dioxide to form bütschliite during the drying of the sample at 105 °C in air.

Three crystal calcium aluminosilicate hydrates, i.e., grossular (hydrogarnet), α-C_2SH, and tobermorite, were formed under the set hydrothermal conditions (Figure 2, Table 1 and Table S2). Higher contents of grossular and tobermorite generated at higher temperatures. Grossular stably existed after its appearance. However, the generation of α-C_2SH and tobermorite was greatly affected by the reaction condition. For example, with the increase in temperature, α-C_2SH only appeared at 190 °C, and tobermorite existed from 190 °C to 250 °C. With the increase of the Ca/Si ratio, α-C_2SH stably existed after its appearance when the Ca/Si ratio reached 1.37. However, tobermorite disappeared when the Ca/Si ratio reached 1.68.

The K-feldspar–lime hydrothermal system is composed of K_2O–CaO–Al_2O_3–SiO_2–H_2O, i.e., K–C–A–S–H, which has the same elements as those of CaO–Al_2O_3–SiO_2–H_2O (C–A–S–H) except K. However, the minerals of tobermorite, α-C_2SH, and hydrogarnet are common phases for both systems [15–17,28–32]. Taylor summarized the phase stability of the CaO–SiO_2–H_2O (C–S–H) system: 11 Å tobermorite stably exists below 200 °C, and α-C_2SH exists below 150 °C [16,17]. Meller et al. [16] determined that the phase stability was radically altered by small quantities of additives. In their study, as the upper temperature limits of α-C_2SH (<250 °C) and 11 Å tobermorite (<300 °C) increased, hydrogarnet was the only aluminum-bearing phase that was stable below 300 °C, and the presence of alumina greatly increased the stability of 11 Å tobermorite. In this study, tobermorite was stable from 190 °C to 250 °C, grossular was stable from 160 °C to 250 °C, and α-C_2SH appeared only at 190 °C. These results are consistent with the results obtained in the studies by Ríos et al. and Meller et al. [15,16]. There was no evidence that Al-substituted tobermorite was formed; however, it is possible that the Al substitution for Si in tobermorite increased its stability [33,34]. Besides, the generation of α-C_2SH and tobermorite was greatly affected by the initial Ca/Si ratio in the starting material (Table 1). Here, the analysis shows that grossular was generated prior to α-C_2SH and tobermorite with an increase in the reaction temperature, revealing that it is easier to generate grossular than α-C_2SH and tobermorite under the hydrothermal conditions. Since the Ca atoms of grossular, α-C_2SH, and tobermorite originate from lime, both Si and Al originate from K-feldspar. Therefore, there is no competition of parallel reactions. The phase transformation is also true for α-C_2SH because α-C_2SH appeared at 190 °C and disappeared at 220 °C, showing that α-C_2SH is metastable and transforms to a more stable phase at higher temperatures [17,35].

As a layered silicate, tobermorite has high exchangeability and selectivity for cations [11], which can alleviate the damage of heavy metals such as Cd, Pb, and Cr on soil [13,36,37]. Tobermorite is also a slow-release reservoir for some nutrients, such as [K^+] and [NH_4^+] [14]. As a primary product of the hydrothermal reaction, the weight percentage of tobermorite can reach 35%–40% by controlling the reaction conditions (Table S2). Tobermorite has a structure similar to that of the 2:1 clay minerals, and this structure plays a key role in the soil K cycle [38] with a positive correlation to the pH buffering capacity [39,40], thereby benefitting the soil.

The general formula of 11 Å tobermorite may be written as $Ca_{4+x}(Al_ySi_{6-y})O_{15+2x-y}(OH)_{2-2x+y}\cdot5H_2O$, with $0 \leq x \leq 1$ and $0 \leq y \leq 1$. The variable x represents the amount of additional calcium hosted in the structural cavities. Consequently, 11 Å tobermorite is actually a series between two endmembers, $Ca_4Si_6O_{15}(OH)_2\cdot5H_2O$ and $Ca_5Si_6O_{17}\cdot5H_2O$ [41]. Therefore, the Ca/Si ratio in the structure of tobermorite is [0.67, 0.83]. Based on a magnified view of XRPD, the CSH gel phase—rather than tobermorite—was formed when the Ca/Si ratio deviated from an ideal one in the structure of tobermorite (the star phase in Figure S3). Wang et al. [42] found that the deviation of the Ca/Si molar ratio from 0.8 to 1.4 impeded the appearance of tobermorite, and this conclusion is in accord with the trend of the tobermorite appearance with the Ca/Si ratio. However, with the increasing of the Ca/Si ratio, it seemed beneficial for the formation of α-C_2SH due to its high Ca/Si ratio. It is well known that Al-substituted tobermorite will lead an increase of the basal spacing (002) between the silicate layers along the *c* axis and the basal spacing increases with Al content [33,43,44]. Compared to tobermorite synthesized without Al, the (002) diffraction peak of samples from L–4 to L–6, from M–1 to M–4, and from Tm–6 to Tm–10 shifted towards lower angles (Figure 3). The content of Al for Si substitution should have been similar because two theta angles of the (002) diffraction peaks of all these samples were almost the same. Considering that the Al/Si ratio was constant due to their shared source from K-feldspar, the conclusion seems reasonable. The EDS data showed that tobermorite contained K and Al atoms. Therefore, we concluded that Al-substituted tobermorite was formed (Table S6).

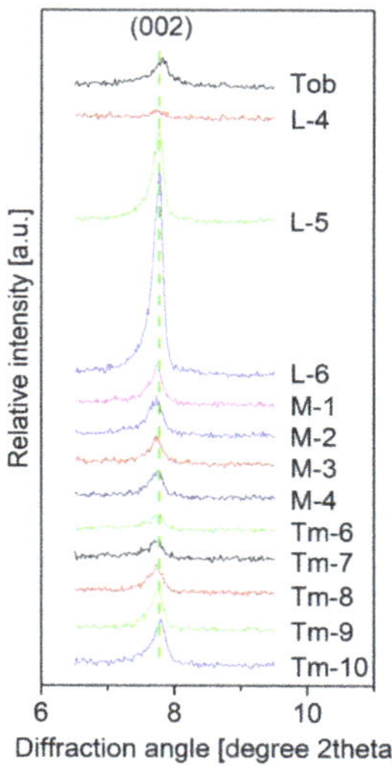

Figure 3. Enlargement of the diffraction peak of tobermorite (002). The sample (Tob) without Al was synthesized with a 0.83 Ca/Si molar ratio under the same temperature, time and water/solid ratio to those of M–1 to M–9, and Ca and Si sources were from lime and fumed silica powder, respectively. Data of Tm–6 to Tm–10 were from the previous publication [22].

The hydrogarnet is an important mineral that is often found in nature and hydrated cement. A calcic hydrogarnet that occurs frequently in the literature is hydrogrossular, $Ca_3Al_2(SiO_4)_{3-x}(OH)_{4x}$ ($0 \leq x \leq 3$), and has intermediate compositions between grossular ($x = 0$) and katoite ($x = 3$). The structure of hydrogarnet (cubic, space group $Ia\bar{3}d$, no. 230) consists of AlO_6 octahedra and SiO_4 tetrahedra linked with oxygen and the hydroxyl component (OH) in a solid solution with both (OH) and (SiO_4) units depending on the composition x [45,46]. A linear relationship between lattice parameter *a* and *x* or *a* and Si content was observed for natural and artificial hydrogarnet [32,47,48], and a good agreement of the Si content estimates from XRD and EPMA was proven [32]. By plotting Si content ($3 - x$) against lattice parameter *a* from the powder diffraction file (PDF) data, a linear equation was obtained for hydrogarnet (Figure 4a):

$$3 - x = 48.4521 - 3.8529a \qquad (1)$$

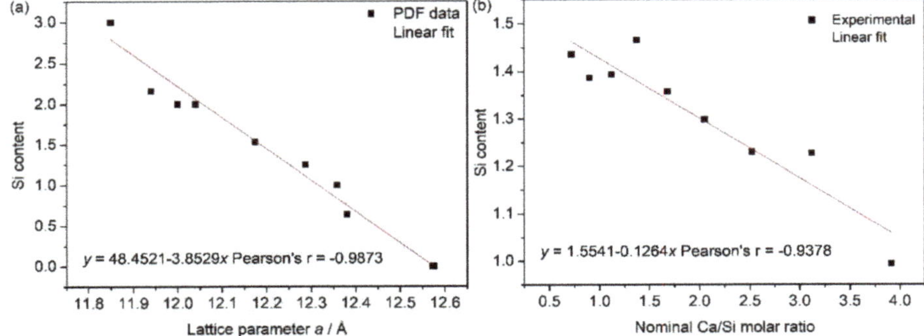

Figure 4. (**a**) Plot of Si content against lattice parameter *a* of hydrogarnet from the powder diffraction file (PDF) data (see Table S4); (**b**) plot of Si content of hydrogarnet calculated from Equation (1) against nominal Ca/Si molar ratio in the starting material (see Table S5).

The Si content of grossular in the samples of M–1 to M–9 was calculated from Equation (1) and was plotted against the Ca/Si ratio in the starting material in Figure 4b. These values of Si content calculated from Equation (1) basically agreed with those obtained in terms of Figure 3 from Kyritsis et al. [32], Figure 1A from Okoronkwo and Glasser [48], and Figure 2 from Adhikari et al. [49]. It is observed from Figure 4b that more hydrogen was incorporated in hydrogarnet through $H_4O_4 \leftrightarrow SiO_4$ with an increase of the Ca/Si molar ratio in the starting material (nominal Ca/Si molar ratio), which is basically consistent with the conclusion of the references [16,32]. Particularly, more hydrogen incorporation proceeded after calcium hydroxide appeared at $N_{Ca/Si} = 1.68$, which agreed with the shift of XRPD towards a lower angle for samples from M–5 to M–9 (a magnified view of the characteristic diffraction peak (420) of hydrogarent in Figure 5b). However, the ratio of $H_4O_4 \leftrightarrow SiO_4$ in hydrogarnet seemed to remain constant due to the same position of the diffraction peak (420) of hydrogarent for the samples of L–2 to L–6 and of Tm–3 to Tm–10. This indicated that neither the reaction temperature nor the reaction time affected the substitution of $H_4O_4 \leftrightarrow SiO_4$ in hydrogarnet if there was the same Ca/Si molar ratio in the starting material.

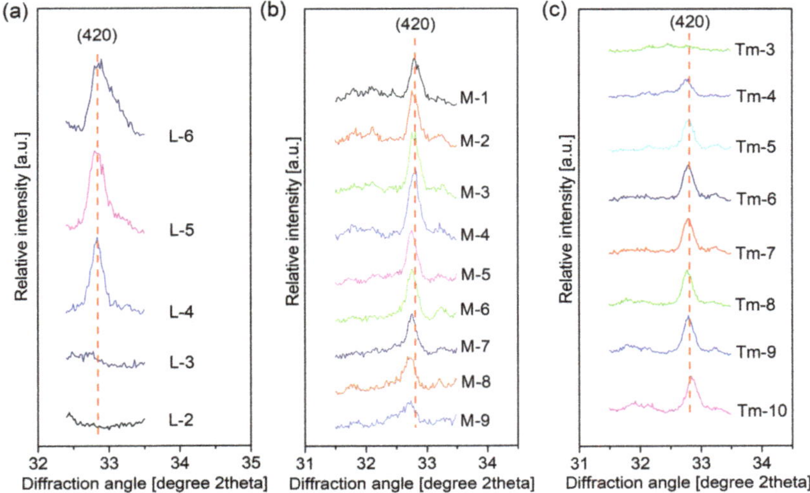

Figure 5. Diffraction peak from hydrogarnet (420) with various hydrothermal conditions: (**a**) reaction temperature; (**b**) Ca/Si molar ratio ($N_{Ca/Si}$); (**c**) reaction time (data of Tm–3 to Tm–10 were from the previous publication [22]).

In order to obtain one single phase, some researchers used the responding oxides to prepare tobermorite and hydrogarnet. For example, tobermorite was synthesized from lime and quartz or lime and fumed silica powder under the hydrothermal condition with the designed Ca/Si molar ratio (0.83) according to the ideal composition of tobermorite ($Ca_5Si_6O_{16}(OH)_2 \cdot 4H_2O$) [13,33, 44,50]. Hydrogarnet was synthesized through silica gel, calcium hydroxide, and γ-alumina [51]. Therefore, the generation of the abovementioned phases was closely related to the atom ratio rather than the sources of Ca, Si, and Al in the starting materials. In this study, Ca was from lime, and Al and Si were both from K-feldspar. In terms of estimated Si content, the Al/Si ratio in hydrogarnet was [1.39, 2.01], which was greater than the ratio of 1:3 in the K-feldspar structure. When one structural unit of K-feldspar was destroyed by Ca^{2+} and OH^- attacking, K^+ was leached into the solution [19]. Released Al and Si combined with Ca to form hydrogarnet, and residual Si reacted with Ca to form crystal or ACSH. Al of K-feldspar mainly entered into the structure of hydrogarnet, and Si distributed in both hydrogarnet and caclium silicate hydrates.

3.1.2. Amorphous

With an increase in the reaction temperature and the Ca/Si ratio, the ACSH diffraction peak was verified by a slight bump at 29° for M–5 to M–9 (* phase in Figure S3) and L–2 to L–3 (* phase in Figure S4). This was also true for the samples of Tm–3 to Tm–5 with a time increase [22]. In fact, the bump corresponded to a diffraction peak of CSH gel and became a stronger diffraction peak of tobermorite when the CSH gel was transformed into tobermorite [52–55]. CSH gel is the principal hydration product and primary binding phase in C–A–S–H or C–S–H systems such as Portland cement. It is generally accepted that the CSH gel phase has a tobermorite-like structure, which can be described as a layer of CaO polyhedra sandwiched between individual silicate chains that are organized in a "dreierketten" structure [10,56]. Some researchers confirmed that lime and Si-containing phases, such as quartz or cement, produced CSH gel with relatively low crystallinity during the hydrothermal reaction and then formed tobermorite with higher crystallinity [57,58]. Shaw et al. employed in situ energy-dispersive X-ray diffraction techniques to study the hydrothermal formation of crystalline tobermorite from 190 °C to 310 °C [57]. First, a poorly crystalline CSH gel phase was formed. The second stage involved the ordering of the CSH gel to form ordered crystalline tobermorite. Based on the phase analysis, ACSH existed in all samples (Table S2). K-feldspar was the only residual after the product was filtrated by 0.5 mol·L^{-1} HCl. Therefore, these amorphous phases should have been acid soluble [22]. All formed phases were concluded to be calcium silicate hydrate or calcium aluminosilicate hydrate by EDS analysis, and the amorphous phases were mainly composed of calcium silicate hydrates or some transition state compounds that were discussed in a previous work [19]. Here, ACSH may have contained other amorphous phases—except calcium silicate hydrates—and all of these phases were judged to be amorphous by XRPD [20]. Therefore, the authors did not further determine the chemical compositions of ACSH because it was still difficult to distinguish different crystal or amorphous phases, even using Electron Probe Micro-Analyzer (EPMA) due to small particles and quantities of intertwining phases [19,20].

3.2. Morphology

Considering the complexity of the products and the inability to identify the phases of different morphology, we do not discuss the SEM images of the samples of M–1 to M–9 and only present the SEM images of the starting materials K-feldspar and lime (Figure 6) and the samples of L–1 to L–6 (Figure 7).

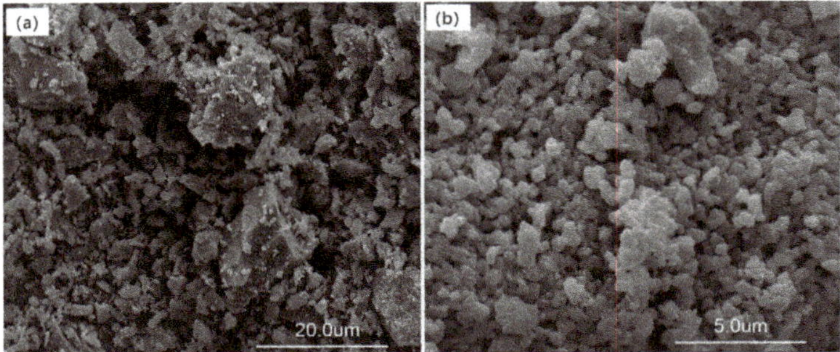

Figure 6. SEM images: (**a**) K-feldspar and (**b**) lime.

Figure 6 shows that the unreacted K-feldspar powder consisted of a few to tens of micron-sized particles and that lime was composed of a few micron-sized particles, which were obviously prepared from a calcining material. When K-feldspar and lime were kept under a closed hydrothermal reactor for 13.6 h at room temperature, no reaction occurred. However, lime reacted with water to form calcium hydroxide and further carbonized to form calcite during the drying of the sample at 105 °C in air. Figure 7a clearly shows that several floccules adhered to the K-feldspar surface, and these floccules were identified as calcium hydroxide and calcite based on the XRPD results (Table S2). The morphology of the hydrothermal product L–2 was different from that of L–1. Large quantities of floccules were deposited on the surface of the K-feldspar particle, and several spherical particles appeared (Figure 7b). When the temperature was increased to 160 °C, many squamiform slices appeared on the K-feldspar particles, and these slices may have been calcium hydroxide, calcite, and potassium carbonate (Figure 7c). At 190 °C, the products were diverse, and it was difficult to distinguish the phases by their different morphologies (Figure 7d). At higher temperatures, the amount of spherical and laminar particles increased, which was in accordance with the QPA results demonstrating that more grossular and tobermorite were formed (Figure 7e,f). The chemical compositions of the laminar and spherical particles were determined using EDS (Figure 8 and Table S6). The Ca/Al and Ca/(Al + Si) atom ratios determined by EDS were 1.63 and 0.62 for the spherical particle (Figure 8a), respectively, and approached the corresponding ratio values of grossular [$N_{Ca/Al}$ = 1.5 $N_{Ca/(Al+Si)}$ = 0.85, $Ca_3Al_2(SiO_4)_{1.53}(OH)_{5.88}$, PDF card number 01-075-1690]. The Ca/Si atom ratio determined by EDS was 0.72 for the laminar particle (Figure 8b) and approached its corresponding ratio value of tobermorite [$N_{Ca/Si}$ = 0.75, $Ca_{2.25}(Si_3O_{7.5}(OH)_{1.5})(H_2O)$, PDF card number 01-083-1520]. The spherical particle contained a K atom and the laminar one contained both K and Al atoms, and this may have been caused by the intertwining of the formed phases with each other. As a result, the ratio determined by EDS slightly deviated from the reference value. Therefore, in this study, the laminar phase was tobermorite, as identified in our previous study and by other researchers [19,59–62], and the spherical particle was grossular, as identified in our previous study and by other researchers [19,60,63].

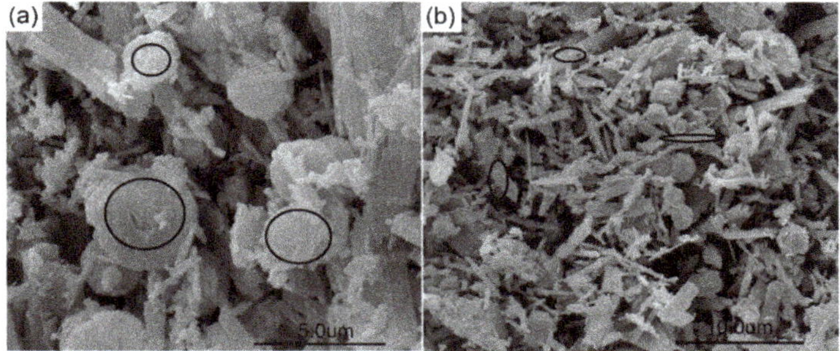

Figure 8. SEM magnified image of L–6: (**a**) spots of energy dispersive X-ray spectroscopy (EDS) for spherical particle; (**b**) spots of EDS for laminar particle. EDS were determined for the spots in the black circles.

3.3. Dissolution Mechanism

For investigating the dissolution mechanism of K-feldspar under hydrothermal condition, we analyzed the percentage of K-feldspar dissolution (Table S7, and the percentage of K-feldspar dissolution was calculated from their corresponding element contents in Table S3; please see the Supplementary Materials) with the reaction temperature. Because K-feldspar did not react with lime at room temperature (25 °C), the authors did not consider the data of K-feldspar dissolution at this temperature. Only five-point data were fitted between the natural logarithm of the dissolution percentage (lnD) and the reciprocal of the temperature ($1/T$, K^{-1}). When lnD versus $1/T$ (K^{-1}) was plotted in Figure 9, a turning point was found rather than a linear relationship. As a result, the plot of lnD vs. $1/T$ could be divided into two regions: I, [130 °C, 190 °C], and II, [190 °C, 250 °C]. Then, two fitted lines were applied to these two regions:

$$I: lnD = 13.4850 - \frac{4346.3643}{T} \quad (403K \leq T \leq 463K;\ \text{Pearson's } r = -0.9914) \tag{2}$$

$$\text{II}: \ lnD = 7.6155 - \frac{1593.4882}{T} \quad (463K \leq T \leq 523K; \ \text{Pearson's } r = -0.9843). \tag{3}$$

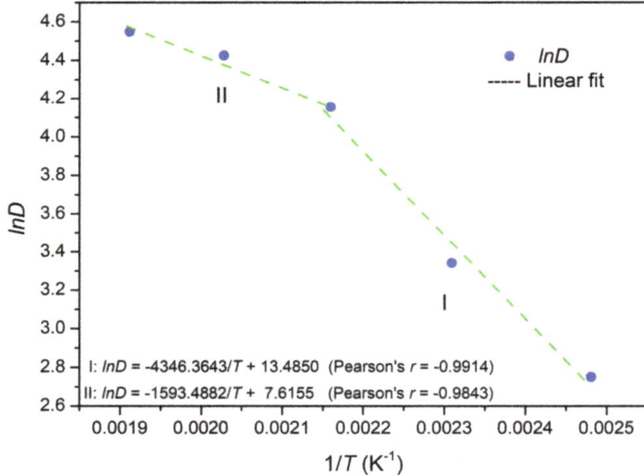

Figure 9. Natural logarithm of the percentage of dissolution (lnD) versus the reciprocal temperature $(1/T, \mathrm{K}^{-1})$. ● lnD; — linear fit.

In terms of the Arrhenius equation, the following represents the relationship between lnD and the dissolution apparent activation energy Ea:

$$lnD = lnA - Ea/RT. \tag{4}$$

When Equation (4) was differentiated, the following equation was obtained:

$$Ea = -R\frac{dlnD}{d\left(\frac{1}{T}\right)}. \tag{5}$$

Based on Equation (5), the activation energies were calculated as 36 kJ/mol at $130\,^{\circ}\mathrm{C} \leq T \leq 190\,^{\circ}\mathrm{C}$ and 13 kJ/mol at $190\,^{\circ}\mathrm{C} \leq T \leq 250\,^{\circ}\mathrm{C}$.

In general, Ea is a constant over a small temperature range, which is true for regions I and II in Figure 9. Using a similar hydrothermal condition at higher temperatures (260–290 °C), Liu et al. [64] found a linear relationship between lnK and $1/T$ and calculated the activation energy for the K-feldspar and slaked lime hydrothermal reaction. However, the activation energy calculated by those authors was 23 kJ/mol, which was larger than 13 kJ/mol at $190\,^{\circ}\mathrm{C} \leq T \leq 250\,^{\circ}\mathrm{C}$ and less than 36 kJ/mol at $130\,^{\circ}\mathrm{C} \leq T \leq 190\,^{\circ}\mathrm{C}$ in this work. It is reasonable to assume that different experimental conditions or approaches can cause a difference in the activation energy calculations [65].

From the activation energy data, we concluded that two different processes controlled the reaction, and a sketch of these processes is illustrated in Figure 10. First, the reaction was controlled by a dissolution-precipitation process (region I). At higher temperatures, the reaction was dominated by a diffusion process due to the quantities of deposited phases (region II). A similar trend was also found for the dissolution of K-feldspar with the reaction time. Through a preliminary interface study, Liu et al. [19] concluded that the reaction between K-feldspar and lime under hydrothermal conditions was a mineral-mineral replacement reaction proceeded by the dissolution-precipitation mechanism. The dissolution-precipitation mechanism was true for a reaction that occurs on the K-feldspar particle surface at medium temperatures, i.e., 160–190 °C. Under this condition, the rate of K-feldspar dissolution was close to the deposition rate of the formed phases. However, at higher temperatures, due to the deposition of large quantities of mineral phases on the K-feldspar surface, the diffusion of K^+ and Ca^{2+} was blocked. The mechanism in Figure 10 is consistent with the SEM results (Figure 7) because more phases were deposited on the K-feldspar surface at higher temperature.

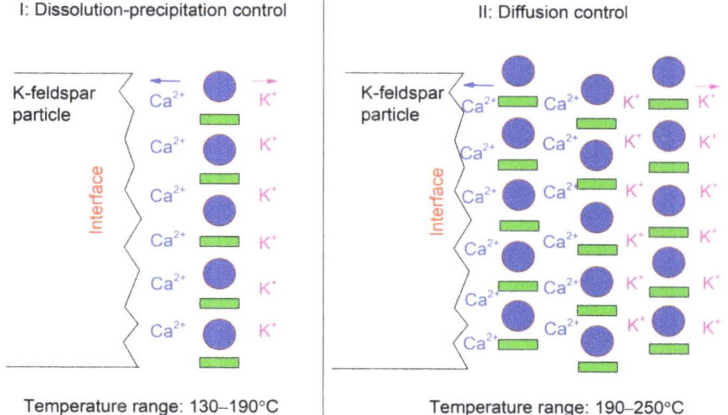

Figure 10. Hydrothermal dissolution mechanism of K-feldspar from 130 °C to 250 °C.

By studying the kinetics and mechanism of the hydrothermal synthesis of barium titanate, Eckert et al. found that multiple reaction mechanisms competed in different regimes [66]. Berger et al. measured the rate of mineral dissolution of sanidine from 100 °C to 300 °C and suggested a competition between two parallel reactions at the mineral surface [67]. In this work, different reaction mechanisms dominated in various temperature regions, which is in agreement with the abovementioned conclusions.

4. Conclusions

Based on an environmentally friendly hydrothermal process, the artificial silicate composite material generated from natural silicate rock, i.e., K-feldspar, exhibited complex mineralogical properties. Except carbonates (namely, potassium carbonate, calcite, and bütschliite), calcium (alumno) silicates—including grossular, tobermorite, alpha-dicalcium silicate hydrate, and amorphous calcium silicate hydrate—were the main products of the K-feldspar and lime hydrothermal reaction in the 130 °C to 250 °C temperature range and in the 7:3 to 3:7 K-feldspar/lime weight ratio range (i.e., the Ca/Si molar ratio from 0.72 to 3.91). Both the temperature and the Ca/Si molar ratio in the starting material greatly affected the formation of phases, especially the generation of tobermorite and α-C_2SH. The substitution of $H_4O_4 \leftrightarrow SiO_4$ proceeded with the increase of the Ca/Si molar ratio rather than the reaction temperature and the reaction time. More hydrogen was incorporated in hydrogarnet through the substitution of $H_4O_4 \leftrightarrow SiO_4$ with the increase of the Ca/Si molar ratio in the starting material.

The dissolution mechanism analysis showed that a turning point between the natural logarithm of the percentage of K-feldspar dissolution and the reciprocal of the temperature (K^{-1}) appeared, implying that different reaction mechanisms controlled K-feldspar dissolution. The apparent activation energies were also calculated.

Based on the results of phase analysis and from a point of environmental and economic views, the mass ratio of K-feldspar and lime should be more than 55:45 to obtain optimized properties of synthesized silicate composite material. More lime in the starting materials means higher product cost due to the higher price compared with K-feldspar. However, because of the properties of tobermorite as a cation exchanger and its potential applications in hazardous waste disposal, experimental parameters should be optimized to obtain better performance of the artificial silicate composite material from the K-feldspar and lime hydrothermal system. A detailed study is worthy of being carried out for investigating the relationship between the mineralogy and the function of the artificial silicate composite material.

As a potential sustainable agromineral, the artificial composite silicate material is particularly appealing, not only because of its green production process, but also because of its excellent performances and multiple functions of remediating the soil [8,9]. Primary studies proved that these excellent performances and functions were closely related to its physicochemical properties and mineral components [8,9]. Therefore, the study yields vital and valuable insights into comprehending the mineralogical properties of the K-feldspar and lime hydrothermal reaction and producing artificial silicate agrominerals from potassium-rich rocks.

Supplementary Materials: The following are available online at http://www.mdpi.com/2075-163X/9/1/46/s1, Element content and QPA; Nomenclature of grossular and hydrogarnet; Definition of the percentage of K-feldspar dissolution (D, %); Table S1: Overview of the hydrothermal reaction with reactant ratio (T = 190 °C, t = 13.6 h, and deionized water = 30 mL); Table S2: Quantitative phase analysis by the Rietveld method (%); Table S3: Element contents measured by ICP–OES and calculated from XRPD; Table S4: PDF data of hydrogarnet used in Figure 4a; Table S5: Estimated Si content, x, and Al/Si ratio of formed hydrogarnet from Equation (1) in this study; Table S6: EDS data of spherical and laminar particle in the sample L–6; Table S7: Percentage of K-feldspar dissolution and original data of Figure 9; Figure S1: Flow chart of mineral phases during the K-feldspar and lime hydrothermal reaction; Figure S2: XRPD of the residue of the filtrated solutions of the samples M–1 to M–9 dissolved in deionized water; Figure S3: A magnified view of XRPD pattern of M–1 to M–9 from 28.2° to 35°; Figure S4: A magnified view of XRPD pattern of L–1 to L–6 from 28.2° to 29.8°; Figure S5: Element contents of ICP–OES vs. XRPD: (a) the hydrothermal reaction with reaction temperature; (b) the hydrothermal reaction with reactant ratio.

Author Contributions: S.L. performed the experiments, analyzed the data and wrote the paper, C.H. and J.L. together presented the research plans and discussed the results.

Funding: This research project was supported by the Open Project Program of Key Laboratory of Mineral Resources, Institute of Geology and Geophysics, Chinese Academy of Sciences (No. KLMR2017-07), Projects in the National Science & Technology Pillar Program during the Eleventh Five-Year Plan Period (China, No. 2006BAD10B04), and the Knowledge Innovation Project of Chinese Academy of Sciences and Spark Program of China (No. 2007EA173003).

Acknowledgments: The authors were appreciated for the precious comments and suggestions of the editor and two reviewers.

Conflicts of Interest: The authors declare no conflict of interest.

References

1. Klein, C.; Philpotts, A.R. Chapter 7 Igneous rock-forming minerals. In *Earth Materials: Introduction to Mineralogy and Petrology*, 2nd ed.; Klein, C., Philpotts, A.R., Eds.; Cambridge University Press: New York, NY, USA, 2017; p. 161.

2. Xu, Y.; Liang, X.; Xu, Y.; Qin, X.; Huang, Q.; Wang, L.; Sun, Y. Remediation of heavy metal-polluted agricultural soils using clay minerals: A review. *Pedosphere* **2017**, *27*, 193–204. [CrossRef]

3. Guo, B.; Liu, B.; Yang, J.; Zhang, S. The mechanisms of heavy metal immobilization by cementitious material treatments and thermal treatments: A review. *J. Environ. Manag.* **2017**, *193*, 410–422. [CrossRef] [PubMed]

4. Misaelides, P. Application of natural zeolites in environmental remediation: A short review. *Microporous Mesoporous Mater.* **2011**, *144*, 15–18. [CrossRef]

5. Liu, J.M.; Liu, S.K.; Han, C.; Sheng, X.B.; Qi, X.; Zhang, Z.L. Mineral technology for soil remediation and improvement: A new applied research direction from mineralogy, petrology and geochemistry. *Bull. Mineral. Petrol. Geochem.* **2014**, *33*, 556–560.

6. Han, C.; Liu, J.M. Method for Preparing Microporous Mineral Fertilizer from Silicates Rock Using Hydrothermal Chemical Reaction. Patent WO2009070953A1, 5 December 2007.

7. Han, C. Method for Producing Micropore Silicon-Potassium-Calcium Mineral Fertilizer. Patent CN101054313B, 26 April 2007.

8. Liu, S.K.; Qi, X.; Han, C.; Liu, J.M.; Sheng, X.B.; Li, H.; Luo, A.M.; Li, J.L. Novel nano-submicron mineral-based soil conditioner for sustainable agricultural development. *J. Clean. Prod.* **2017**, *149*, 896–903. [CrossRef]

9. Liu, S.K.; Li, H.; Han, C.; Sheng, X.B.; Liu, J.M. Cd inhibition and pH improvement via a nano-submicron mineral-based soil conditioner. *Environ. Sci. Pollut. Res.* **2017**, *24*, 4942–4949. [CrossRef] [PubMed]

10. Richardson, I.G. The calcium silicate hydrates. *Cem. Concr. Res.* **2008**, *38*, 137–158. [CrossRef]
11. Komarneni, S.; Roy, D.M. Tobermorites: A new family of cation exchangers. *Science* **1983**, *221*, 647–648. [CrossRef] [PubMed]
12. Zou, J.J.; Guo, C.B.; Zhou, X.Q.; Sun, Y.J.; Yang, Z. Sorption capacity and mechanism of Cr^{3+} on tobermorite derived from fly ash acid residue and carbide slag. *Colloids Surf. A* **2018**, *538*, 825–833. [CrossRef]
13. Guo, X.L.; Shi, H.S. Microstructure and heavy metal adsorption mechanisms of hydrothermally synthesized Al-substituted tobermorite. *Mater. Struct.* **2017**, *50*, 10. [CrossRef]
14. Yao, Z.D.; Tamura, C.; Matsuda, M.; Miyake, M. Resource recovery of waste incineration fly ash: Synthesis of tobermorite as ion exchanger. *J. Mater. Res.* **1999**, *14*, 4437–4442. [CrossRef]
15. Rios, C.A.; Williams, C.D.; Fullen, M.A. Hydrothermal synthesis of hydrogarnet and tobermorite at 175 °C from kaolinite and metakaolinite in the $CaO-Al_2O_3-SiO_2-H_2O$ system: A comparative study. *Appl. Clay Sci.* **2009**, *43*, 228–237. [CrossRef]
16. Meller, N.; Kyritsis, K.; Hall, C. The mineralogy of the $CaO-Al_2O_3-SiO_2-H_2O$ (CASH) hydroceramic system from 200 to 350 °C. *Cem. Concr. Res.* **2009**, *39*, 45–53. [CrossRef]
17. Meller, N.; Hall, C.; Kyritsis, K.; Giriat, G. Synthesis of cement based $CaO-Al_2O_3-SiO_2-H_2O$ (CASH) hydroceramics at 200 and 250 °C: Ex-situ and in-situ diffraction. *Cem. Concr. Res.* **2007**, *37*, 823–833. [CrossRef]
18. Liu, S.K.; Han, C.; Liu, J.M.; Li, H. Research of extracting potassium, silica and aluminum from potassium feldspar by hydrothermal chemical reaction. *Acta Mineral. Sin.* **2009**, *29*, 320–326.
19. Liu, S.K.; Han, C.; Liu, J.M.; Li, H. Hydrothermal decomposition of potassium feldspar under alkaline conditions. *RSC Adv.* **2015**, *5*, 93301–93309. [CrossRef]
20. Ciceri, D.; de Oliveira, M.; Allanore, A. Potassium fertilizer via hydrothermal alteration of K-feldspar ore. *Green Chem.* **2017**, *19*, 5187–5202. [CrossRef]
21. Skorina, T.; Allanore, A. Alkali Metal Ion Source with Moderate Rate of Ion Release and Methods of Forming. U.S. Patent US9340465B2, 6 May 2013.
22. Liu, S.; Han, C.; Liu, J. Study of K-feldspar and lime hydrothermal reaction at 190 °c: Phase, kinetics and mechanism with reaction time. *ChemistrySelect* **2018**, *3*, 13010–13016. [CrossRef]
23. Larson, A.C.; Von Dreele, R.B. *General Structure Analysis System (GSAS)*; Los Alamos National Laboratory Report LAUR; Los Alamos National Laboratory: New Mexico, NM, USA, 2004; pp. 86–748.
24. Toby, B.H. EXPGUI, a graphical user interface for GSAS. *J. Appl. Crystallogr.* **2001**, *34*, 210–213. [CrossRef]
25. Rietveld, H.M. A profile refinement method for nuclear and magnetic structures. *J. Appl. Crystallogr.* **1969**, *2*, 65–71. [CrossRef]
26. Reardon, E.J.; Fagan, R. The calcite/portlandite phase boundary: Enhanced calcite solubility at high pH. *Appl. Geochem.* **2000**, *15*, 327–335. [CrossRef]
27. Pabst, A. Synthesis, properties, and structure of $K_2Ca(CO_3)_2$, Bütschliite. *Am. Mineral.* **1974**, *59*, 353–358.
28. Al-Wakeel, E.I.; El-Korashy, S.A. Reaction mechanism of the hydrothermally treated $CaO-SiO_2-Al_2O_3$ and $CaO-SiO_2-Al_2O_3-CaSO_4$ systems. *J. Mater. Sci.* **1996**, *31*, 1909–1913. [CrossRef]
29. Klimesch, D.S.; Ray, A. DTA-TG study of the $CaO-SiO_2-H_2O$ and $CaO-Al_2O_3-SiO_2-H_2O$ systems under hydrothermal conditions. *J. Therm. Anal. Calorim.* **1999**, *56*, 27–34. [CrossRef]
30. Klimesch, D.S.; Ray, A. DTA-TGA evaluations of the $CaO-Al_2O_3-SiO_2-H_2O$ system treated hydrothermally. *Thermochim. Acta* **1999**, *334*, 115–122. [CrossRef]
31. Watanabe, O.; Kitamura, K.; Maenami, H.; Ishida, H. Hydrothermal treatment of a silica sand complex with lime. *J. Am. Ceram. Soc.* **2001**, *84*, 2318–2322. [CrossRef]
32. Kyritsis, K.; Meller, N.; Hall, C. Chemistry and morphology of hydrogarnets formed in cement-based CASH hydroceramics cured at 200 to 350 °C. *J. Am. Ceram. Soc.* **2009**, *92*, 1105–1111. [CrossRef]
33. Guo, X.L.; Meng, F.J.; Shi, H.S. Microstructure and characterization of hydrothermal synthesis of Al-substituted tobermorite. *Constr. Build. Mater.* **2017**, *133*, 253–260. [CrossRef]
34. Organova, N.I.; Koporulina, E.V.; Ivanova, A.G.; Trubkini, N.V.; Zadov, A.E.; Khomyakov, A.P.; Marcille, I.M.; Chukanov, N.V.; Shmakov, A.N. Structure model of Al,K-substituted tobermorite and structural changes upon heating. *Crystallogr. Rep.* **2002**, *47*, 950–956. [CrossRef]
35. Hu, X.L.; Yanagisawa, K.; Onda, A.; Kajiyoshi, K. Stability and phase relations of dicalcium silicate hydrates under hydrothermal conditions. *J. Ceram. Soc. Jpn.* **2006**, *114*, 174–179. [CrossRef]
36. Pena, R.; Guerrero, A.; Goni, S. Hydrothermal treatment of bottom ash from the incineration of municipal solid waste: Retention of Cs(I), Cd(II), Pb(II) and Cr(III). *J. Hazard. Mater.* **2006**, *129*, 151–157. [CrossRef] [PubMed]

37. Coleman, N.J. Interactions of Cd(II) with waste-derived 11 angstrom tobermorites. *Sep. Purif. Technol.* **2006**, *48*, 62–70. [CrossRef]

38. Hinsinger, P. Potassium. In *Encyclopedia of Soil Science*; Lal, R., Ed.; Marcel Dekker Inc.: New York, NY, USA, 2002; pp. 1354–1358.

39. Xu, R.K.; Zhao, A.Z.; Yuan, J.H.; Jiang, J. pH buffering capacity of acid soils from tropical and subtropical regions of China as influenced by incorporation of crop straw biochars. *J. Soils Sediments* **2012**, *12*, 494–502. [CrossRef]

40. Weaver, A.R.; Kissel, D.E.; Chen, F.; West, L.T.; Adkins, W.; Rickman, D.; Luvall, J.C. Mapping soil pH buffering capacity of selected fields in the coastal plain. *Soil Sci. Soc. Am. J.* **2004**, *68*, 662–668. [CrossRef]

41. Biagioni, C.; Merlino, S.; Bonaccorsi, E. The tobermorite supergroup: A new nomenclature. *Mineral. Mag.* **2015**, *79*, 485–495. [CrossRef]

42. Wang, Z.H.; Ma, S.H.; Zheng, S.L.; Wang, X.H. Incorporation of Al and Na in Hydrothermally Synthesized Tobermorite. *J. Am. Ceram. Soc.* **2017**, *100*, 792–799. [CrossRef]

43. Matsui, K.; Kikuma, J.; Tsunashima, M.; Ishikawa, T.; Matsuno, S.-Y.; Ogawa, A.; Sato, M. In situ time-resolved X-ray diffraction of tobermorite formation in autoclaved aerated concrete: Influence of silica source reactivity and Al addition. *Cem. Concr. Res.* **2011**, *41*, 510–519. [CrossRef]

44. Maeda, H.; Abe, K.; Ishida, E.H. Hydrothermal synthesis of aluminum substituted tobermorite by using various crystal phases of alumina. *J. Ceram. Soc. Jpn.* **2011**, *119*, 375–377. [CrossRef]

45. Grew, E.S.; Locock, A.J.; Mills, S.J.; Galuskina, I.O.; Galuskin, E.V.; Halenius, U. Nomenclature of the garnet supergroup. *Am. Mineral.* **2013**, *98*, 785–810. [CrossRef]

46. Cohenaddad, C.; Ducros, P.; Bertaut, E.F. Etude de la substitution du groupement SiO_4 par $(OH)_4$ dans les composes $Al_2Ca_3(OH)_{12}$ et $Al_2Ca_3(SiO_4)_{2,16}(OH)_{3,36}$ de type grenat. *Acta Crystallogr.* **1967**, *23*, 220–230. [CrossRef]

47. Oneill, B.; Bass, J.D.; Rossman, G.R. Elastic properties of hydrogrossular garnet and implications for water in the upper-mantle. *J. Geophys. Res. Solid Earth* **1993**, *98*, 20031–20037. [CrossRef]

48. Okoronkwo, M.U.; Glasser, F.P. Compatibility of hydrogarnet, $Ca_3Al_2(SiO_4)_{(x)}(OH)_{(4(3-x))}$, with sulfate and carbonate-bearing cement phases: 5–85 °C. *Cem. Concr. Res.* **2016**, *83*, 86–96. [CrossRef]

49. Adhikari, P.; Dharmawardhana, C.C.; Ching, W.Y. Structure and properties of hydrogrossular mineral series. *J. Am. Ceram. Soc.* **2017**, *100*, 4317–4330. [CrossRef]

50. Galvankova, L.; Masilko, J.; Solny, T.; Stepankova, E. Tobermorite synthesis under hydrothermal conditions. In *Ecology and New Building Materials and Products 2016*; Drdlova, M., Kubatova, D., Bohac, M., Eds.; Elsevier Science Bv: Amsterdam, The Netherlands, 2016; Volume 151, pp. 100–107.

51. Maeda, H.; Kurosaki, Y.; Nakamura, T.; Nakayama, M.; Ishida, E.H.; Kasuga, T. Control of chemical composition of hydrogrossular prepared by hydrothermal reaction. *Mater. Lett.* **2014**, *131*, 132–134. [CrossRef]

52. Grangeon, S.; Fernandez-Martinez, A.; Baronnet, A.; Marty, N.; Poulain, A.; Elkaim, E.; Roosz, C.; Gaboreau, S.; Henocq, P.; Claret, F. Quantitative X-ray pair distribution function analysis of nanocrystalline calcium silicate hydrates: A contribution to the understanding of cement chemistry. *J. Appl. Crystallogr.* **2017**, *50*, 14–21. [CrossRef] [PubMed]

53. Hou, D.S.; Ma, H.Y.; Li, Z.J. Morphology of calcium silicate hydrate (C-S-H) gel: A molecular dynamic study. *Adv. Cem. Res.* **2015**, *27*, 135–146. [CrossRef]

54. Grangeon, S.; Claret, F.; Lerouge, C.; Warmont, F.; Sato, T.; Anraku, S.; Numako, C.; Linard, Y.; Lanson, B. On the nature of structural disorder in calcium silicate hydrates with a calcium/silicon ratio similar to tobermorite. *Cem. Concr. Res.* **2013**, *52*, 31–37. [CrossRef]

55. Houston, J.R.; Maxwell, R.S.; Carroll, S.A. Transformation of meta-stable calcium silicate hydrates to tobermorite: Reaction kinetics and molecular structure from XRD and NMR spectroscopy. *Geochem. Trans.* **2009**, *10*, 14. [CrossRef] [PubMed]

56. Lothenbach, B.; Scrivener, K.; Hooton, R.D. Supplementary cementitious materials. *Cem. Concr. Res.* **2011**, *41*, 1244–1256. [CrossRef]

57. Shaw, S.; Clark, S.M.; Henderson, C.M.B. Hydrothermal formation of the calcium silicate hydrates, tobermorite ($Ca_5Si_6O_{16}(OH)_{(2)} \cdot 4H_{(2)}O$) and xonotlite ($Ca_6Si_6O_{17}(OH)_{(2)}$): An in situ synchrotron study. *Chem. Geol.* **2000**, *167*, 129–140. [CrossRef]

58. Mitsuda, T.; Sasaki, K.; Ishida, H. Phase evolution during autoclaving process of aerated concrete. *J. Am. Ceram. Soc.* **1992**, *75*, 1858–1863. [CrossRef]

59. Zhang, P.; Ma, H.W. The synthesis of tobermorite from potassium feldspar powder: An experimental study. *Acta Petrol. Et Mineral.* **2005**, *24*, 333–338.

60. Qiu, M.Y.; Ma, H.W.; Nie, Y.M.; Zhang, P.; Liu, H. Experimental study on synthesis of tobermorite by decomposing potassium feldspar. *Geoscience* **2005**, *19*, 348–354.

61. Klimesch, D.S.; Ray, A. Effects of quartz particle size and kaolin on hydrogarnet formation during autoclaving. *Cem. Concr. Res.* **1998**, *28*, 1317–1323. [CrossRef]

62. Klimesch, D.S.; Ray, A. Hydrogarnet formation during autoclaving at 180 °C in unstirred metakaolin-lime-quartz slurries. *Cem. Concr. Res.* **1998**, *28*, 1109–1117. [CrossRef]

63. Klimesch, D.S.; Ray, A. Autoclaved cement-quartz pastes with metakaolin additions. *Adv. Cem. Based Mater.* **1998**, *7*, 109–118. [CrossRef]

64. Liu, Y.; Xia, H.; Ma, H. Kinetics of hydrothermal decomposition of potassium feldspar with calcium hydroxide. *Adv. Mater. Res.* **2012**, *549*, 65–69. [CrossRef]

65. Skorina, T.; Allanore, A. Aqueous alteration of potassium-bearing aluminosilicate minerals: From mechanism to processing. *Green Chem.* **2015**, *17*, 2123–2136. [CrossRef]

66. Eckert, J.O.; HungHouston, C.C.; Gersten, B.L.; Lencka, M.M.; Riman, R.E. Kinetics and mechanisms of hydrothermal synthesis of barium titanate. *J. Am. Ceram. Soc.* **1996**, *79*, 2929–2939. [CrossRef]

67. Berger, G.; Beaufort, D.; Lacharpagne, J.C. Experimental dissolution of sanidine under hydrothermal conditions: Mechanism and rate. *Am. J. Sci.* **2002**, *302*, 663–685. [CrossRef]

Article

Evaluation of Magnetic Separation Efficiency on a Cassiterite-Bearing Skarn Ore by Means of Integrative SEM-Based Image and XRF–XRD Data Analysis

Markus Buchmann [1,*,†], Edgar Schach [2,*,†], Raimon Tolosana-Delgado [2], Thomas Leißner [1], Jennifer Astoveza [2], Marius Kern [2], Robert Möckel [2], Doreen Ebert [2], Martin Rudolph [2], Karl Gerald van den Boogaart [2] and Urs A. Peuker [1]

[1] Institute of Mechanical Process Engineering and Mineral Processing, TU Bergakademie Freiberg, Agricolastraße 1, 09599 Freiberg, Germany; thomas.leissner@mvtat.tu-freiberg.de (T.L.); urs.peuker@mvtat.tu-freiberg.de (U.A.P.)

[2] Helmholtz-Zentrum Dresden-Rossendorf, Helmholtz Institute Freiberg for Resource Technology, Chemnitzer Straße 40, 09599 Freiberg, Germany; r.tolosana@hzdr.de (R.T.-D.); jennifer.astoveza@kerneos.com (J.A.); m.kern@hzdr.de (M.K.); r.moeckel@hzdr.de (R.M.); d.ebert@hzdr.de (D.E.); m.rudolph@hzdr.de (M.R.); k.van-den-boogaart@hzdr.de (K.G.v.d.B.)

* Correspondence: markus.buchmann@mvtat.tu-freiberg.de (M.B.); e.schach@hzdr.de (E.S.); Tel.: +49-3731-39-3023 (M.B.); +49-351-260-4480 (E.S.)

† These authors contributed equally to this work.

Received: 8 June 2018; Accepted: 31 August 2018; Published: 6 September 2018

Abstract: Image analysis data obtained from scanning electron microscopy provided data for a detailed evaluation of the separation efficiency for various processes involving the beneficiation of particulate materials. A dry magnetic separation by a drum type magnetic separator served as a case study to visualize effects of processing of a skarn ore with a high content of cassiterite as ore mineral (~4 wt%). For this material, iron oxides and silicates are the main gangue mineral groups. Based on the obtained data, partition curves were generated with the help of local regression. From the partition curves, the separation efficiency was evaluated and the relevant particle properties deduced. A detailed analysis of the bias of the quantitative mineralogical data is presented. This bias was monitored and further analyzed in detail. Thorough analysis of feed and products of magnetic separation enabled identification of the most important factors that control losses of cassiterite to the magnetic product, namely the association with iron oxides and particle sizes below ~40 μm. The introduced methodology is a general approach applicable for the optimization of different separation processes and is not limited to the presented case study.

Keywords: SEM-based image analysis; MLA (Mineral Liberation Analyzer); magnetic separation; cassiterite; partition curve; local regression

---

1. Introduction

Complex skarn ores containing considerable amounts of tin, zinc, indium, and other metals are located in the Ore Mountains (Erzgebirge) region of Germany. Due to the finely disseminated nature and the complex mineralogy of these skarn ores, past beneficiation attempts failed, and the ore was considered economically unprocessable. Yet, increasing resource demand has resulted in renewed interest in these deposits [1]. In this study, a skarn ore from the Hämmerlein deposit, one of the prime examples of skarn ores from the Ore Mountains, was investigated. Cassiterite is the main tin-bearing mineral in this ore. Due to the high density of cassiterite (~7 g/cm^3), gravity separation is considered as an important beneficiation technique [2]. Yet, the investigated skarn ore contains about ~30 wt% of iron oxides (magnetite and hematite). The effective separation of these iron oxides is important for the following reasons:

- Low magnetic separation is usually implemented as a pre-concentration step to reduce the mass streams in the downstream flow sheet [3];
- A by-product of high grade iron oxides will increase the feasibility of the planned mining operations;
- Subsequent density separation steps will become more efficient due to a higher contrast in the separation properties;
- Iron oxides are a source for multivalent ions, which have a detrimental effect on the froth flotation of cassiterite [2,4];

Because of the brittle nature of cassiterite [2,4] this mineral is easily enriched in the fine particle size fractions, reducing the effectivity of separation processes. It is thus beneficial to apply a first magnetic separation step to remove iron oxides at a relatively coarse particle size (<250 μm) [5]. However, evidence from prior studies at Hämmerlein [1,6] showed that cassiterite is intergrown with iron oxides, thus raising the question of how much value can potentially be lost in the magnetic products. The first aim of this paper is to quantify these potential losses, evaluate possible reasons, and discuss feasible remedies.

To answer this sort of questions, the feed and output streams are often analyzed by one or more of X-ray fluorescence (XRF), X-ray diffractometry (XRD) and scanning electron microscopy-based image analysis (such as QEMSCAN (FEI Company, Hillsboro, OR, USA) or MLA—Mineral Liberation Analyzer (FEI Company, Hillsboro, OR, USA)). With SEM-based image analysis, it is possible to identify characteristic mineral specific effects that help to achieve a deeper understanding of the different processes. For example Little et al. [7,8] describe specific shape effects during grinding as a function of different minerals. Similarly, Leißner et al. (2016a) [9] were able to quantify selective breakage effects. Various authors have utilized SEM-based image analysis in their studies for detailed characterization of their materials [10–12]. However, most of these studies disregard the existence of uncertainties (material heterogeneity, sample representativity, sample preparation issues, and counting errors) despite the fact that tools for monitoring them have been proposed [13,14]. The second aim of this paper is thus to test some tools for an assessment of the uncertainty on the fate of value and waste minerals through processing, here illustrated with the behavior of cassiterite and iron oxides in magnetic separation.

The application of SEM-based image analysis data to magnetic separation was inter alia carried out by Leißner et al. (2016b) [15]. The magnetic susceptibility of different minerals is one major determining parameter influencing the results of magnetic separation. The iron content in minerals shows a big influence on their response in magnetic separation [16]. As these contents are different for each deposit, a practical testing of this magnetic separation for the different materials is necessary. For ferromagnetic substances, different from paramagnetic and diamagnetic materials [17], the susceptibility shows strong dependence on the applied magnetic field, hence for materials containing such ferromagnetic minerals (i.e., magnetite) this property is difficult to access [18,19].

2. Materials and Methods

A complex tin-bearing skarn ore from the Hämmerlein deposit was processed with a magnetic separator. The feed material and the different products were analyzed by a Mineral Liberation Analyzer (MLA), a commercially available SEM-based image analysis system [20].

To evaluate the SEM-based image analysis, special focus was on the statistical representativity of the obtained results. Due to the 2D structure of the image data, a stereological bias arises [21]. Numerous authors have addressed this topic and given certain relatively complex tools at hand to remedy this problem [22,23]. Furthermore, the sampling of the material is of high significance for the validity of the obtained results but also an adequate sample preparation is crucial for obtaining representative results [24].

2.1. Materials

Table 1 represents the modal mineralogy of the sample characterized by MLA technique. The modal mineralogy of the material shows high contents of silicates, phyllosilicates, and iron oxides. Additionally, the cassiterite content with approximately 4 wt% (Table 1), (corresponding to 3.44 wt% Sn measured by XRF) is rather high compared to many other tin ores [2].

Table 1. Modal mineralogy defined by the Mineral Liberation Analyzer (MLA) of the input material.

Mineral Groups	Content in wt%
Silicates	47
Phyllosilicates	16
Iron oxides	29
Fluorite	2
Sulfides	2
Cassiterite	4

It is worth noting that an appropriate grouping of the identified minerals is a crucial step for determination of the significant effects for the investigated system. For the grouping, various approaches are possible. The mineral groups can be based on the significance of the applied process (e.g., magnetic separation), on important properties for the applied measurement method (e.g., atomic mass), or the geological classification. The mineral grouping shown in Table 1 was applied to the investigated separation process and was focused on the main mineral groups that influence magnetic separation (additionally there are carbonates and others, but these groups represent only small amounts below 1 wt%). The silicates group mainly consists of quartz, feldspar, garnet, and members of the amphibole group. The group phyllosilicates is dominated by mica and chlorite. The main constituents of the group sulfides are sphalerite and minor pyrite, arsenopyrite, and galena. The iron oxides are comprised of magnetite and hematite. Figueroa et al. [25] reported that it is possible to discriminate hematite and magnetite with MLA by their backscattered electron (BSE) values. This approach is inapplicable with the Hämmerlein ore because the BSE values of hematite and magnetite are subject to fluctuations caused by varying amounts of Sn and other elements in the crystal lattice. There are typical ranges regarding susceptibility for the most important minerals of the present study (Table 2).

Table 2. Susceptibility ranges for some chosen minerals [26].

Mineral	Susceptibility in 10^{-9} m^3kg^{-1}
Magnetite	>6505
Hematite	720–6505
Cassiterite	<22.5

Table 2 explicitly represents that the susceptibility ranges are significantly different for iron oxides and cassiterite. As already stated, the mineral susceptibility is highly influenced by Fe content and varies strongly for different deposits. This fact necessitates a practical testing of the individual materials.

The feed material was comminuted via a cone crusher and a subsequent screen ball mill (screen aperture w = 250 µm). Figure 1 shows the particle size distribution (measured by laser diffraction: HELOS, Sympatec, Germany) after comminution of the feed material for the magnetic separation test. Additionally, the mineral grain size distribution of cassiterite (based on MLA data) is shown.

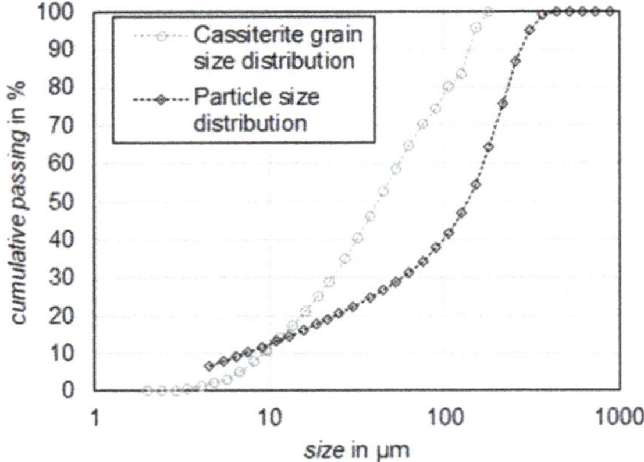

Figure 1. Particle size distribution (from laser diffraction measurements) and grain size distribution of cassiterite (from Mineral Liberation Analyzer—MLA data) of the feed material for magnetic separation.

2.2. Magnetic Separation

For the experimental tests, a drum type magnetic separator according to Figure 2 was utilized. The separator operated in a dry magnetic environment. An installed permanent magnet generated a maximum magnetic flux density of approx. 160 mT on the surface of the drum.

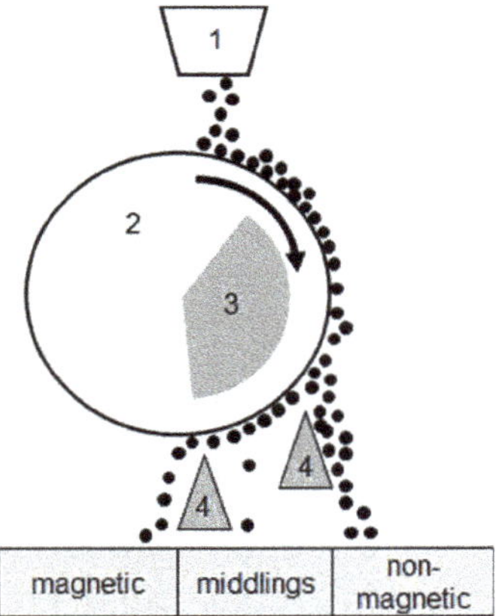

Figure 2. Rotating drum type magnetic separator (1 feeder; 2 drum; 3 permanent magnet and 4 splitter).

The drum operated at a rotational speed of approx. 33 revolutions per minute (rpm) and the diameter of the drum was approx. 600 mm. The magnetic field of this specific magnetic separator was described in detail by Madai et al. (1998) [27]. Magnetic separation tests were carried out in a batch condition feeding 4 kg while the feed rate was adjusted manually to generate a mono-particle layer on the surface of the drum.

Since the magnetic separator was operated in a low magnetic field, the main task was to remove the ferromagnetic components and to concentrate them in the magnetic product. The main ferromagnetic components in the feed material were identified as the magnetite.

The different products (magnetic, middlings and non-magnetic) were collected below the drum and representative samples for subsequent analysis were obtained using rifle- and rotator-splitters.

2.3. X-ray Diffraction and X-ray Fluorescence Measurements

X-ray powder diffractometry (XRD) analysis were undertaken to distinguish between the iron oxides. A PANalytical Empyrean diffractometer (PANalytical, Almelo, Netherlands) was used, equipped with a PIXcel3D-Medipix area detector (in combination with a Fe-filter) and a proportional counter (with monochromator) as well as a Co X-ray tube. It was operated at 35 kV and 35 mA, measurements were done in the 2θ range of $5°–80°$, with step size of at least $0.013°$ 2Θ and an overall measurement time of 2.5 h. The irradiated area was kept constant (10×12 mm) by means of an automated divergence slit. Quantifications were implemented utilizing the Rietveld method by the BGMN/Profex software package v. 3.9.2 [28].

Bulk chemical assays were carried out by X-ray fluorescence spectrometry at a commercial laboratory (ALS, Ireland) using an analytical method reported elsewhere [29].

2.4. Mineral Liberation Analysis (MLA)

The particulate samples were characterized by a Mineral Liberation Analyzer (MLA). Grain mounts of the particulate samples were prepared. For this purpose, aliquots of 3 g of material were mixed with the same volume of graphite and epoxy resin. The grain mounts were polished and then carbon coated with Leica (Baltec) MED 020 vacuum evaporator (Leica Microsystems, Wetzlar, Germany).

The MLA consists of a FEI Quanta 650F field emission SEM (FE-SEM) (FEI Company, Hillsboro, OR, USA) equipped with two Bruker Quantax X-Flash 5030 energy dispersive X-ray detectors (EDX). Identification of mineral grains by MLA was based on backscattered electron (BSE) image segmentation and collection of EDX-spectra of the particles and grains distinguished in BSE imaging mode. Collected EDX-spectra were further classified using a list of mineral spectra collected by the user. More detailed information with regard to the functionality of the MLA system can be found in the literature [20,30]. The GXMAP measurement mode [20] was applied to all samples. Data processing and evaluation was exanimated with the software package MLA Suite 3.1.4.686 using the modified approach for automated mineralogy as described by Kern et al. (2018) [6].

2.5. Bootstrap Resampling

The bootstrapping algorithm was developed by Efron and Tibshirani [31]. The algorithm enabled information to be gained about the statistical reliability of the results. In the processing stage, a bootstrap resampling with replacement method was used as described by Evans and Napier-Munn (2013) [13]. Bootstrap resampling created M virtual subsets of the sample by randomly taking N particles from the measured particle population. Particles once selected by the algorithm were directly replaced in the population to be again available for sampling. The potential of error analysis based on the bootstrap resampling method was tested and compared to other methods available in the literature by Mariano and Evans (2015) [32]. It was shown that bootstrap resampling indicated reliable estimates of sample statistics without being based on any assumptions.

2.6. Tromp Curves

The partition curve so-called Tromp curve ($T(\xi)$) for a separation property ξ, is a suitable instrument to evaluate the quality of selectivity of a separation process. Basically, the separation feature is either particle size or another physical property. Values for the partition curve were calculated for all discrete classes using Equation (1) [33].

$$T_i(\xi) = \frac{u_{i,\,mag}}{u_{i,relcalc.\ feed}} R_{m,mag} \tag{1}$$

where $R_{m,mag}$ represents the recovery of mass into the magnetic product in wt%.

The mass portion μ_{ij} of a certain property class i in a certain material stream j was calculated with the mass in this certain class and the material fraction m_{ij} and the total mass of this fraction m_j:

$$\mu_{ij} = \frac{m_{ij}}{m_j} \tag{2}$$

With respect to the partition curves, the feed material represents the recalculated material based on magnetic product, middlings, and non-magnetic products. The results from MLA enable partition curves to be generated for various separation properties. Therefore, a reasonable classification for the investigated properties was applied and T_i was calculated for each class.

For fitting smooth curves to the values of the partition curve based on the empirical data for each class, local regression ("loess") was applied. It represents a simple nonparametric fitting for smooth curves to empirical data [34], similar to a moving window regression. The method parameters are α and λ. The parameter α represents a smoothing parameter and determines the portion of nearest observations that are used for estimating the regression at each point. It is specified between 0 and 1, with higher values producing smoother curves. The parameter λ defines the degree of the polynomial regression that is fitted to the data in that window. Here, α and λ are set to 0.75 and 1, respectively. Linear equations are fitted to the data points in each window. Further, the confidence interval of 95% for each curve is indicated.

An ideal separation is as a vertical line, whereas an ideal splitting is characterized as a horizontal line for T. For a real separation process, the performance will be between these two cases as a function of the particular investigated feature. For the magnetic separation, the magnetic susceptibility may be used as separation property.

3. Results and Discussion

In this section, the plausibility of the MLA data is discussed by comparing the recalculated feed with the real feed sample. Further, the quality of the magnetic separation is evaluated with a special focus on possible reasons for the significant amounts of cassiterite lost into the magnetic and middlings product. For the following considerations, the focus is on cassiterite and the iron oxides as these minerals show high relevance for magnetic separation processes. The particle size (xparticle) is defined as the maximum length of the minimum bounding rectangle (MLMR) around a 2D particle image from MLA.

3.1. Plausibility of MLA Data

A detailed analysis of the data quality is presented. Thereby, the feed for the magnetic separation is compared with a recalculated feed based on the products of magnetic separation. The mass distribution over different particle size classes is compared and a critical evaluation of the mineral contents is given. Furthermore, XRF and XRD analyses complement the assessment of the established mass balance.

3.1.1. Comparison of Feed and Recalculated Feed

A comparison of the real and recalculated feed obtained from the products of the magnetic separation is presented to evaluate the quality of the MLA-data precisely as suggested by Grant et al. (2018) [35]. It enables deviations to be understood and for these to be categorized in the context of assessing various effects of magnetic separation. For that reason, Figure 3 displays mass fractions (μ) in the different particle size (MLMR) classes by comparing them with the real and recalculated feeds.

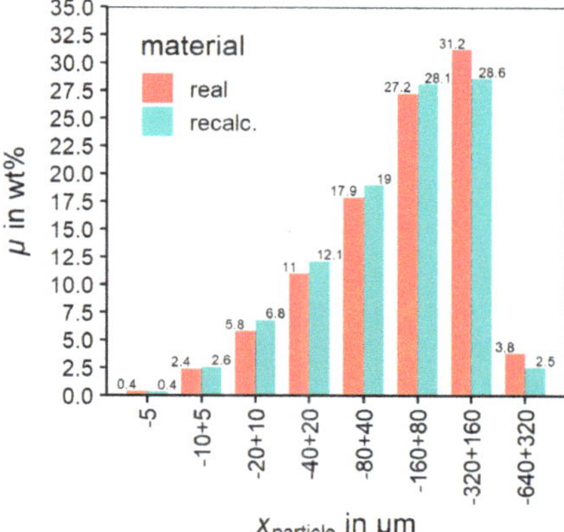

Figure 3. Comparison of mass fractions in different particle size classes for the real feed (real) and recalculated feed (recalc.). The latter is recalculated from the MLA-data of the three products of magnetic separation.

There are deviations for μ between the real and recalculated feed visible in Figure 3. For particle size classes—160 µm—μ is overestimated for the recalculated feed, whereas it is underestimated for sizes +160 µm compared to the real feed. The deviations for the particle size fractions +160 µm are explained by the comparative smaller number of particles in these coarser size fractions that cause a larger statistical uncertainty. As mass fractions have to total 100%, the uncertainty for coarse particle sizes also influences the bias for fine particle sizes.

A similar approach can also be applied to mineral contents. Figure 4 illustrates the contents of the main mineral groups (Table 1) in the real and recalculated feeds from the magnetic product stream. For the estimation of mineral contents, bootstrap resampling was applied [32]. The bootstrap resampling methodology allows for error estimation in quantitative textural analysis [13]. In addition, it enables the calculation of the confidence interval of the mineral content of the real and recalculated feed samples. The results of the bootstrap resampling are visualized via boxplots in Figure 4.

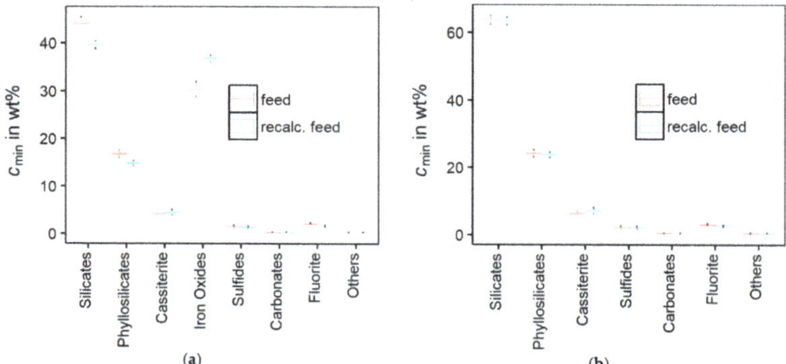

Figure 4. Comparison of mineral contents of real feed and recalculated feed based on the three products from magnetic separation (**a**) for all mineral groups; (**b**) without iron oxides closed to 100 wt%) (N = particle number of MLA sample; M = 1000). The median marks the mid-point of the data and is shown by the line that divides the box into two parts. Half the values are larger than or equal to this value and half are less. The middle "box" represents the middle 50% of values for the group. The range of scores from lower to upper quartile is referred to as the inter-quartile range. The middle 50% of values fall within the inter-quartile range. Outliers that fall out of the boxplot range are marked by dots.

Figure 4a illustrates obvious discrepancies between the mineral contents of the real and recalculated feeds. These deviations are especially prominent for silicates and iron oxides. The recalculated feed overestimates the content of iron oxides and underestimates the silicates content compared to the real feed. If the iron oxides are excluded and the content of the remaining minerals is close to 100 wt% (Figure 4b), the mineral contents of the recalculated feed are in good agreement with the actual feed. This illustrates that the content of iron oxides carries an error to all other mineral components in the middlings, non-magnetic or magnetic product.

The bulk chemical composition of feed and products of magnetic separation can be used as an external monitor for the validity of the MLA data. In Table 3, the concentrations of Sn and Fe as determined by XRF are compared to concentrations calculated from the mineral chemistry and contents of all minerals contained in the feed sample as well as the recalculated feed based on the results for the three products of magnetic separation. Compared with deviations for the mineral contents from MLA, the element contents from XRF show only minor discrepancies. It has to be stated that the elemental contents for Sn and Fe cannot be easily assigned to the mineral contents of cassiterite and iron oxides respectively, due to the complex character of the investigated ore and the elements also being present in various other minerals.

Table 3. X-ray fluorescence (XRF) analysis of c_{Sn} and c_{Fe} measured for real feed and the recalculated feed based on the three products from magnetic separation.

	XRF		MLA	
	Real Feed	**Recalc. Feed**	**Real Feed**	**Recalc. Feed**
c_{Sn}	3.44	3.21	3.42	3.64
c_{Fe}	29.75	28.69	27.87	31.74

All values in wt%.

Possible reasons for the discrepancies of the real feed and the recalculated feed are pointed out below under Section 3.1.3.

3.1.2. Hematite–Magnetite Ratio

For a more detailed analysis of the iron oxides content (c_{ironox}), XRD analyses were conducted for the feed and the products of magnetic separation. From these XRD measurements, it is possible to distinguish between hematite and magnetite. However, this is just a bulk value, not particle wise as by MLA. In Table 4, the values for c_{ironox} and the proportions for magnetite and hematite are shown. In comparison to the MLA data, the mass balance for the iron oxides shows fewer deviations for the XRD-results and also the ratio of magnetite to hematite for the feed is in good agreement with the recalculated feed.

Table 4. c_{ironox}, magnetite and hematite proportions for the real feed, recalculated feed and the products measured by X-ray diffractometry (XRD). All values for c_{ironox} in wt%, for magnetite and hematite in rel.%.

	Real Feed	**Magnetic**	**Middling**	**Non-Magnetic**	**Recalc. Feed**
c_{ironox}	32.5	53.1	16.7	4.8	33.2
Magnetite	48	51	56	not detected	48
Hematite	52	49	44	100	52

From Table 4 it is obvious that the XRD measurement indicates the absence of magnetite (e.g., concentrations below the detection limit of 1 wt%) in the non-magnetic product. All of the iron oxides recovered in this product were found to be hematite. For the feed, the magnetic and the middlings, the ratio for hematite and magnetite is in a similar range. Here the mass ratio of hematite to magnetite for the materials is approx. 50:50, with only slight deviations. For the feed, the hematite proportion is higher than that for magnetite. In contrast to that for the magnetic product and the middlings, magnetite slightly dominates. The non-magnetic product contains only 4.8 wt% of total iron oxides, whereas an enrichment to 53.1% can be found in the magnetic product.

From the findings, it can be concluded that particles even containing only small amounts of iron oxides are recovered into the magnetic product or the middlings. It follows, that the applied magnetic separation is a suitable method to remove the iron oxides from the feed material.

3.1.3. Assessment of Data Quality

As shown above the discrepancies of the mass balance in MLA data can be mainly assigned to the iron oxide contents. However, as shown in Tables 3 and 4 the consistency of XRF and XRD results indicate only minor deviations. Therefore, it can be concluded that the overall mass balance is correct. The bias in MLA data for the mineral content of feed and recalculated feed can be explained with the following points.

As differentiation of magnetite and hematite by MLA is not possible for this material, the mean values of Fe-content and mineral density are applied. In the products of magnetic separation the ratio of the iron oxide phases varies from the feed, as is shown in Table 4, leading to an error in the mineral content estimation in these fractions.

The resolution limit of MLA makes a detailed analysis of agglomerates of fine particles in the sample problematic. An inaccurate determination of the minerals in the agglomerates causes an error in the overall mineral contents, which can be selective for specific phases. Further, different effects such as a nugget effect or segregation during sampling or sample preparation contribute to deviations of mineral contents.

According to the obtained results, the following partition curves were calculated based on the recalculated feed, as a closed mass balance is necessary. Hence, the presented partition curves only consider the distribution of the products. To assess the cassiterite distribution over the different property classes for the products of magnetic separation, the real feed is always shown for comparison.

3.2. Overview of Results of Magnetic Separation

According to Figure 2, there are three products resulting from the magnetic separation process. An integral view on the content and recovery of the magnetic separation gives a first impression about the enrichment of cassiterite and iron oxides in the products as shown in Table 5. The values presented in Table 5 were calculated using the following Equations:

$$R_{\mathrm{m}} = \frac{m_p}{m_{\text{total}}} \times 100 \text{ in wt\%} \tag{3}$$

$$R_{\text{cas}} = \frac{c_{\text{cas, }p}}{c_{\text{cas, f}}} R_{\mathrm{m}} \text{ in wt\%} \tag{4}$$

$$R_{\text{ironox}} = \frac{c_{\text{ironox, }p}}{c_{\text{ironox, f}}} R_{\mathrm{m}} \text{ in wt\%} \tag{5}$$

where the index f represents the feed material and p represents the three different products of magnetic separation including magnetic, middlings and non-magnetic materials. R_{m} is the mass recovery of the products, whereas R_{cas} and R_{ironox} represent the mineral recovery in the different products for cassiterite and iron oxides, respectively.

Table 5. Mass recovery, contents and mineral recovery with correlated error (standard deviation) estimated by bootstrap resampling and error propagation of the different products and feed.

	Magnetic	**Middlings**	**Non-Magnetic**	**Feed**
R_m	56.6 ± 2.1	1.7 ± 0.2	41 ± 2.1	-
c_{cas}	2.5 ± 0.1	5.4 ± 0.3	6.2 ± 0.4	3.8 ± 0.2
c_{ironox}	60.2 ± 0.4	12.6 ± 0.5	3.8 ± 0.2	30.4 ± 0.5
R_{cas}	39.2 ± 3.2	2.42 ± 0.5	66.1 ± 7.8	-
R_{ironox}	112.1 ± 4.8	0.7 ± 0.1	5.2 ± 0.5	-

All values in wt%—concentrations determined with MLA.

The values of R_m in Table 5 represent the mean value of mass recovery of six individual tests of magnetic separation with identical parameter settings and material. For c_{cas} and c_{ironox} (concentration of cassiterite and iron oxide) the values represent the mean of 1000 virtual samples from bootstrap resampling (according to chapter 2.5, N = particle number of MLA sample; M = 1000) based on the original data from MLA for the products and the feed. R_{cas} and R_{ironox} (recovery of cassiterite and iron oxide) were calculated with these values. The standard deviations for R_{cas} and R_{ironox} were estimated by error propagation. The calculated recovery for iron oxides into the magnetic product of 112.1%, which is obviously not possible, is caused by the bias in the iron oxide content obtained by MLA (refer to Section 3.1.1). A correction of this bias is not possible, as its origin cannot be clearly assigned to either the feed data or the data of the magnetic product. Therefore, the following results and considerations are only based on calculation methods, which are independent of the found bias in the data. For the partition curve, this is valid, as the calculation was done based on the recalculated feed (Section 3.3). Due to the fact, that the combination of the feed data and the product data is not possible, the distribution of cassiterite bearing particles within the feed and the products is shown individually (Section 3.4). The presented assessment of the uncertainty in the data allows for the decision of which calculations and methods are applicable to the data to obtain reliable information.

In the case of the product mass distribution for R_m, it is found that approx. 60 wt% of the material accumulated in the magnetic product and 40 wt% in the non-magnetic product. However, only a small proportion reports to the middlings. The majority of the iron oxides report to the magnetic product; for cassiterite, however, the allocation is less pronounced. More than 50% of the cassiterite is recovered in the non-magnetic product, whereas approximately 40 and 2 to 3 wt% respectively report to the magnetic and middlings products. This fact justifies a more detailed review of the cassiterite behavior on a particle basis to identify the effects that determine the cassiterite recovery in the different products.

3.3. Evaluation of Quality of Magnetic Separation

Leißner et al. (2016) [15] established partition curves on the mixed susceptibility of the separated material from isodynamic separator experiments. This was performed because the material investigated by Leißner et al. (2016) predominantly consisted of minerals showing paramagnetic character. In the present case, the feed material contains a high proportion of ferromagnetic minerals (e.g., magnetite). In the case of ferromagnetic material, the susceptibility is a function of the applied magnetic field. It is thus not constant in the variable external field of a magnetic separator and difficult to calculate. Therefore, it is necessary to generate partition curves based on other relevant parameters. One reasonable parameter is the content of iron oxides (c_{ironox}) of the particles as these minerals show the most pronounced ferromagnetic character compared to other minerals present in the studied skarn ores [36]. Although magnetite and hematite show different magnetic properties it can be stated, based on the presented results, that the two minerals behave rather similarly during the applied magnetic separation step—and because it was difficult to separate these two iron oxides by MLA [37], it was deemed appropriate to consider them together as iron oxides for further analysis via partition curves.

The partition curve as function of the mineral content in the particles is shown in Figure 5a for the iron oxides, which illustrates the behavior of a particle in the presence of the magnetic field. In addition, another partition curve is plotted for the cassiterite content in the particle. This example shows how the partition curve behaves in dependency of a particle property that is not determent for the separation behavior. To generate the data points for $T(c)$, a grouping of the mineral content for each particle from MLA data is applied. The recalculated feed is chosen as a reference point due to the findings reported in Section 3.1. It follows that the generated partition curves define the distribution between the magnetic product, middlings and non-magnetic product. The classes vary from 0 to 100 wt% for each mineral content with a step width of 5 wt%. The values of $T(c)$ are plotted against the mean of each content class and are represented as points in the diagram. For both minerals, a smoothed curve is added via the "loess" method as explained above in Section 2.6.

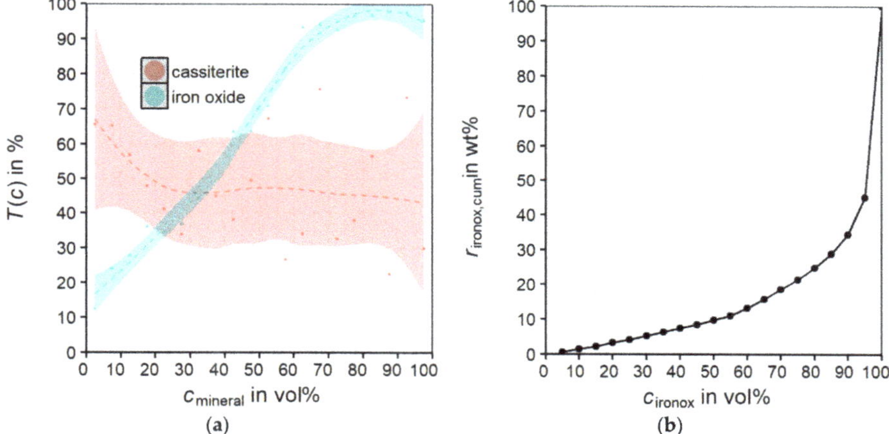

Figure 5. (a) $T(c)$ vs. mineral content in the particles for cassiterite (red) and iron oxides (blue); (b) cumulative distribution of iron oxides ($r_{\text{ironox,cum}}$) in the content classes (c_{ironox}) of the feed material.

From Figure 5a, it is obvious that the higher the iron oxides content of a particle (c_{ironox}), the more likely it reports to the magnetic product. For c_{ironox} above 65 vol%, the particles accumulate to >90% in the magnetic product. Below approx. 35 vol% c_{ironox}, the particles do not accumulate predominantly in the magnetic product ($T(c) < 50\%$). The partition curve of Figure 5a for iron oxides has to be understood as qualitative representation and should be seen in context of the mass distribution of the iron oxides in the described content classes to get a more detailed understanding of the specific material behavior. Therefore, the mass distribution of iron oxides is calculated by Equation 6 and represented in Figure 5b for the feed material of the magnetic separation.

$$r_{\text{ironox,cum}} = \sum_{i=1}^{n} \frac{m_{\text{ironox},i}}{m_{\text{ironox,tot}}} \times 100 \text{ in wt\%} \tag{6}$$

where $m_{\text{ironox,tot}}$ represents the total mass of iron oxides in the feed material and $m_{\text{ironox,i}}$ denotes the mass of iron oxides in the iron oxides content classes (c_{ironox}). A steeper gradient for the partition curve is expected over the iron oxides content as nearly all iron oxides report to the magnetic product. The relatively flat slope of the partition curve might be related to the mass distribution of the iron oxides over the volume liberation classes as shown in Figure 5b. It can be found that only small portions of iron oxides are present in the classes with a low volume liberation. Most of the iron oxides are highly liberated, explaining the discrepancy between the cumulative recovery for the iron oxides and the partition curve. This is in accordance with the high recovery of iron oxides in the magnetic product from Table 5. As seen, the trend of the separation curve for the iron oxides content is changed by the magnetic field. By reducing the magnetic force (F_{mag}), the separation process is possibly more selective to c_{ironox} in the way that particles with lower c_{ironox} deplete in the magnetic product.

For comparison, the $T(c)$ for cassiterite is shown in Figure 5a. The shape of the curve for cassiterite is distinctly different from that for iron oxides. In case of cassiterite, the appearance of the curve shows unselective behavior regarding mineral content compared to the iron oxides.

In Figure 6a, the $T(c)$ is shown as a function of c_{ironox} for three different particle size fractions. It is found that $T(c)$ increases with increasing particle size. For finer particles, the minimum c_{ironox} is higher enriched in the magnetic product compared to the coarser particles. Feeding from the top should support the separation of fine particles relative to coarse particles due to higher specific magnetic forces. One possible reason for the behavior of particles finer than 40 µm shown in Figure 6a is an agglomeration phenomenon with respect to ultrafine particles. This agglomeration is not visible in MLA due to desagglomeration during preparation of the samples for measurement. Agglomerates which are separated based on the agglomerate properties might be seen as individual particles. Further, misplaced liberated iron oxides and misplaced locked particles might influence the separation results. Another possible reason for the behavior in Figure 6a is an enrichment of hematite and magnetite in different particle size ranges. This cannot be identified by the MLA technique owing to the problematic distinction of these iron oxides for this material. The stereological effect must not be neglected, which also has an influence on the shape of the partition curves.

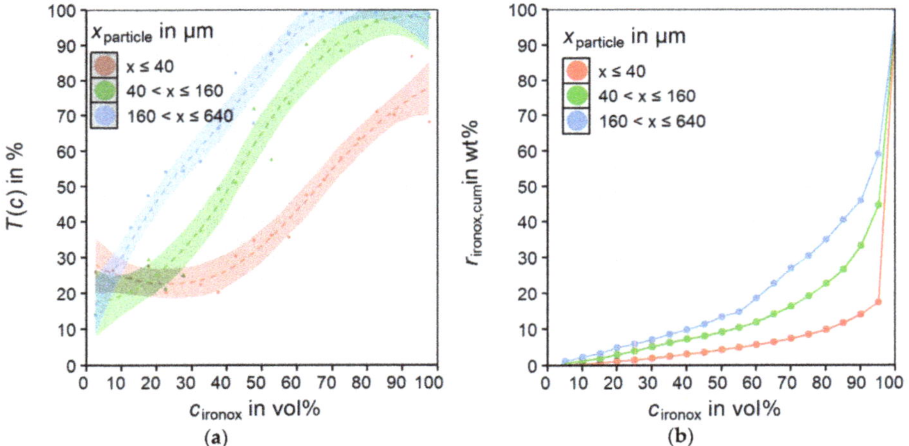

Figure 6. $T(c)$ vs. c_{ironox} (**a**) for three different particle size classes; (**b**) cumulative distribution of iron oxides ($r_{ironox,cum}$) in the content classes (c_{ironox}) of the feed material for three different particle size classes.

It can be also found in Figure 6a that the particle size has a significant influence on the separation quality. In Figure 6b, the mass distribution of iron oxides in the different classes for c_{ironox} is shown for the three size classes. It reveals that the finer the particle size range, the more liberated are the iron oxides.

The data presented above points out that c_{ironox} is a significant parameter for magnetic separation. There is a minimum c_{ironox}, which is defined by the separation cut-point ($T(c_{ironox})$ = 50%), necessary for particles to be enriched in the magnetic product.

3.4. Cassiterite-Bearing Particles

The following considerations focus on cassiterite and its distribution in the feed and the three products of magnetic separation. Special attention is paid to the association of cassiterite with iron oxides due to the fact that the iron oxide content is a crucial factor for magnetic separation.

3.4.1. Liberation of Cassiterite

In this section, the feed represents the results of the MLA of the real feed material. For the analysis of MLA data, the number of particles N always has to be considered. Table 6 represents the total and cassiterite-bearing number of particles in the feed and the three products. In Table 6, the total number of particles, N_{total}, for the analysis and the number of cassiterite bearing particles, N_{cas}, are indicated.

Table 6. Number of particles (N_{total}: all particles; N_{Cas}: particles containing cassiterite).

	Feed	Magnetic	Middlings	Non-Magnetic
N_{total}	193,476	299,510	143,411	91,558
N_{cas}	7391	12,081	6777	4505

In Figure 7 a typical particle image from SEM-based image analysis is shown, which is binarized in cassiterite and the gangue area.

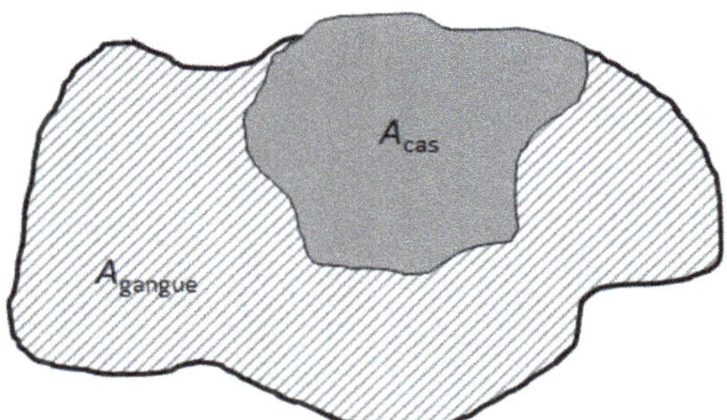

Figure 7. Typical 2D projection of a particle in the cutting plane from MLA measurement.

The cassiterite volume liberation is calculated for each particle according to Equation (7):

$$L_{cas} = \frac{A_{cas}}{A_{cas} + A_{gangue}} \times 100 \text{ in vol\%} \tag{7}$$

where A_{cas} and A_{gangue} denote the proportion of the area of cassiterite and other gangue minerals, respectively. As the data given in the MLA is 2D, the volume liberation for cassiterite is calculated based on obtained area information, considering that the bias for this transformation is in an acceptable range. This is a reasonable assumption given the near isometric shape of most cassiterite grains and the lack of any preferred orientation of mineral grains in the fine-grained skarn ore.

The grouping of the cassiterite-containing particles as a function of particle size is shown in Figure 8. In this diagram, only the particles N_{cas} (Table 6) are utilized. For the feed and each product p (magnetic, middlings and non-magnetic), $r_{cas,p,i}$ sums up to 100 wt% and i represents the particle size and liberation classes for which r_{cas} is calculated. Therefore, the colors give an indication of the cassiterite distribution within the product with regard to the particle size and cassiterite liberation. For the feed and the three products of magnetic separation, the distribution of cassiterite over the particle size classes is shown in Figure 8.

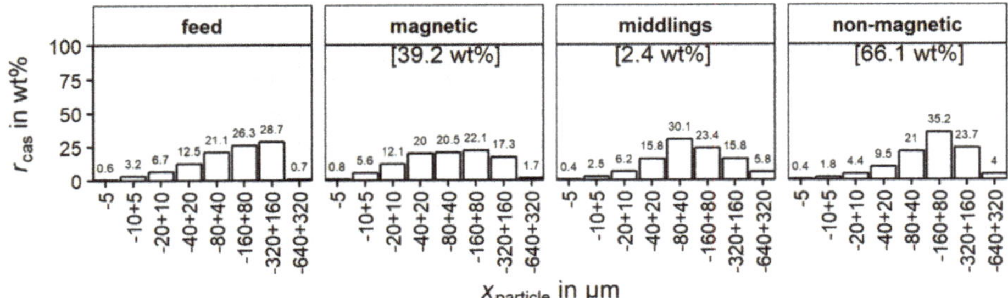

Figure 8. Distribution of cassiterite (r_{cas}) in different size classes for feed and the products (values in brackets indicate distribution of cassiterite into the products from Table 5).

The distribution of cassiterite in the non-magnetic product shows a clear maximum in the $-160 + 80$ µm class. However, the maximum for the middlings is in a lower class at $-80 + 40$ µm. In contrast, the magnetic product shows no clear maximum and a more uniform distribution of the cassiterite between 20–320 µm. The feed material is added for comparison and indicates left skewed distribution with a maximum at $-320 + 160$ µm.

The distribution of cassiterite in the feed is similar to the cassiterite distribution in the non-magnetic product, also showing a left skewed distribution. The recovery of cassiterite into the middlings and the magnetic product is caused by intergrown iron oxides, especially magnetite which reports completely to these products. Finely, disseminated cassiterite is intergrown with magnetite, which also explains the greater amounts of cassiterite in the finer-grained fractions compared to the non-magnetic and middlings fraction. For a better understanding of this behavior, the liberation of cassiterite in the different size fractions has to be investigated. For this purpose, r_{cas} is studied as a function of liberation together with the particle size classes in Figure 9. The r_{cas} for every product sums up to 100% for a better visualization. Further, the color scale displays the cassiterite recovery into the property classes (r_{cas}) for each product individually calculated by Equation (8).

$$r_{cas,\,p,\,i} = \frac{m_{cas,p,i}}{m_{cas,p}} \times 100 \text{ in wt\%} \tag{8}$$

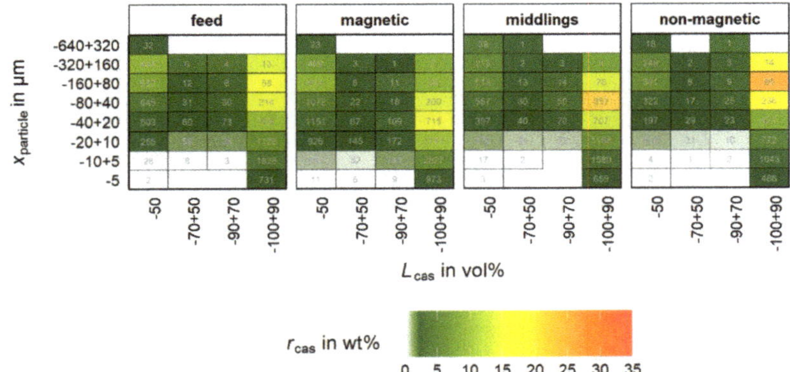

Figure 9. Distribution of cassiterite (r_{cas}) as a function of $x_{particle}$ classes and L_{cas} classes (recovery in each product sums up to 100%; green: 1%, yellow: 17.5%, red: 35%). The numbers in the boxes indicate the amount of particles in each bin.

The number of particles is added for each property class. These values are added for review of the statistical representative status of the data, where a high value of r_{cas} together with a low number of particles is problematic and must be noted.

For all four samples, the cassiterite accumulates in the liberation class $-100 + 90\%$. Further, there is a significant amount of cassiterite in the -50% class. It is worth noting that the depicted class is comprised of a much higher range than the other classes, which influences the optical perception in this diagram. Nevertheless, there is a significant number of particles in all three products with low cassiterite liberation.

Regarding the particle size, the maximum cassiterite recovery is found in the non-magnetic product in the range of $-160 + 80$ µm, followed by $-80 + 40$ µm. In comparison, for the magnetic product the maximum cassiterite recovery is in the range of $-40 + 20$ µm. The maximum is less pronounced than in the non-magnetic product and the cassiterite recovery shows a broader distribution over the $-100 + 90\%$ class. Cassiterite with a high degree of liberation predominantly accumulates in the non-magnetic product due to its paramagnetic nature. A discharge into the middlings or magnetic product either must be of statistical nature or caused by additional particle properties beside the content of iron oxides, e.g., mineral association, particle size, shape or density. The influence of these properties cannot be assessed in Figure 9.

Summarizing Figure 9, the cassiterite containing particles are divided into three main liberation groups including low (-50%), intermediate (50%–90%) and high (90%–100%) liberated cassiterite. Another important group represents the fully (100%) liberated cassiterite. This particle section shows no other interconnected mineral in the particle and is therefore of high significance. This group is not visible in Figure 9 as it is included in the $-100 + 90\%$ group.

Based on bootstrap resampling (refer to chapter 2.5, N = particle number of MLA sample; $M = 1000$) the coefficient of variation (CV) was calculated for the bins of Figure 9. The CV for each bin is shown in Figure 10.

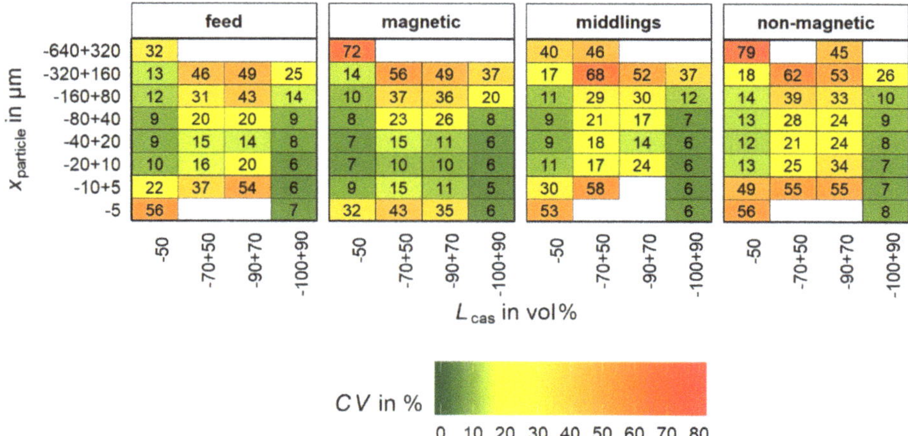

Figure 10. Coefficient of variation (CV) for the bins of $x_{particle}$ classes and L_{cas} based on data from bootstrap resampling; green: 0%, yellow: 20%, red: 80%). The numbers in the boxes indicate the value of CV for each bin.

From Figure 10 in combination with Figure 9 it is seen that for bins with a low number of particles in the original sample for MLA the value for CV is high. According to Lamberg et al. [22] the maximum acceptable value of CV is in the range of 20%. If the value is higher, bins will be merged. This approach reduces the mean error for the following calculation methods. However, it does not allow a tracking of the bias along the analysis process.

For a better overview of the distribution of cassiterite regarding these described liberation classes, r_{cas} is shown for these groups in Figure 11 for the three products and the feed.

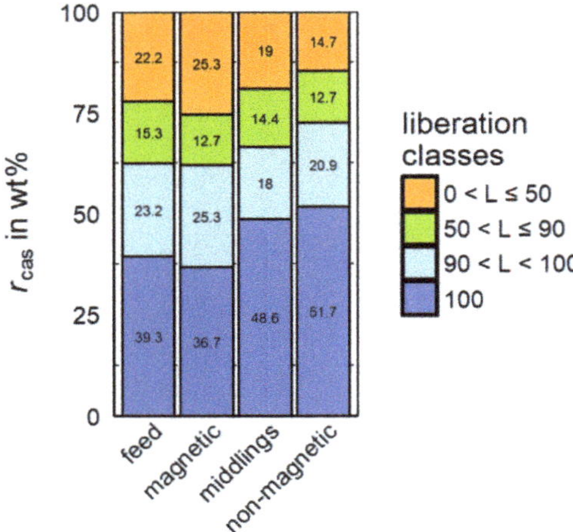

Figure 11. Distribution of cassiterite (r_{cas}) in the different liberation classes for the three products.

In Figure 11, the 100% liberated cassiterite is the predominant proportion of all products. It should be mentioned that MLA slightly overestimates the liberation degree by only cutting the particles in the cutting plane of the grain mounts and the related 2D character of the data. The described stereological effect leads to the fact that 100% liberated cassiterite can be intergrown with other minerals, which are not visible in MLA as these minerals are not positioned in the cutting plane of the grain mount. Nevertheless, the non-magnetic product shows the highest amount of fully liberated cassiterite compared to all other products and the feed.

3.4.2. Association with Iron Oxides

The association of cassiterite with iron oxides is of high importance. The content of cassiterite together with iron oxides in one particle is relevant for its recovery into the magnetic product. Table 7 shows the particle population that is utilized for the analysis of cassiterite in connection with iron oxides.

Table 7. Number of particles in the feed and three products classifying, $N_{cas,\ ironox}$, particles containing cassiterite and iron oxides and $N_{cas,\ no\ ironox}$ as particles includes cassiterite and no iron oxides.

	Feed	Magnetic	Middlings	Non-Magnetic
$N_{cas,\ ironox}$	1508	3561	797	447
$N_{cas,\ no\ ironox}$	5883	8520	5980	4057

From Figure 12 it follows, that the highest portion of cassiterite intergrown with iron oxides can be found in the magnetic product. Nevertheless, a significant amount of cassiterite (17.5 wt%) is also associated with iron oxides in the non-magnetic product. As no magnetite is present in the non-magnetic product, the cassiterite is only intergrown with hematite. For these particles, a more detailed analysis is necessary.

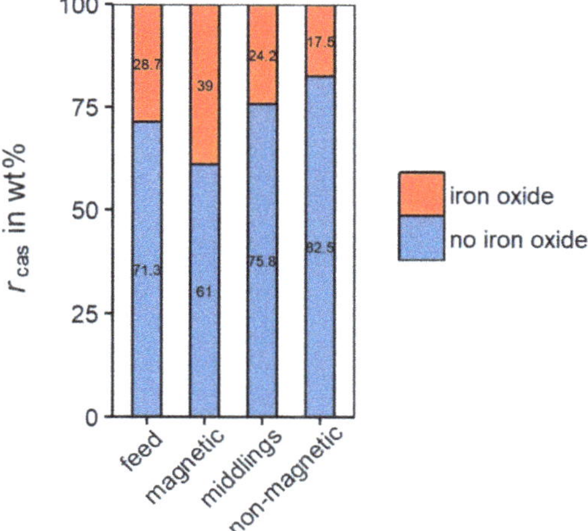

Figure 12. Distribution of cassiterite (r_{cas}) in particles with iron oxides and without iron oxides.

Figure 12 implies the necessity of a more detailed analysis of the particles that are comprised of cassiterite and iron oxides. In Figure 13, the distribution of these particles is shown as a function of particle size and c_{ironox} classes.

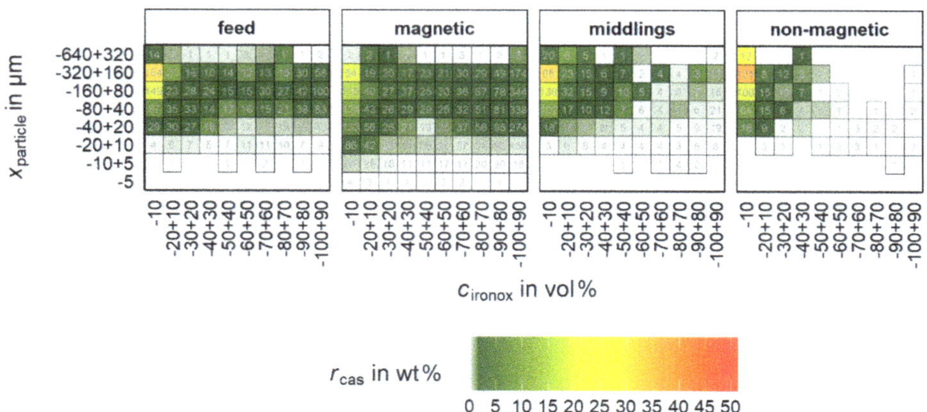

Figure 13. Distribution of cassiterite (r_{cas}) as function of $x_{particle}$ and c_{ironox} in the feed and the products (recovery in each product sums up to 100%; green: 1%, yellow: 25%, red: 50%). The numbers in the boxes indicate the amount of particles in each bin.

For r_{cas}, the non-magnetic product shows a clear accumulation at low c_{ironox}. The maximum r_{cas} is for c_{ironox} at -10 wt % and $x_{particle}$ at $-320 + 160$ µm. For the non-magnetic product, the majority of the cassiterite shows c_{ironox} below 30 vol%. The magnetic product shows an obviously broader distribution of cassiterite over the c_{ironox} classes. Despite that, the maximum is also in the class -10 vol% but significantly less pronounced than in the non-magnetic product. Furthermore, the optical impression in Figure 13 implies a slight shift of r_{cas} to lower particle size classes for the magnetic product compared to the non-magnetic product. The behavior of the middlings shows a clear maximum in the fraction c_{ironox} -10 vol% but also a broader distribution of r_{cas} over c_{ironox} than in the non-magnetic product.

It can be concluded that already small proportions of iron oxides (most probably magnetite) in the particle suffice to recover it in the magnetic product or the middlings.

For the magnetic and the non-magnetic products, respectively two typical particle images that contain cassiterite and iron oxides are shown in Figure 14.

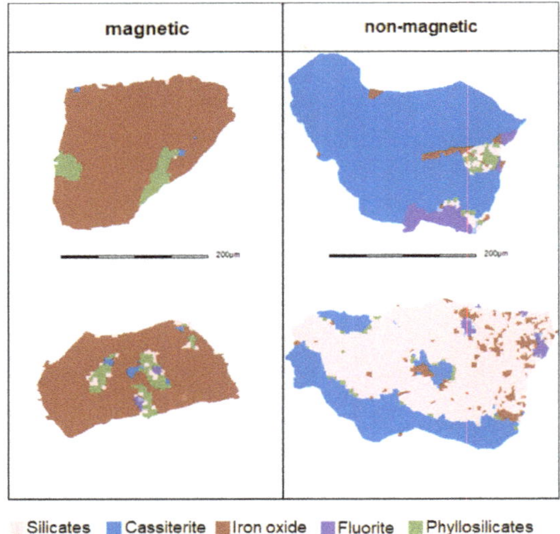

Figure 14. Examples of particle images for particles containing cassiterite and iron oxides in the magnetic and the non-magnetic products.

As seen in Figure 14, the magnetic product exhibits the finely disseminated cassiterite grains within the matrix of iron oxides. Conversely, in the case of the non-magnetic product, the iron oxides are finely disseminated within matrices of other minerals. These characteristics of the ore material represent one reason for discharge of minerals with iron oxides into the non-magnetic product and the loss of finely intergrown cassiterite into the magnetic product.

3.4.3. Fully Liberated Cassiterite

For an ideal magnetic separation, the assumption is that a particle, solely consisting of cassiterite, can be recovered in the non-magnetic fraction. Table 8 shows, however, that a considerable number of particles that show an apparent (2D) fully liberation of cassiterite are recovered in the magnetic product.

Table 8. Number of particles (N_{cas}: particles containing cassiterite; $N_{cas,100\%}$: particles with 100% volume liberated cassiterite).

	Feed	Magnetic	Middlings	Non-Magnetic
N_{cas}	7391	12,081	6777	4505
$N_{cas,100\%}$	4362	5953	4437	2990

To compare the particle size range of fully-liberated cassiterite, r_{cas} is shown as a function of the particle size classes for the feed and the products of magnetic separation in Figure 15. The distribution for r_{cas} of the magnetic product is shifted to a lower particle size range compared to the non-magnetic product. For the non-magnetic product, the maximum of r_{cas} can be found for the particle size class $-160 + 80$ µm, whereas for the magnetic product, r_{cas} shows a maximum value in the class $-40 + 20$ µm.

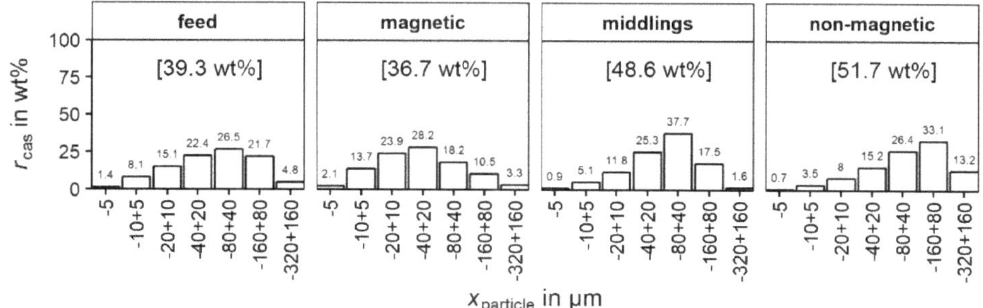

Figure 15. Distribution of cassiterite (r_{cas}) in 100% liberated class as function of $x_{particle}$ size for the different products (values in brackets indicate proportion of 100% liberated cassiterite of total cassiterite in feed and each product).

The described observations lead to the conclusion that one reason for the relatively high recovery of 100% (apparently) liberated cassiterite in the magnetic product is agglomeration of fine particles and their attachment to coarse and high magnetic particles. Nevertheless, as already stated, due to stereological effects the liberation based on MLA data is overestimated. Associated minerals might not be visible in the cutting plane and a proportion of the 100% liberated cassiterite can still be intergrown with some other minerals regarding the three dimensional structure.

4. Conclusions and Outlook

The present study presents the characterization of the magnetic properties and potential magnetic separation in a complex tin-bearing skarn of the Erzgebirge. It is impossible to over-emphasize the application of MLA data coupled with XRD and XRF for the evaluation of such magnetic separation unit operations in mineral processing. Indeed, based on the insights provided by the discussed parameters a relatively simple, efficient work flow for process analysis can be established based on a process mineralogical approach for an ore body which was thought to be uneconomic for many years. It was shown that for an evaluation of the separation efficiency via partition curves, it is also important to take the mass distribution over the volume liberation of iron oxides into account. This helps to explain the high discrepancy of the overall recovery of the iron oxides into the magnetic fraction and the relatively small slope of the partition curve. The main findings of this article can be concluded in the following statements:

- Besides the iron oxide content, the particle size had a significant influence on the magnetic separation process. The distinction of iron oxides in hematite and magnetite is possible with an additional characterization technique like XRD. All iron oxides in the non-magnetic product were found to be hematite.
- Partition curves for the evaluation of the separation process were calculated. The overall iron oxide content was used instead of the susceptibility, which is difficult to measure for ferromagnetic materials. The obtained results provided a precise assessment of separation efficiency.
- The applied magnetic separation was able to reduce the amount of iron oxides to below 5%. However, this caused significant losses of cassiterite into the magnetic product.
- The main reasons for cassiterite losses into the magnetic product were determined. The most important one is the finely disseminated character of cassiterite in the matrix of iron oxides. Losses of fully liberated cassiterite can be due to agglomeration with iron oxide bearing particles.
- Losses of cassiterite could be significantly decreased by a further comminution and refeeding of the upper particle size fraction in the magnetic product and by a better dispersion during the separation (e.g., by using a wet magnetic separator).

The study illustrates that the interpretation of MLA data to evaluate the separation efficiency should always contain an assessment of uncertainties. In the presented study, these uncertainties originate from two different sources. First, unbalanced mass streams for single or multiple minerals such as the group of iron oxides reported in the present work. Second, a statistically unrepresentative number of particles in property classes (bins) during particle tracking which were used to track the cassiterite through the process in this case study. In addition to the particle tracking approach by Lamberg et al. [22] the error of the data was monitored and further analyzed in detail. Nevertheless, new approaches urgently need to overcome the problem of discrete bins in particle tracking.

Author Contributions: E.S. and M.B. conceived the paper. M.K. was responsible for sample analysis with a Mineral Liberation Analyzer (MLA) including quality control, data processing and data evaluation. R.T.-D. developed the basis for evaluation of the MLA data in the R statistical environment and interpreted results. R.M. and D.E. were responsible for sample preparation for XRD, data acquisition and evaluation of the measurements using the Rietveld method. E.S., M.B. and J.A. analyzed, interpreted, and joined together all results. T.L., M.R., K.G.v.d.B., as well as U.P., interpreted results, supervised the work and supported preparation of the manuscript.

Funding: This research was funded by BMBF in the r^4 program grant number: 033R128.

Acknowledgments: We would like to thank Saxore Bergbau GmbH for providing the material samples and UVR-FIA GmbH for preprocessing of the material.

Conflicts of Interest: The authors declare no conflict of interest.

References

1. Schuppan, W.; Hiller, A. *Die Komplexlagerstätten Tellerhäuser und Hämmerlein, Uranbergbau und Zinnerkundung in der Grube Pöhla der SDAG Wismut*; Bergbaumonografie, Landesamt für Umwelt, Landwirtschaft und Geologie, Sächsisches Oberbergamt: Freiberg, Germany, 2012; p. 17.
2. Angadi, S.I.; Sreenivas, T.; Jeon, H.-S.; Baek, S.-H.; Mishra, B.K. A review of cassiterite beneficiation fundamentals and plant practices. *Min. Eng.* **2015**, *70*, 178–200. [CrossRef]
3. Xu, C.; Zhang, Y.; Liu, T.; Huang, J. Characterization and Pre-Concentration of Low-Grade Vanadium-Titanium Magnetite Ore. *Minerals* **2017**, *7*, 137. [CrossRef]
4. Bulatovic, S.M. *Handbook of Flotation Reagents. Chemistry, Theory and Practice*, 1st ed.; Elsevier: Amsterdam, The Netherlands, 2010.
5. Falcon, L.M. The gravity recovery of cassiterite. *J. S. Afr. Inst. Min. Met.* **1982**, *4*, 112–117.
6. Kern, M.; Möckel, R.; Krause, J.; Teichmann, J.; Gutzmer, J. Calculating the deportment of a fine-grained and compositionally complex Sn skarn with a modified approach for automated mineralogy. *Min. Eng.* **2018**, *116*, 213–225. [CrossRef]
7. Little, L.; Becker, M.; Wiese, J.; Mainza, A.N. Auto-SEM particle shape characterisation: Investigating fine grinding of UG2 ore. *Min. Eng.* **2015**, *82*, 92–100. [CrossRef]
8. Little, L.; Mainza, A.N.; Becker, M.; Wiese, J.G. Using mineralogical and particle shape analysis to investigate enhanced mineral liberation through phase boundary fracture. *Powder Technol.* **2016**, *301*, 794–804. [CrossRef]
9. Leißner, T.; Hoang, D.H.; Rudolph, M.; Heinig, T.; Bachmann, K.; Gutzmer, J.; Schubert, H.; Peuker, U.A. A mineral liberation study of grain boundary fracture based on measurements of the surface exposure after milling. *Int. J. Min. Process.* **2016**, *156*, 3–13. [CrossRef]
10. Alsafasfeh, A.; Alagha, L. Recovery of Phosphate Minerals from Plant Tailings Using Direct Froth Flotation. *Minerals* **2017**, *7*, 145. [CrossRef]
11. Liu, X.; Zhang, Y.; Liu, T.; Cai, Z.; Sun, K. Characterization and Separation Studies of a Fine Sedimentary Phosphate Ore Slime. *Minerals* **2017**, *7*, 94. [CrossRef]
12. Philander, C.; Rozendaal, A. A process mineralogy approach to geometallurgical model refinement for the Namakwa Sands heavy minerals operations, west coast of South Africa. *Min. Eng.* **2014**, *65*, 9–16. [CrossRef]
13. Evans, C.L.; Napier-Munn, T.J. Estimating error in measurements of mineral grain size distribution. *Min. Eng.* **2013**, *52*, 198–203. [CrossRef]
14. Lotter, N.O.; Evans, C.L.; Engström, K. Sampling—A key tool in modern process mineralogy. *Min. Eng.* **2018**, *116*, 196–202. [CrossRef]

15. Leißner, T.; Bachmann, K.; Gutzmer, J.; Peuker, U.A. MLA-based partition curves for magnetic separation. *Min. Eng.* **2016**, *94*, 94–103. [CrossRef]
16. Jeong, S.; Kim, K. Pre-Concentration of Iron-Rich Sphalerite by Magnetic Separation. *Minerals* **2018**, *8*, 272. [CrossRef]
17. Chehreh Chelgani, S.; Leißner, T.; Rudolph, M.; Peuker, U.A. Study of the relationship between zinnwaldite chemical composition and magnetic susceptibility. *Min. Eng.* **2015**, *72*, 27–30. [CrossRef]
18. Waters, K.E.; Rowson, N.A.; Greenwood, R.W.; Williams, A.J. Characterising the effect of microwave radiation on the magnetic properties of pyrite. *Sep. Purif. Technol.* **2007**, *56*, 9–17. [CrossRef]
19. Tripathy, S.K.; Singh, V.; Rama Murthy, Y.; Banerjee, P.K.; Suresh, N. Influence of process parameters of dry high intensity magnetic separators on separation of hematite. *Int. J. Min. Process.* **2017**, *160*, 16–31. [CrossRef]
20. Fandrich, R.; Gu, Y.; Burrows, D.; Moeller, K. Modern SEM-based mineral liberation analysis. *Int. J. Min. Process.* **2007**, *84*, 310–320. [CrossRef]
21. Gottlieb, P.; Wilkie, G.; Sutherland, D.; Ho-Tun, E.; Suthers, S.; Perera, K.; Jenkins, B.; Spencer, S.; Butcher, A.; Rayner, J. Using quantitative electron microscopy for process mineralogy applications. *JOM* **2000**, *52*, 24–25. [CrossRef]
22. dLamberg, P.; Vianna, S. *A Technique for Tracking Multiphase Mineral Particles in Flotation Circuits*; XXII ENTMME/VII MSHMT: Ouro Preto, MG, Brazil, 2007.
23. Ueda, T.; Oki, T.; Koyanaka, S. Statistical effect of sampling particle number on mineral liberation assessment. *Min. Eng.* **2016**, *98*, 204–212. [CrossRef]
24. Heinig, T.; Bachmann, K.; Tolosana-Delgado, R.; van den Boogaart, G.; Gutzmer, J. *Monitoring Gravitational and Particle Shape Settling Effects on MLA Sampling Preparation*; International Association for Mathematical Geosciences: Freiburg, Germany, 2015.
25. Figueroa, G.; Moeller, K.; Buhot, M.; Gloy, G.; Haberla, D. Advanced Discrimination of Hematite and Magnetite by Automated Mineralogy. In Proceedings of the 10th International Congress for Applied Mineralogy (ICAM), Trondheim, Norway, 1–5 August 2012; pp. 197–204.
26. Rosenblum, S.; Brownfield, I.K. *Magnetic Susceptibility of Minerals*; U.S. Department of Interior, U.S. Geological Survey: Washington, DC, USA, 1999; p. 10.
27. dMadai, E. Limitations of magnetic separation in relation to particle size and susceptibility. *Aufbereit. Tech. Min. Process.* **1998**, *39*, 394–406.
28. Doebelin, N.; Kleeberg, R. Profex: A graphical user interface for the Rietveld refinement program BGMN. *J. Appl. Crystallogr.* **2015**, *48*, 1573–1580. [CrossRef] [PubMed]
29. ALS Geochemistry. *Shedule of Services and Fees: Refractory Ores and Geological Materials*; ALS Ltd.: Brisbane, Australia, 2017.
30. Gu, Y. Automated Scanning Electron Microscope Based Mineral Liberation Analysis an Introduction to JKMRC/FEI Mineral Liberation Analyzer. *J. Min. Mater. Charact. Eng.* **2003**, *2*, 33–41.
31. Efron, B.; Tibshirani, R. *An Introduction to the Bootstrap*; Chapman & Hall: Boca Raton, FL, USA, 1998.
32. Mariano, R.A.; Evans, C.L. Error analysis in ore particle composition distribution measurements. *Min. Eng.* **2015**, *82*, 36–44. [CrossRef]
33. Leißner, T. Beitrag zur Kennzeichnung von Aufschluss- und Trennerfolg am Beispiel der Magnetscheidung. Ph.D. Thesis, Freiberg University of Mining and Technology, Freiberg, Germany, 2015.
34. Jacoby, W.G. Loess: A nonparametric, graphical tool for depicting relationships between variables. *Elect. Stud.* **2000**, *19*, 577–613. [CrossRef]
35. Hunt, C.P.; Moskowitz, B.M.; Banerjee, S.K. Magnetic properties of rocks and minerals: A Handbook of Physical. AGU Reference Shelf 3. *Rock Phys. Phase Relat.* **1995**, *3*, 189–204.
36. Grant, D.C.; Goudie, D.J.; Voisey, C.; Shaffer, M.; Sylvester, P. Discriminating hematite and magnetite via Scanning Electron Microscope–Mineral Liberation Analyzer in the −200 mesh size fraction of iron ores. *Appl. Earth Sci.* **2018**, *127*, 30–37. [CrossRef]
37. Parian, M.; Lamberg, P.; Rosenkranz, J. Developing a particle-based process model for unit operations of mineral processing—WLIMS. *Int. J. Min. Process.* **2016**, *154*, 53–65. [CrossRef]

Article

Determination of Dissolution Rates of Ag Contained in Metallurgical and Mining Residues in the $S_2O_3{}^{2-}$-O_2-Cu^{2+} System: Kinetic Analysis

Julio C. Juárez Tapia [1], Francisco Patiño Cardona [2], Antonio Roca Vallmajor [3], Aislinn M. Teja Ruiz [1,*], Iván A. Reyes Domínguez [4], Martín Reyes Pérez [1], Miguel Pérez Labra [1] and Mizraim U. Flores Guerrero [5]

[1] Área Académica de Ciencias de la Tierra y Materiales, Universidad Autónoma del Estado de Hidalgo (UAEH), Pachuca de Soto, Hidalgo 42184, Mexico; jcjuarez@uaeh.edu.mx (J.C.J.T.); mar_77_mx@hotmail.com (M.R.P.); MIGUELABRA@hotmail.com (M.P.L.)

[2] Ingeniería en Energía, Universidad Politécnica Metropolitana de Hidalgo, Tolcayuca, Hidalgo 43860, Mexico; franciscopatinocardona@gmail.com

[3] Facultat de Química, Departament de Ciència dels Materials I Enginyeria Metallúrgica, Universitat de Barcelona, Provincia de Barcelona, 08028 Barcelona, Spain; roca@ub.edu

[4] Catedrático CONACYT-Instituto de Metalurgia, Universidad Autónoma de San Luis Potosí, San Luis Potosí, San Luis Potosí 78210, Mexico; iareyesdo@conacyt.mx

[5] Área de Electromecánica Industrial, Universidad Tecnológica de Tulancingo, Tulancingo, Hidalgo 43642, México; uri.fg@hotmail.com

*   Correspondence: aislinn_teja@uaeh.edu.mx; Tel.: +52-771-7172000 (ext. 2279)

Received: 28 June 2018; Accepted: 18 July 2018; Published: 23 July 2018

Abstract: The materials used to conduct kinetic study on the leaching of silver in the $S_2O_3{}^{2-}$-O_2-Cu^{2+} system were mining residues (tailings) from the Dos Carlos site in the State of Hidalgo, Mexico, which have an estimated concentration of Ag = 71 g·ton^{-1}. The kinetic study presented in this paper assessed the effects of the following variables on Ag dissolution rate: particle diameter (d_0), temperature (T), copper concentration [Cu^{2+}], thiosulfate concentration [$S_2O_3{}^{2-}$], pH, [OH^-], stirring rate (RPM), and partial pressure of oxygen (PO_2). Temperature has a favorable effect on the leaching rate of Ag, obtaining an activation energy (Ea) = 43.5 kJ·mol^{-1} in a range between 288 K (15 °C) and 328 K (55 °C), which indicates that the dissolution reaction is controlled by the chemical reaction. With a reaction order of n = 0.4, the addition of [Cu^{2+}] had a catalytic effect on the leaching rate of silver, as opposed to not adding it. The dissolution rate is dependent on [$S_2O_3{}^{2-}$] in a range between 0.02 mol·L^{-1} and 0.06 mol·L^{-1}. Under the studied conditions, variables d_0, [OH^-] and RPM did not have an effect on the overall rate of silver leaching.

Keywords: silver leaching; thiosulfate; mining residues; kinetic analysis

1. Introduction

In Mexico, mining accounts for 4% of gross domestic product, which places this country among the top ten world producers of sixteen minerals, among which silver stands out. In 2015, 49% of mining production in Mexico was based on the extraction of precious metals, which represented 7.5 trillion dollars. However, in the last three years there has been a decrease in silver production indices due to the massive exploitation of Au-Ag-Cu-Zn deposits [1]. One way to increase the production of silver is by reprocessing scrap, slag and mining waste that contain economically reasonable amounts of this metal. In the State of Hidalgo, Mexico, due to the extensive mining activity that has lasted approximately 480 years in the mining districts of Pachuca and Mineral del Monte, millions of tons of mining waste have been accumulated, and are nowadays posing a serious environmental problem [2,3].

In addition, reprocessing these residues presents a challenge for the metallurgical industry because of the presence of quartz and pyrite ores, where the metal values of interest are encapsulated [4].

During the last 30 years, research on the extraction of metal values of gold and silver has been focused on the search for leaching processes alternative to cyanidation, which is a process that has been used for over a century and is considered highly toxic and with a strong impact on the environment [5]. Leaching agents that can substitute cyanide include compounds such as thiocyanate, thiourea and thiosulfate, the last two being the most promising processes, as their application has a lower environmental impact [6–11].

Aylmore and Muir [10] reported that thiosulfate has the advantage of being capable of recovering precious metals and increasing the dissolution rate of metals contained in hard-to-process ores. Since higher dissolution rates mean smaller leaching tanks, investment costs and energy consumption can be reduced [12,13]. The use of thiourea and thiosulfate has been mainly applied to refractory gold and silver ores, in addition to other metals of interest, such as Cu, Pb and Zn, as traditional processes have shown certain deficiencies when treating such minerals [14–17]. As regards to silver leaching, the available published works have been limited, and the kinetic aspects approached show certain inconsistencies. Some of the first research works on silver leaching were conducted by Mohammadi et al. [12], who developed a leaching process at atmospheric pressure using ammonium thiosulfate to recover gold and silver from residues generated during ammoniacal leaching of copper sulfide concentrates. Other studies discussed the chemistry of silver sulfide leaching with solutions of thiosulfates, among which stands out the use of copper sulfate, emphasizing that copper catalyzes the dissolution reaction of the metal values, and assuming that certain equilibrium between the solution and the Cu^+ and Cu^{2+} ions is needed for the extraction to occur [13–19]. Since then several studies have had the aim of explaining how the leaching process with these compounds occurs. Lampinen et al. [20] reported that thiosulfate is a good alternative for leaching gold, noting that the process is favored with the presence of copper ions, increased temperature and by injecting oxygen into the system. Zipperian et al. [21] showed that temperature has a drastic influence on the extraction of silver. At room temperature, they obtained dissolutions of 18% of the precious metal, which increased up to 60% at 60 °C within a period of 3 h.

Thus, it would seem that the thiosulfate system is the most promising for metal extraction, as it can increase the dissolution rate of metals, and because, compared to cyanide, it is a selective method for refractory ores [10,22]. Another alternative that has been studied (in the laboratory) in previous works on the leaching of silver in sulfide or metal form with thiosulfate, is the addition of metal ions Zn^{2+} and Cu^{2+}, where these two oxidize the noble metals while thiosulfate forms stable complexes. In this regard, Juárez et al. [23] and Hernández et al. [24] have used these metal ions in their research, and they found that the process can be a good alternative in the recovery of silver contained in mining residues. They obtained dissolutions higher than 95%, and showed that Zn^{2+} and Cu^{2+} have a catalytic effect on the process by considerably reducing leaching times. In the case of research conducted on the leaching of metal silver using the ammoniacal thiosulfate system, was reported that the process appears to be affected by the formation of copper oxides and sulfides on the surface of the silver particles, thus favoring the precipitation of a considerable amount of this metal in the form of sulfide. Most of the leaching systems with thiosulfate described above have been found to be controlled by the chemical reaction [25]. However, Rivera et al. showed that stirring rate had an effect on the leaching of a silver plate, which was indicative of a strong influence of the oxygen mass transfer in the solid-liquid interface, affecting the reaction rate. This was due to the injection of oxygen and to the geometry of the silver sample [26].

As these examples show, the leaching kinetics of silver with thiosulfate and metal ions as oxidizers has not been widely approached. Moreover, regarding the effect of temperature, optimal concentration of thiosulfate, and effect of catalysts and dissolution atmosphere, several dissimilarities have been published. It is also important to point out that this process has been scarcely used to recover precious metals contained in mining waste. For this reason, we conducted a kinetic study on the leaching of silver contained in mining residues in the $S_2O_3^{2-}$-O_2-Cu^{2+} system. We assessed the effect of temperature, reagent concentration, and also of oxygen injection into the system, with the aim of reaching recoveries higher than those obtained with conventional processes, such as cyanidation.

2. Materials and Methods

For this research work we used the Dos Carlos mine dumps from the State of Hidalgo, Mexico, as it is one of the oldest yet least studied areas. It has a total volume of 14.3 million tons, produced in three technologically different periods: grinding-amalgamation, followed by grinding-cyanidation and grinding-flotation-cyanidation [27]. The samples were digested with aqua regia and analyzed using Perkin Elmer Optima 3200 RL ICP-OES spectrometer in order to identify and quantify the elements present in the mine dumps. The concentration of silver was determined by dry route, using the crucible melting cupellation technique, in which the substance is dissolved in a crucible with a reducing flow, to recover the precious metal inside a lead nugget. To determine separately the laws of precious metals in the silver and gold buttons, it was necessary to dissolve them in nitric acid and hydrochloric acid and finally the solutions were analyzed using the ICP-OES technique.

The samples extracted on the field were homogenized and then characterized for identification of oxides by X-ray fluorescence (XRF) using a X-ray sequential spectrophotometer by wavelength dispersion (WDXRF), Philips model PW2400 (Labexchange, Burladingen, Germany). The mineralogical characterization was performed by the X-Ray Diffraction (XRD) technique using a powder diffractometer, Philips model X'Pert and the Scanning Electron Microscopy-Energy-Dispersive X-Ray Spectroscopy (SEM-EDS) technique using a scanning electron microscope, JOEL Model JSM-840 (JEOL, Peabody, MA, USA), which has coupled an oxford dispersive energy spectrometer. The leaching experiments were carried out in a conventional 500 mL glass reactor mounted on a heating plate equipped with stirring rate and temperature controls.

pH was continually measured with an OAKTON pH meter equipped with a gel filled ROSS Ultra pH/ATC Triode electrode that can operate in the 0–14 range. pH adjustments of each experiment were performed by adding a 0.2 mol·L^{-1} NaOH solution directly into the reactor. The system's temperature was controlled through a thermometer coupled to the heating plate.

Samples extracted at different times during the leaching experiments were analyzed by atomic absorption spectrophotometry (AAS) using a Perking Elmer AAnalyst 200 spectrometer to determine the concentration of silver in solution at a given time t.

The fraction of silver was calculated according to the following expression:

$$X_{Ag} = \frac{[Ag]_{sol}}{[Ag]_T} \tag{1}$$

where: X_{Ag} = Fraction of silver in solution, $[Ag]_{sol}$ = concentration of silver at a given time t, $[Ag]_T$ = total silver concentration, which value corresponds to the amount of silver retained in the particle size selected by the granulometric analysis as the most viable for the kinetic study.

The reagents used for the present work were the following: $Na_2S_2O_3$ (99.9% purity), $CuSO_4$ (98.9% purity), O_2 (99.9% purity), NaOH (98.8% purity) all Sigma-Aldrich (St. Louis, MO, USA). In order to determine the particle size distribution of the mining residues, a particle size analysis was conducted using a Ro-Tap sieve shaker (Cole-Parmer, Vernon Hills, IL, USA) for 10 min with Taylor sizing sieves and the following mesh sizes: 149, 106, 75, 56, 44, 37, and 25 μm. Sample density was calculated with a pycnometer, using water as immersion liquid. Experimental conditions used in the kinetic study are shown in Table 1.

Table 1. Experimental conditions of the kinetic study.

Parameters	Experimental Conditions
d_0 (μm)	149, 106, 75, 56, 44, 37,25
$[Cu^{2+}]$ (mol·L^{-1})	0.001, 0.002, 0.003, 0.004, 0.006
$[S_2O_3{}^{2-}]$ (mol·L^{-1})	0.02, 0.04, 0.06, 0.08, 0.16, 0.40
T (K)	288, 298, 308, 318, 328
RPM (min^{-1})	250, 350, 450, 550, 650, 750
PO$_2$	1 atm (excess)
$[OH^-]$ (mol·L^{-1})	$1 \times 10^{-9}, 1 \times 10^{-7}, 1 \times 10^{-5}, 1 \times 10^{-3}, 1 \times 10^{-2}$

3. Results

3.1. Chemical Characterization

By using XRF determined the elements present in the residues as oxides, obtaining the following chemical composition: SiO$_2$ (75%), Al$_2$O$_3$ (8%), Fe$_2$O$_3$ (5%), SO$_3$ (5%), K$_2$O (3.4%), CaO (1.5%), Na$_2$O (0.4%), MgO (0.4%) TiO$_2$ (0.3%), MnO (0.15%), and P$_2$O$_5$ (0.1%).

3.2. Mineralogical Characterization

The X-ray spectrum in Figure 1 shows that silica is the main species (JCPDS No. 01-074-3485), representing the mineralization matrix of these mining residues [27]. Ag appears to be absent in the diffractogram because silver concentration in the residues is under the detection limit of this technique.

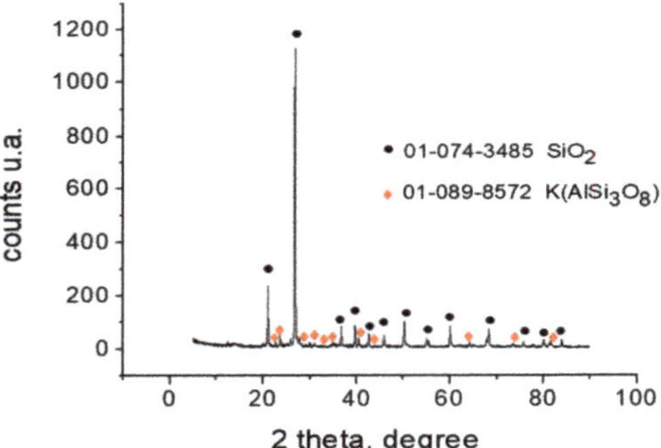

Figure 1. X-ray diffractogram showing the mineralization matrix of the Dos Carlos mine dumps of the State of Hidalgo, Mexico.

Figure 2 shows a micrograph obtained by Reflected light microscopy where the matrix of the mineral sample and the presence of iron sulfides were identified.

The micrograph shown in Figure 3 was obtained by scanning electron microscopy (SEM), in this the main species of quartz and silicates (also obtained through XRD) can be noticed. Figure 4 shows a micrograph obtained by backscattered electrons, whose gray contrast allowed to identify metal particles corresponding to silver sulfide.

Figure 2. Optical micrograph of the mining residues showing the presence of metal inclusions of pyrite in a quartz matrix.

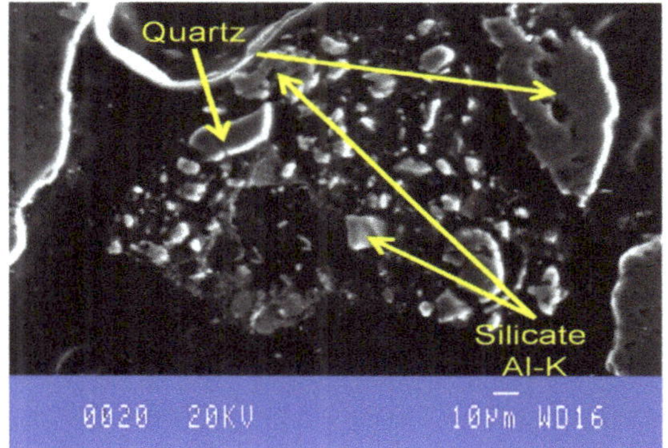

Figure 3. Mixed particles of quartz and silicates. scanning electron microscopy (SEM), secondary electrons.

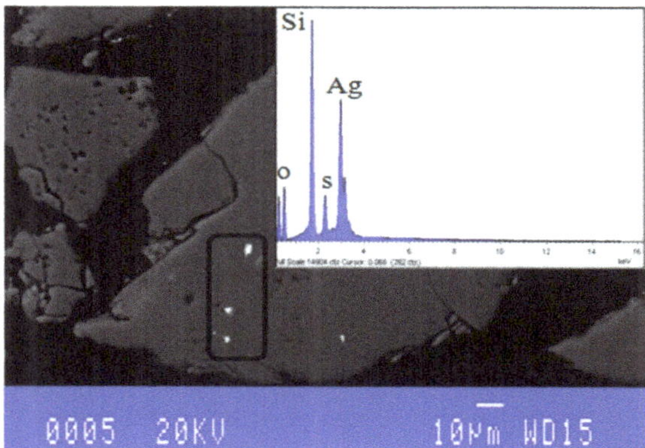

Figure 4. Backscattered electron image and energy dispersive X-ray spectrometry (EDS) spectrum showing the presence of silver sulfide encrusted in a quartz matrix.

3.3. Physical Characterization

The density of the mineral was of 2.57 g/cm^3, and the particle size analysis results of the mine dumps are shown in Table 2.

Table 2. Granulometric analysis of the Dos Carlos mine dumps (tailings).

Mesh	Opening (μm)	Ag Retained (g/ton^{-1})	Weight (wt %)	Distribution of Ag (wt %)
<100	149	50.13	40	43.51
100–140	106	24.36	53	28
140–200	75	2.81	66	4.01
200–270	56	1.27	63	1.73
270–325	44	1.18	73	1.86
325–400	37	1.20	82	2.12
>400	25	1.65	101	3.53

3.4. Dissolution Curves of Ag and Kinetic Model

In heterogeneous reactions where there is a solid-liquid interface, it is important to consider that the process consists of several stages that include mass transport and the chemical reaction. The variable that allows describing the progress of the reaction is defined as conversion factor (Equation (1)). It represents the amount of reacted mass in relation to the initial mass. In the case of the dissolution of silver, the reaction was monitored by observing the fraction of Ag that passed into the solution (by forming a complex with thiosulfate) in relation to the amount of silver present in the concentrate.

The development of a kinetic study for this kind of non-catalytic reaction, where solid particles are suspended in an active solution, comprises the application of two main models: the progressive conversion model and the unreacted shrinking core model. When the chemical stages occur in shorter times, one can assume that the reagent in heterogeneous reactions is rapidly running out on the solid's surface. The flow of the leaching solution diffuses by unit of time in a perpendicular direction on the surface of the solid particle. In this case, the applied kinetic expression is shown in Equation (2).

$$t/\tau = 1 - 3(1 - X_{Ag})^{2/3} + 2(1 - X_{Ag}) = k_{exp} \cdot t \tag{2}$$

where t is time in min, τ is the total time of the reaction, X_{Ag} is the fraction of silver in solution at a given time t.

When the chemical reaction takes more time compared to phenomena of hydrodynamic nature, no significant concentration gradients appear on the fluid film, so the mechanism of dissolution is controlled by the chemical stages where a low dependence on matter transport, a high sensitivity towards temperature increase, and reaction orders different from the unit are observed [28]. The kinetic equation that describes this process is presented in Equation (3).

$$t/\tau = 1 - (1 - X_{Ag})^{1/3} = k_{exp} \cdot t \tag{3}$$

Figure 5 is an example of how the leaching process of silver with thiosulfates in the proposed medium evolves. The plot shows a rapid dissolution of silver of X_{Ag} = 0.4 in the first 12 s, as well the absence of an induction period, which can be attributed to two factors: (1) part of this silver is found released in the mining residues; and [2] selectivity of the complexing agent in relation to silver. Thus, we only observed a progressive conversion period followed by a stabilization period, which indicates the end of the reaction.

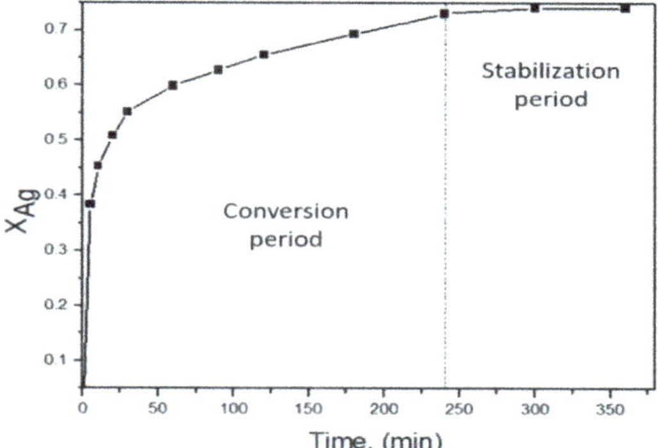

Figure 5. Dissolution curve of silver in the $S_2O_3^{2-}$-Cu^{2+} system (V = 0.5 L, 40 g L^{-1} Mineral, PO_2 = 1 atm, pH = 10, leaching time = 360 min, $[S_2O_3^{2-}]$ = 0.16 mol·L^{-1}, $[Cu^{2+}]$ = 0.006 mol·L^{-1}, 298 K (25 °C), RPM = 750 min^{-1}).

The data representing X_{Ag} vs. t in Figure 5 were used to assess Equations (4) and (5) (corresponding to control by matter transport and chemical control), which should have a linear behavior, and the obtained slope represents the experimental constant k_{exp} during the progressive conversion period. As seen on Figure 6, the chemical control model is consistent with the linear requirement; therefore, this model can be accepted to describe the dissolution process of silver.

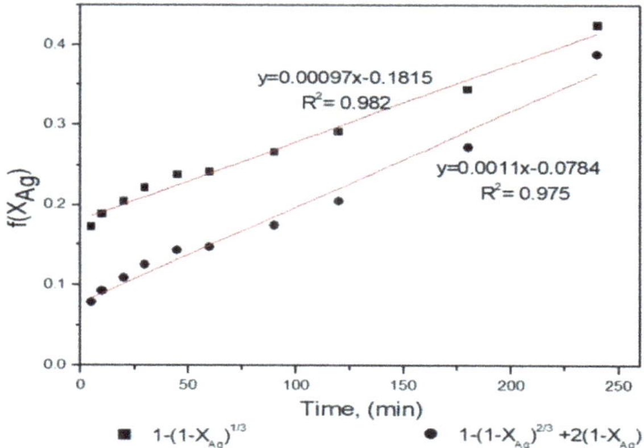

Figure 6. Comparison between the chemical control model and the control-by-transport model for the leaching of silver in the $S_2O_3^{2-}$-Cu^{2+} system.

3.4.1. Effect of Partial Pressure of Oxygen

For the kinetic study of the leaching of silver in the proposed medium, we assessed the following effects on the dissolution rate of Ag: d_0, $[Cu^{2+}]$, $[S_2O_3^{2-}]$, $[OH^-]$, T, RPM and PO_2. Two experiments were conducted at 0.2 and 1.0 atm of pressure in order to assess the best leaching conditions. Figure 7 shows that the leaching rate of silver is proportional to the partial pressure of oxygen in the system.

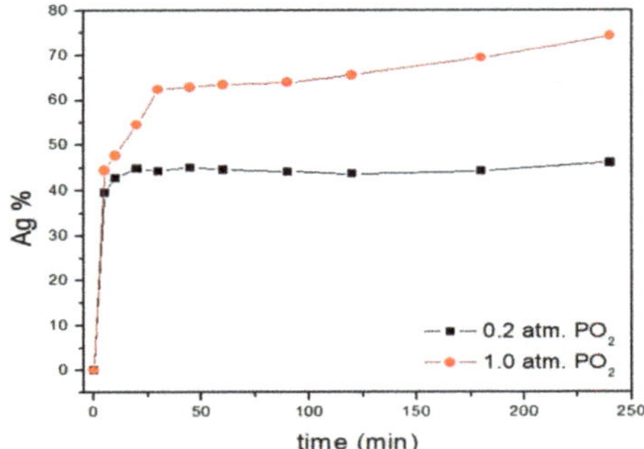

Figure 7. Effect of partial pressure of oxygen on the leaching reaction of silver (V = 0.5 L, 40 g·L^{-1} Mineral, pH = 10, leaching time = 240 min, [S$_2$O$_3{}^{2-}$] = 0.16 mol·L^{-1}, [Cu^{2+}] = 0.006 mol·L^{-1}, 298 K (25 °C), RPM = 750 min^{-1}).

3.4.2. Particle Size Effect

For the kinetic study of the leaching of silver in the proposed medium, we assessed the following effects on the dissolution rate of Ag: d$_0$, [Cu^{2+}], [S$_2$O$_3{}^{2-}$], [OH$^-$], T, RPM and PO$_2$. In order to determine the effect of the initial particle size, we conducted a series of experiments where the initial particle size was varied, while the other variables were kept constant. According to Equation (4) corresponding to the shrinking core model with chemical control, the constant k$_{exp}$ is defined as follows:

$$k_{exp} \;=\; \frac{2V_m K_q C_A^n}{d_0}. \tag{4}$$

where V$_m$ is molar volume, k$_q$ is the chemical rate constant, C$_A$ is the reagent concentration, *n* is the order of reaction, and d$_0$ is the initial particle diameter.

Thus, according to Equation (6), a representation of the experimental constants determined at constant concentration, temperature and stirring rate, vs the inverse of particle radius should be linear and pass through the origin, as show the Figure 8.

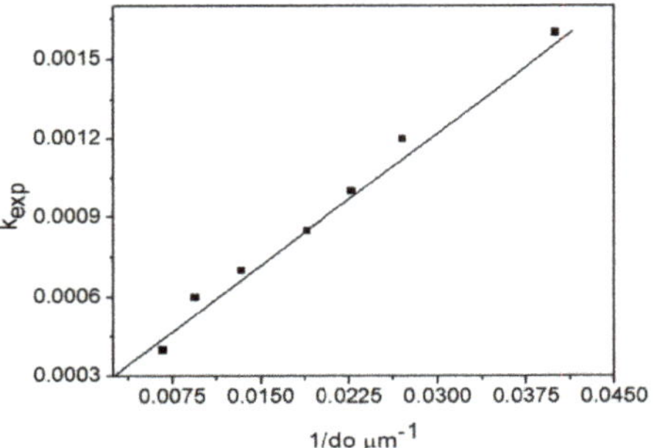

Figure 8. k$_{exp}$ versus the inverse of particle diameter. (V = 0.5 L, 40 g·L^{-1} Mineral, PO$_2$ = 1 atm, pH = 10, leaching time = 240 min, [S$_2$O$_3{}^{2-}$] = 0.16 mol·L^{-1}, [Cu^{2+}] = 0.006 mol·L^{-1}, 298 K, RPM = 750 min^{-1}).

3.4.3. Thiosulfate Concentration Effect

The concentration of thiosulfate is another very important variable for this kinetic study, as it is the complexing agent of silver. This variable was studied in an [S$_2$O$_3{}^{2-}$] interval = 0.02–0.40 mol·L^{-1}. At the beginning and for the whole reaction time of each experiment we injected industrial-grade oxygen in excess. The reaction order was determined by linear regression analysis, plotting the log $[S_2O_3{}^{2-}]$ vs the log of k_{exp} to obtain a straight slope (m). Figure 9 clearly shows two reaction orders for the studied thiosulfate concentrations, i.e., between 0.02, and 0.06 mol·L^{-1}. A pseudo- order of reaction of $\alpha = 0.4$ was obtained, which indicates that the dissolution rate of silver is dependent on the concentration of thiosulfate. However, for the concentration range between 0.06, and 0.40 mol·L^{-1}, the order of reaction changed drastically, obtaining in this case a pseudo- order of reaction of $\alpha = 0$, which indicates that, at this concentration range of thiosulfate, the leaching rate of Ag is independent from the concentration of the complexing agent.

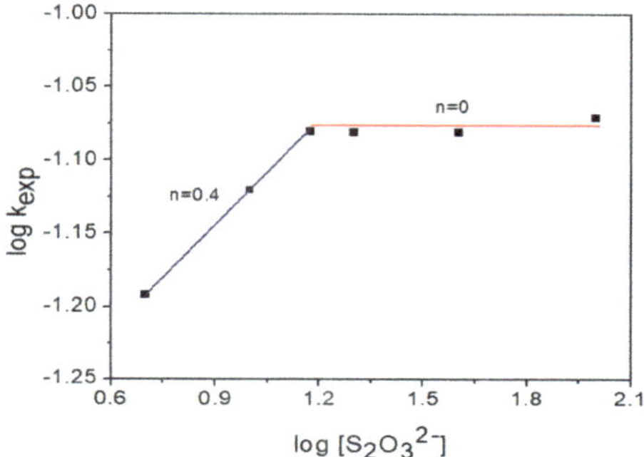

Figure 9. k_{exp} dependence vs. $[S_2O_3{}^{2-}]$. Reaction order in the concentration range of 0.02 to 0.06 mol·L^{-1} (V = 0.5 L, 40 g·L^{-1} Mineral, PO$_2$ = 1 atm, pH = 10, leaching time = 240 min, $[Cu^{2+}]$ = 0.006 mol·L^{-1}, 298 K (25 °C), RPM = 750 min^{-1}).

3.4.4. Cu^{2+} Concentration Effect

It has been reported that the presence of Cu ions favors the dissolution rate of silver, as well as its maximum release in a leaching solution with sodium thiosulfate [8]. In order to confirm this, we conducted several experiments where concentration of copper was modified in a range between 0.001 mol·L^{-1} and 0.006 mol·L^{-1}, while keeping the other parameters constant. Figure 10 shows log k_{exp} vs. log $[Cu^{2+}]$; the straight slope gives a pseudo-order of reaction of $\beta = 0.41$, which indicates that there is a significant dependence of the leaching rate of silver on the presence and increase of the concentration of copper ions in the solution.

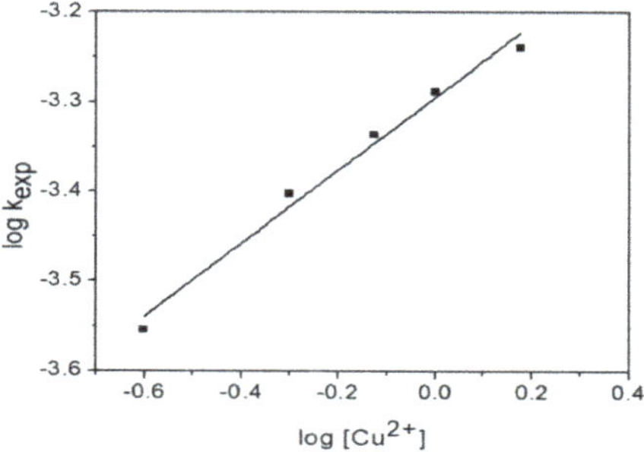

Figure 10. k_{exp} dependence in function of $[Cu^{2+}]$. Reaction order $n = 0.4$. (V = 0.5 L, 40 g·L^{-1} Mineral, PO$_2$ = 1 atm, pH = 10, leaching time = 240 min, $[S_2O_3{}^{2-}]$ = 0.16 mol·L^{-1}, 298 K (25 °C), RPM = 750 min^{-1}).

3.4.5. pH Effect

The effect of [OH$^-$] on the leaching rate of silver was assessed in a range of pH between 8 and 12, since thiosulfate is more stable in this pH range [25]. As seen on Figure 11.

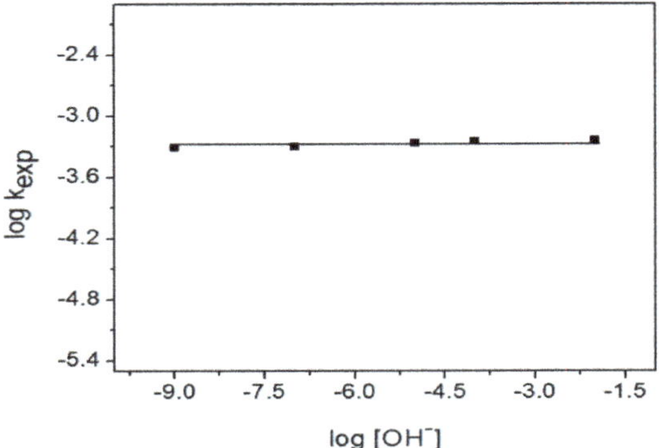

Figure 11. Dependence of log k_{exp} in function of log [OH$^-$], obtained reaction order *n* = 0 (V = 0.5 L, 40 g·L^{-1} Mineral, PO$_2$ = 1 atm, leaching time = 240 min, [S$_2$O$_3{}^{2-}$] = 0.16 mol·L^{-1}, [Cu^{2+}] = 0.006 mol·L^{-1}, 298 K (25 °C), RPM = 750 min^{-1}).

3.4.6. Temperature Effect

The effect of temperature was studied in a range between 288 K (15 °C) and 328 K (55 °C), while the other parameters were kept constant. Figure 12 represents the fraction of leached silver; it was assessed using the shrinking core model with chemical control in function of time for all of the temperatures used.

In order to determine the activation energy of the system, we used the Arrhenius equation [28] shown in Equation (5), where k_q represents the rate constant, k_0 is the frequency factor, Ea is the activation energy, R is the universal gas constant (8.314 kJ·mol^{-1}), and *T* is temperature in *K*.

$$k_q = k_0 e^{-Ea/RT} \tag{5}$$

Although it is known that the ionization constant of water (k_w) varies considerably along with temperature, this dependence of [OH$^-$] on temperature was eliminated with the $k_{exp}/[\text{OH}^-]^n$ relation according to Equation (4) [29]. By substituting this relation in Equation (6) and applying the natural logarithm, we obtained Equation (6). Equation (12) represents a straight line whose slope corresponds to the –Ea/R value, and the ordinate to the origin corresponds to k_0.

$$\ln \frac{k_{exp}}{[\text{OH}^-]^n} = \ln \frac{2V_m k_0}{d_0} - \frac{Ea}{R} \cdot \frac{1}{T} \tag{6}$$

where V_m represent molar molume, k_0 is the frequency factor and d_0 is particle diameter. The activation energy value is used to obtain the control mechanism. Therefore, we plotted the Napierian logarithm of $k_{exp}/[\text{OH}^-]^n$ vs. 1/T, allowing us to determine the activation energy of the system, as shown in Figure 13.

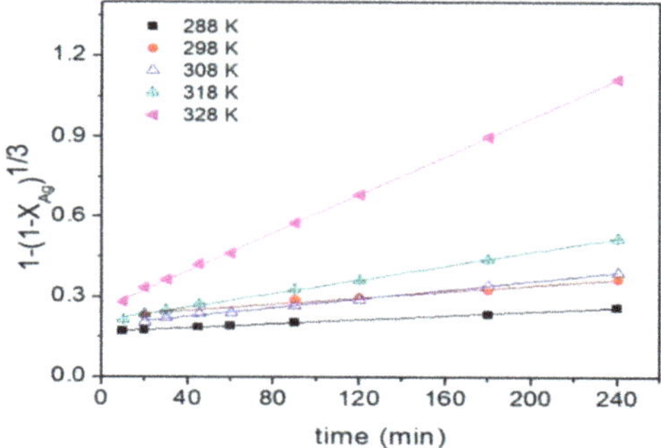

Figure 12. Assessment of the application of the chemical control model on the conversion values of Ag in function of time for the study of the temperature effect in the $S_2O_3^{2-}$-O_2 system, (V = 0.5 L, 40 g·L^{-1} Mineral, PO_2 = 1 atm, pH = 10, leaching time = 240 min, $[S_2O_3^{2-}]$ = 0.16 mol·L^{-1}, $[Cu^{2+}]$ = 0.006 mol·L^{-1}, 298 K (25 °C), RPM = 750 min^{-1}).

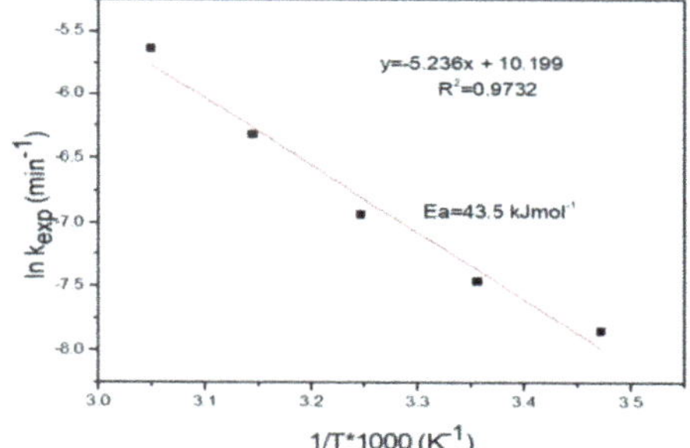

Figure 13. Dependence of ln k_{exp} vs temperature. Calculated activation energy, Ea = 43.5 kJ·mol^{-1}.

3.4.7. Stirring Rate Effect

In order to know if there are changes in the leaching rate of silver when the stirring rate is modified, we conducted a number of experiments in an interval between 250 min^{-1} and 750 min^{-1}, keeping the other variables constant. In Figure 14, one can observe that the global rate of the reaction does not depend on the stirring rate, since the values of k_{exp} were constant in all the experiments ($k_{exp} \approx 0.0585$ min^{-1}).

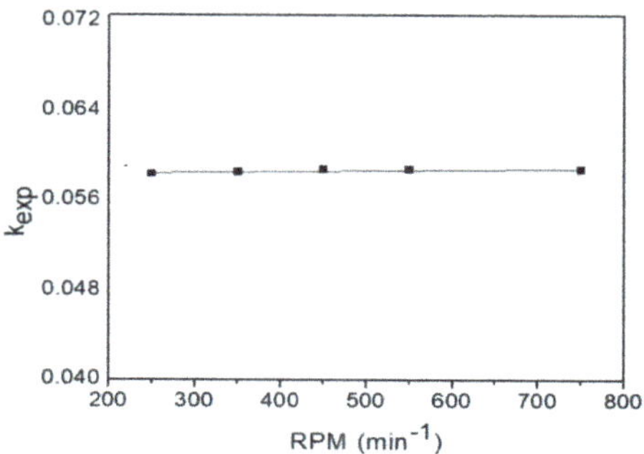

Figure 14. Dependence of k_{exp} in function of stirring rate.

4. Discussion

Results of the chemical characterization, which was carried out by X-ray fluorescence (XRF) and by the cupellation technique, showed similar silver contents. We could establish that the obtained concentrations of Ag in the mine dumps were of the order of 75 g·ton^{-1} and 71 g·ton^{-1} respectively. The Figure 1 is consistent with the results obtained by XRF. It is also possible to notice the presence of some feldspars, such as potassium silicate (JCPDS No. 01-089-8572), which is one of the secondary minerals formed by supergene oxidation. It is common in small amounts in veins associated with silver-rich ores [4,30].

As regards the mineralogical characterization, the technique of Reflected light microscopy identified abundant fragments of quartz crystals, silicates and scarce metal mineralization. Such mineralization is composed mainly of base metal sulfides that make up the ore [31–33] such as pyrite (FeS_2) associated with quartz (SiO_2), as shown in the micrograph on Figure 2. To complete the characterization study, the mineral was analyzed by scanning electron microscopy (SEM) along with energy dispersive X-ray spectrometry (EDS). Microanalyses performed on particles with a coarse aspect showed a mixed composition, as observed in Figure 3. Upon sample analysis with backscattered electrons, we were able to see that silver is present in the form of argentite (Ag_2S) encrusted in a quartz matrix, as observed in Figure 4. This has previously been reported by Geyne A.R. and Fries C. [4]. According to the results obtained by granulometric analysis of the Dos Carlos mine dumps, it can be observed that the highest retained weight percentage (50.13%) and distribution of silver (43.51%) were obtained with mesh size 100 (149 µm). Therefore, this particle size (100) was used in all of the experiments for the kinetic study.

Determining the stoichiometry of the reaction is challenging due to the complexity of the mineralization contained in the Dos Carlos tailings. Thus, we used theoretical stoichiometry based on the characterization results in order to estimate the presence of silver in the mine dumps as silver sulfide. The proposed equations are the following:

$$Ag_2S + \tfrac{1}{2}O_2 + H_2O \rightarrow 2AgOH + S°\downarrow + 2e^- \tag{7}$$

$$AgOH + Na_2S_2O_3 \rightarrow AgNa[S_2O_3] + NaOH \tag{8}$$

The kinetic study of silver leaching in $S_2O_3^{2-}$-O_2-Cu^{2+} medium revealed that the recovery of Ag increases, from 46% at 0.2 atm, to 74% when using 1 atm of gas pressure at the same reaction time of 240 min (Figure 7). Since there were no significant changes in the dissolution rates at more than 1 atm of O_2 pressure (O_2 is clearly in excess), we decided to use 1 atm pressure in all the experiments for practical reasons. In the study of Particle size effect can be noted that the experimental constant obtained in experiments with different initial particle sizes is inversely proportional to particle diameter ($k_{exp}·\alpha·(1/d_0)$); thus, the leaching reaction of silver is consistent with the proposed model (Figure 8). In addition, we could confirm that, as particle size decreases, the leaching rate of silver increases. This is due to the presence of a larger surface area on the silver particles that are in contact with the leaching agent [26]. Moreover, the thiosulfate concentration effect reported an apparent change in the reaction occurred because, at low thiosulfate concentrations, the leaching rate of silver depends on the partial pressure of oxygen, and on a high consumption of complexing agent, which, possibly, is not present in enough quantities to sustain the dissolution reaction (Figure 9). Nonetheless, at concentrations ranging from 0.06 mol·L^{-1} to 0.40 mol·L^{-1}, a saturation of the complexing agent in the solution was observed (it was found in excess) that keeps the leaching rate constant. Thus, the leaching rate depends only on the concentration of dissolved oxygen.

The increase of the leaching rate in the presence of Cu^{2+} is due to a parallel reaction to the oxidation of silver. The leaching reaction of Ag in the O_2-$S_2O_3^{2-}$ system without a catalyst basically depends on the oxygen dissolved in the system, as shown in Equation (9):

$$Ag_2S_{(s)} + 4(S_2O_3)^{2-}_{(aq)} + 4H^+_{(aq)} + O_2 \rightarrow 2Ag(S_2O_3)_2^{3-}_{(aq)} + S^{2+} + 2H_2O_{(aq)} \tag{9}$$

This is different from the reaction in the presence of $[Cu^{2+}]$, where the reduction of Cu^{2+} to Cu^+ causes the Ag° to oxidize to Ag^+ as described in Equations (10)–(12).

$$2Cu^{2+} + 2S_2O_3{}^{2-} \leftrightarrow 2Cu^+ + S_4O_6{}^{2-} \tag{10}$$

$$6OH^- + 4S_4O_6{}^{2-} \leftrightarrow 2S_3O_6{}^{2-} + 3H_2O + 5S_2O_3{}^{2-} \tag{11}$$

$$Cu^{2+} + Ag^\circ + 5S_2O_3{}^{2-} \rightarrow Ag(S_2O_3)_2{}^{3-} + Cu(S_2O_3)_3{}^{5-} \tag{12}$$

This confirms that, in the presence of oxygen and copper ions, leaching with thiosulfate has a favorable effect on the reaction (Figure 10), with up to 30% increase compared to the same process without using copper [26,34,35]. The maximum recovery of silver was 74.4% with the addition of 0.25 $(g \cdot L^{-1})$ Cu^{2+} at room temperature, while without copper the Ag recoveries were of 43% at the same conditions of the reaction. While the study of pH effect allowed the calculation of a reaction order of $n = 0$ was obtained, which indicates that the leaching rate does not depend on the $[OH^-]$ values (Figure 11); this is consistent with previous works by Rivera et al. [26], who explained that the degradation of thiosulfate occurs preferably in the pH range of 8 < pH > 12.

The Figure 12 shows the strong influence of temperature on the dissolution rate. The plot demonstrates straight lines whose slopes represent the experimental rate constants (k_{exp}), where, as temperature is increased, the fraction of silver in solution rises as well. There is a moderate increase from 288 K (15 °C) to 318 K (45 °C), while upon reaching the highest temperature of 323 K (50 °C), the fraction of leached silver increases drastically until reaching 0.96 in a time of 240 min. The obtained Ea value was 43.5 $kJ \cdot mol^{-1}$ (Figure 13) which indicates that the chemical reaction is the controlling mechanism of the leaching process of silver in the studied system [36]. Finally, the study of Stirring rate effect the transport of matter in the solid-liquid interface has no significant effect on the dissolution rate when modifying the range of revolutions per minute (Figure 14). It can be confirmed that the slow, controlling stage, is the chemical reaction.

Due to the obtained value of >40 $kJ \cdot mol^{-1}$ for the activation energy, which established a control by chemical reaction, and to the fact that k_{exp} was inversely proportional to the initial particle diameter and passes through the origin, it was possible to confirm that the shrinking core model with chemical control is the one that satisfactorily describes the leaching process of silver in $S_2O_3{}^{2-}$-Cu^{2+}. Therefore, by combining and arranging Equations (3)–(5), it is possible to obtain a general kinetic model that describes the Ag dissolution of the metallurgical residues in the $S_2O_3{}^{2-}$-Cu^{2+} system, taking into consideration that the O_2 concentration in the system remains constant:

$$1 - (1 - X_{Ag})^{\frac{1}{3}} = \frac{V_m K_o}{r_o} e^{-\frac{Ea}{RT}} \left[S_2O_3^{2-} \right]^\alpha [Cu^{2+}]^\beta t \tag{13}$$

where α and β correspond to pseudo- orders of reaction related to $[S_2O_3{}^{2-}]$ and $[Cu^{2+}]$ respectively, and $\alpha + \beta = n$ is the global order of reaction for the studied system. By substituting the obtained kinetic parameters of k_0, Ea, and n in Equation (13), the following expression was obtained:

$$1 - (1 - X_{Ag})^{\frac{1}{3}} = 1.6 \times 10^4 e^{-\frac{43.5000}{RT}} \left[S_2O_3^{2-} \right]^{0.4} [Cu^{2+}]^{0.4} t \tag{14}$$

Equation (14) shows the effect of time, and of $S_2O_3{}^{2-}$ and Cu^{2+} concentrations on the fraction of silver that is transformed into product, which is valid in the temperature range between 288 K (15 °C) and 328 K (55 °C), and in concentrations ranging from 0.02 to 0.06 $mol \cdot L^{-1}$ $S_2O_3{}^{2-}$ and 0.001 to 0.006 $mol \cdot L^{-1}$ Cu^{2+}.

5. Conclusions

Results of the characterization of the mine dumps showed that silver is found in the form of sulfide encrusted in quartz and pyrite matrixes. The Ag concentration is of the order of 71 $g \cdot ton^{-1}$. The main non-metal mineral phase (determined by XRD) was quartz, while the reflected light microscope and EDS microanalysis revealed that the most abundant metal species in the mine dumps is pyrite.

The dissolution rate is influenced by concentrations of thiosulfate in the 0.02–0.06 mol·L^{-1} range, with saturation of the complexing agent observed from 0.06 mol·L^{-1}, which causes the leaching rate to stay constant at higher concentration values. The pseudo- order of reaction with respect to the concentration of copper ions was $n = 0.4$, which suggests a catalytic effect, since an increase in the leaching rate of silver was observed. In addition, the partial pressure of oxygen dissolved in the system with thiosulfates has a favorable effect on the leaching rate, with an increase of up to 30% when increasing gas pressure from 0.2 atm to 1.0 atm.

Since temperature was found to have a drastic effect on the leaching rate of silver ($E_a = 43.5$ kJ·mol^{-1}), while modification of the stirring rate was not found to have such an effect, it was demonstrated that the matter chemical stages evolve in shorter time intervals, compared to the transport stages. This was confirmed through the adjustment made on the shrinking core model with chemical control ($k_{exp} \cdot t = 1 - (1 - X)^{1/3}$). The maximum silver recovery was of (96.8%) under the following experimental conditions: $[S_2O_3^{-2}] = 0.08$ mol·L^{-1}, $[Cu^{2+}] = 0.006$ mol·L^{-1}, RPM = 650 min^{-1}, T = 328 K (55 °C), $PO_2 = 1$ atm., pH = 10 and t = 240 min.

According to the results obtained, the present work suggests that there is a real possibility of using the $S_2O_3^{2-}$-O_2 system, adding copper ions (II) as an oxidizing reagent, to improve the kinetic conditions of the leaching process. The proposed system represents a reasonably viable alternative, since silver is present in real refractory waste from the mining industry, making this an innovative method.

Author Contributions: J.C.J.T. designed and performed the experiments and wrote the paper; F.P.C. realized the chemical characterization; A.R.V. performed the cupellation technique and the characterization by XRF and MEB-EDS; A.M.T.R. and I.A.R.D. conducted the discussion of results and wrote the paper; M.R.P and M.P.L. performed the characterization by DRX and the granulometric analysis; M.U.F.G. realized the characterization by MOP.

Funding: The authors would like to thank the Universidad Autónoma del Estado de Hidalgo (Autonomous University of the State of Hidalgo) and the Consejo Nacional de Ciencia y Tecnología, CONACyT (National Council for Science and Technology), of the Mexican government for the funding of the present work.

Acknowledgments: The authors would like to thank the University of Barcelona for their technical-scientific services and for their collaboration in this piece of work.

Conflicts of Interest: The authors declare no conflicts of interest.

References

1. Lopez, C. México: Política en Materia Minera. (Diálogo Unión Europea-América Latina Sobre Materias Primas 2014). Available online: europa.eu/geninfo/query/index.do?queryText=plata&query_source=GROWTH&summary=summary&more_options_source=restricted&more_options_date=*&more_options_date_from=&more_options_date_to=&more_options_language=es&more_options_f_formats=*&swlang=en (accessed on 14 July 2018).

2. Patiño, F.; Hernández, J.; Flores, M.U.; Reyes, I.A.; Reyes, M.; Juárez, J.C. Characterization and Stoichiometry of the Cyanidation Reaction in NaOH of Argentian Waste Tailings of Pachuca, Hidalgo, México. In *Characterization of Minerals, Metals, and Materials*; Springer: Cham, Switzerland, 2016; Volume 145, pp. 355–362, ISBN 978-1-119-26439-2.

3. Flores, J.; Hernández, J.; Rivera, I.; Reyes, M.I.; Cerecedo, E.; Reyes, M.; Salinas, E.; Guerrero, M. Study of the Effect of Surface Liquid Flow during Column Flotation of Mining Tailing of the Dos Carlos Dam. In *Characterization of Minerals, Metals, and Materials*; Springer: Cham, Switzerland, 2017; Volume 146, pp. 787–798.

4. Teja Ruiz, A.M.; Juarez Tapia, J.C.; Reyes Perez, M.; Hernandez Cruz, L.E.; Flores, M.U.; Reyes, I.A.; Perez Labra, M.; Moreno, R. Characterization and Leaching Proposal of Ag(I) from a Zn Concentrate in an $S_2O_3^{2-}$-O_2 Medium. In *Characterization of Minerals, Metals, and Materials*; Springer: Cham, Switzerland, 2017; Volume 146, pp. 567–575.

5. Hilson, G.; Monhemius, J. Alternatives to cyanide in the gold mining industry: What prospects for the future. *J. Clean. Prod.* **2015**, *14*, 1158–1167. [CrossRef]

6. Hernandez, J.; Patino, F.; Rivera, I.; Reyes, I.A.; Flores, M.U.; Juarez, J.C.; Reyes, M. Leaching kinetics in cyanide media of Ag contained in the industrial mining-metallurgical wastes in the state of Hidalgo, Mexico. *Int. J. Min. Sci. Technol.* **2014**, *24*, 689–694. [CrossRef]

7. Briones, R.; Lapidus, G.T. The leaching of silver sulfide with the thiosulfate–ammonia–cupric ion system. *Hydrometallurgy* **1998**, *50*, 243–260. [CrossRef]

8. Alvarado-Macías, G.; Fuentes-Aceituno, J.C.; Nava-Alonso, F. Study of silver leaching with the thiosulfate-nitrite-copper alternative system: Effect of thiosulfate concentration and leaching temperature. *Miner. Eng.* **2016**, *86*, 140–148. [CrossRef]

9. Alonso, A.R.; Lapidus, G.T.; González, I. A strategy to determine the potential interval for selective silver electrodeposition from ammoniacal thiosulfate solutions. *Hydrometallurgy* **2007**, *85*, 144–153. [CrossRef]

10. Aylmore, M.G.; Muir, D.M. Thiosulfate leaching of gold—A review. *Miner. Eng.* **2001**, *14*, 135–174. [CrossRef]

11. Abbruzzese, C.; Fornari, P.; Massidda, R.; Veglio, F.; Ubaldini, S. Thiosulphate leaching for gold hydrometallurgy. *Hydrometallurgy* **1995**, *39*, 265–276. [CrossRef]

12. Mohammadi, E.; Pourabdoli, M.; Ghobeiti-Hasab, M.; Heidarpour, A. Ammoniacal thiosulfate leaching of refractory oxide gold ore. *Int. J. Miner. Process.* **2017**, *164*, 6–10. [CrossRef]

13. Ubaldini, S.; Fornari, P.; Massidda, R.; Abbruzzese, C. An innovative thiourea gold leaching process. *Hydrometallurgy* **1998**, *48*, 113–124. [CrossRef]

14. Alonso-Gómez, A.; Lapidus, G. Inhibition of lead solubilization during the leaching of gold and silver in ammoniacal thiosulfate solutions (effect of phosphate addition). *Hydrometallurgy* **2009**, *99*, 89–96. [CrossRef]

15. Deutsch, J.L.; Dreisinger, D.B. Silver sulfide leaching with thiosulfate in the presence of additives part I: Copper–ammonia leaching. *Hydrometallurgy* **2013**, *137*, 156–164. [CrossRef]

16. Feng, D.; Van Deventer, J.S.J. Ammoniacal thiosulphate leaching of gold in the presence of pyrite. *Hydrometallurgy* **2006**, *82*, 126–132. [CrossRef]

17. Solis-Marcial, O.J.; Lapidus, G.T. Chalcopyrite leaching in alcoholic acid media. *Hydrometallurgy* **2014**, *147–148*, 54–58. [CrossRef]

18. Xu, B.; Yang, Y.; Jiang, T.; Li, Q.; Zhang, X.; Wang, D. Improved thiosulfate leaching of a refractory gold concentrate calcine with additives. *Hydrometallurgy* **2015**, *152*, 214–222. [CrossRef]

19. Grosse, A.C.; Dicinoski, G.W.; Shaw, M.J.; Haddad, P.R. Leaching and recovery of gold using ammoniacal thiosulfate leach liquors (a review). *Hydrometallurgy* **2003**, *69*, 1–21. [CrossRef]

20. Lampinen, M.; Laari, A.; Turunen, I. Ammoniacal thiosulfate leaching of pressure oxidized sulfide gold concentrate with low reagent consumption. *Hydrometallurgy* **2015**, *151*, 1–9. [CrossRef]

21. Zipperian, D.; Raghavan, S.; Wilson, J.P. Gold and silver extraction by ammoniacal thiosulfate leaching from a rhyolite ore. *Hidrometallurgy* **1988**, *19*, 361–378. [CrossRef]

22. Patiño, F.; Reyes, I.A.; Flores, M.U.; Pandiyan, T.; Roca, A.; Reyes, M.; Hernández, J. Kinetics modeling and experimental design of the sodium arsenojarosite descomposition in alkaline media: Implications. *Hydrometallurgy* **2013**, *137*, 115–125. [CrossRef]

23. Juárez, J.C.; Rivera, I.; Patiño, F.; Reyes, M. Efecto de la Temperatura y Concentración de Tiosulfatos sobre la Velocidad de Disolución de Plata contenida en Desechos Mineros usando Soluciones $S_2O_3^{2-}$-O_2-Zn^{2+}. *Inf. Tecnol.* **2012**, *18*, 1–12. [CrossRef]

24. Hernández, J.; Rivera, I.; Patiño, F.; Juárez, J.C. Estudio Cinético de la Lixiviación de Plata en el Sistema $S_2O_3^{2-}$-O_2-Cu^{2+} Contenido en Residuos Minero-Metalúrgicos. *Inf. Tecnol.* **2012**, *16*, 1–10. [CrossRef]

25. Skoog, A.D.; West, D.M. *Analytical chemistry an Introduction*, 1st ed.; Reverté: Barcelona, Spain, 2002; pp. 444–445, ISBN 8429175113.

26. Rivera, I.; Patiño, F.; Roca, A.; Cruells, M. Kinetics of metallic silver leaching in the O_2–thiosulfate system. *Hydrometallurgy* **2015**, *156*, 63–70. [CrossRef]

27. Teja-Ruiz, A.M.; Juárez-Tapia, J.C.; Reyes-Domínguez, I.A.; Hernández- Cruz, L.E.; Reyes-Pérez, M.; Patiño-Cardona, F.; Flores-Guerrero, M.U. Kinetic Study of Ag Leaching from Arsenic Sulfosalts in the S2O32--O2-NaOH System. *Metals* **2017**, *7*, 411. [CrossRef]

28. Levenspiel, O. *Ingeniería de las Reacciones Químicas*, 2nd ed.; Reverté: Barcelona, Spain, 2002; pp. 397–441, ISBN 968-18-5860-3.

29. Tanda, B.C.; Eksteen, J.J.; Oraby, E.A. Kinetics of chalcocite leaching in oxygenated alkaline glycine solutions. *Hydrometallurgy* **2018**, *178*, 264–273. [CrossRef]

30. Moreno-Tovar, R.; Téllez-Hernández, J.; Monroy-Fernández, M.G. Evaluación geoquímica de residuos mineros (jales o colas) de mineralización de tipo epitermal, Hidalgo, México. *Rev. Geol. Am. Cent.* **2009**, *41*, 79–98, ISSN: 0256-7024.

31. Moreno-Tovar, R.; Monroy, M.G.; Castañeda, E.P. Influencia de los minerales de los jales en la bioaccesibilidad de arsénico, plomo, zinc y cadmio en el distrito minero Zimapán, México. *Rev. Geol. Am. Cent.* **2012**, *28*, 3, ISSN: 0188-4999.

32. Asamoah, R.A.; Skinner, W.; Addai-Mensah, J. Alkaline cyanide leaching of refractory gold flotation concentrates and bio-oxidised products: The effect of process variables. *Hydrometallurgy* **2018**, *179*, 79–93. [CrossRef]

33. Melo, P.; Halmenschlager, P.; Veit, H.M.; Bernardes, A.M. Leaching of gold and silver from printed circuit board of mobile phones. *Rev. Esc. Minas* **2015**, *68*, 61–68. [CrossRef]

34. Sitando, O.; Senanayake, G.; Dai, X.; Nikoloski, A.N.; Breuer, P. A review of factors affecting gold leaching in non-ammoniacal thiosulfate solutions including degradation and in-situ generation of thiosulfate. *Hydrometallurgy* **2018**, *178*, 151–175. [CrossRef]

35. Langhans, J.W., Jr.; Lei, K.P.V.; Carnahan, T.G. Copper-catalyzed thiosulfate leaching of low-grade gold ores. *Hydrometallurgy* **1992**, *29*, 191–203. [CrossRef]

36. Ballester, A.; Felipe Verdeja, L.; Sancho, J. *Metalurgia Extractiva Vol. I Fundamentos*, 2nd ed.; Síntesis: Barcelona, Spain, 2000; pp. 37–67, ISBN 9788477388029.

Article

Heat-Assisted Batch Settling of Mineral Suspensions in Inclined Containers

Cristian Reyes [1,2], **Christian F. Ihle** [1,2,*], **Fernando Apaz** [3] **and Luis A. Cisternas** [4]

[1] Laboratory for Rheology and Fluid Dynamics, Department of Mining Engineering, Universidad de Chile, Beauchef 850, 8370448 Santiago, Chile; cristian.reyes@ug.uchile.cl
[2] Advanced Mining Technology Center, Universidad de Chile, Tupper 2007, 8370451 Santiago, Chile
[3] Department of Mechanical Engineering, Universidad de Antofagasta, Av. Angamos 601, 1240000 Antofagasta, Chile; fernando.apaz@uantof.cl
[4] Departamento de Ingeniería Química y Procesos de Minerales, Universidad de Antofagasta, Av. Angamos 601, 1240000 Antofagasta, Chile; luis.cisternas@uantof.cl
* Correspondence: cihle@ing.uchile.cl; Tel.: +56-2-2978-4991

Received: 2 February 2019; Accepted: 4 April 2019; Published: 12 April 2019

Abstract: In mineral processing, the common requirement for progressively finer milling due to the decreasing of ore grades implies the need for more challenging water recovery conditions in thickeners. Several mining operations exist in arid areas, where water recovery becomes critical. The present paper explores the process of particle separation in batch inclined settlers where the downward facing wall is subject to heating. To this purpose, two-dimensional numerical simulations using a mixture model have been run for a number of combinations of temperature jumps at the downward facing fall, particle diameters, and concentrations. Results show that, for particle sizes on the order of 10 μm, heating has a significant effect on the particle settling velocity at the bottom, but it also promotes particle resuspension, affecting the particle concentration at the supernatant layer. The initial concentration also affects settling: for the concentration range tested (8%–15% by volume), when re-normalized by the average concentration, particle accumulation rates at the bottom were found to be lower for higher average concentrations, thus suggesting that the separation process is more efficient at lower concentrations.

Keywords: thickening; water recovery; Boycott effect; tailings; settling velocity; natural convection

1. Introduction

Solid-liquid separation in mining operations is a critical unit operation occurring in commonly adverse conditions such as water scarcity, energy cost [1–3], or a naturally fine particle size range [4–6]. While there are several means to recycle water, including centrifugal separation and filtration [7], due to its considerably low cost per recovered m^3 water, gravity settling [8] is still the most commonly-used technique in the sector. Among this type of technology are lamellar settlers, which are a type of process equipment consisting of a vessel fitted with a set of parallel inclined plates immersed in a slurry vessel. Differently from other industries such as water treatment, this kind of equipment has been historically scarcely used in the mining sector, possibly because their capacity to separate fine solids at small particle loadings was not such an important requirement in the past; this has changed significantly in recent years due to the progressively decreasing particle sizes in comminution products. In the present paper, this working principle is explored, for the first time, in combination with the use of heat, considerably abundant from natural sources in arid areas, to further enhance the particle settling process. To this purpose, a set of two-dimensional numerical simulations for the case of batch settling within a single convection cell consisting of two parallel inclined plates is presented and analyzed.

The present paper uses a two-phase CFD model using the OpenFOAM library [9], adapted for the purposes of this problem.

Lamellar equipment [10] uses the Boycott effect [11] either to enhance the settling process in comparison with upright settlers such as conventional thickeners and clarifiers or to separate fine from coarse particles in a slurry [12]. In batch settling, as the suspension flows within the inclined plates, particle settling near the upward facing wall induces the upward flux of a thin, clear liquid below the downward facing wall. As a result, the velocity of the horizontal boundary between the suspension and the clear fluid region resulting from the existence of a settling process is considerably larger than the particle settling velocity. Indeed, when the plate spacing is small—and thus the suspension height-to-plate spacing, H/b, is large—then $dH/dt \sim v_0 H/b$, where v_0 is the vertical settling velocity of a particle subject or not to hindrance effects due to the concentration [13].

The same arid environment that challenges water supply in several mining operations can offer, in exchange, very generous solar radiation conditions. The potential for solar energy harvesting in arid countries is significant, and governments have already started to put incentives and goals toward the use of these energy sources [14,15]. In the absence of particles, a heated vertical plate induces an upward vertical flow, whose characteristics will depend on the relative importance of inertial heat advection and diffusion (given by the Grashof number) and the relative thicknesses of the momentum and the heat boundary layer (Prandtl number, both described in [16]). When laminar flow is present, natural convection below a downward facing boundary has been found to share features with its vertical plate counterpart [16], except by the projection of the gravity term along the angle of the plate [17]. The paper is organized as follows: Section 2 describes the governing equations used in the numerical approach; Section 3 indicates the specific cases to be analyzed, recalling the present focus on fine fraction recovery from process water; Section 4 presents results and a discussion of them, and final remarks are given in Section 5.

2. Governing Equations

We consider an initially homogeneous, solid-liquid suspension confined in an inclined, slender batch container of overall height h (Figure 1). According to this figure, the plate spacing is $b = W \cos\theta$, where θ is the cell angle measured with respect to the vertical. The length of the plates is $h/\cos\theta$.

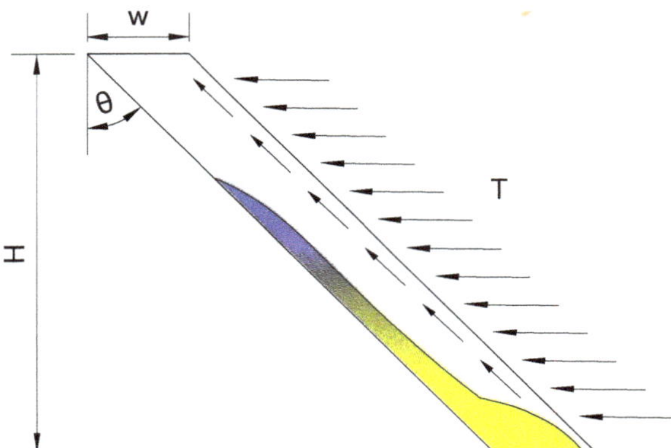

Figure 1. Cell. The heated downward-facing, heated wall corresponds to the right side boundary of the domain. The horizontal arrows denote the heat flux on the downward facing wall, while the diagonal arrows show a scheme of the liquid (and partially fine solid) fraction flowing upwards near it.

This configuration mimics the batch operation mode of lamella settlers, where normally the upper and lower plates are closely spaced (a few centimeters' distance, as reviewed by

Leung and Probstein [10]). This configuration promotes an accelerated settling of particles, which implies an upward flow near the downward-facing top plate when compared to the case of particle settling in an upright settler.

The suspension is modeled as a two-dimensional continuum with solid velocity, density and volume fraction fields $\mathbf{u}_s$, ρ_s, and ϕ_s, respectively, superimposed on the liquid with velocity ($\mathbf{u}_l$), density ρ_l, and volume fraction $\phi_l = 1 - \phi_s$, with the initial average concentration $\phi_s(\mathbf{x}, t = 0) = \phi_0$, a constant value.

The mass transport equation for the solid phase is:

$$\frac{\partial(\phi_s\rho_s)}{\partial t} + \nabla \cdot (\phi_s\rho_s\mathbf{u}_s) = 0, \tag{1}$$

while the liquid phase transport equation is given by:

$$-\frac{\partial(\phi_s\rho_l)}{\partial t} + \nabla \cdot [(1 - \phi_s)\rho_l\mathbf{u}_l] = 0. \tag{2}$$

From Equations (1) and (2), it is concluded that the mean velocity, defined as $\langle \mathbf{u} \rangle = \phi_s\mathbf{u}_s + (1 - \phi_s)\mathbf{u}_l$, verifies $\nabla \cdot \langle \mathbf{u} \rangle = 0$.

The momentum conservation equations are computed in standard form as [18,19]:

$$\frac{\partial(\phi_i\rho_i\mathbf{u}_i)}{\partial t} + \nabla \cdot (\phi_i\rho_i\mathbf{u}_i\mathbf{u}_i) + \nabla \cdot (\phi_i\tau_i) = -\phi_i\nabla p + \nabla \cdot (\phi_i p_{s,i}) + \phi_i\rho_i\mathbf{g} + \mathbf{f}_i, \tag{3}$$

where the sub-index i stands for solid s or liquid l. The pressure is assumed common to both phases. An additional solid pressure contribution $p_s\phi_s$, which sharply increases around the maximum packing fraction ϕ_{max}, is added on the momentum equation for the solid in order to bound the volume fraction of solid ϕ_s below ϕ_{max}. This extra pressure term in the liquid equation is zero $p_{s,l} = 0$. The liquid density or water density was calculated as a function of temperature only, from Maidment et al. [20]:

$$\rho_l(T) = 1000 \times \frac{1 - (T + 288.9414)}{508929.2(T + 68.12963)}(T - 3.9863)^2, \tag{4}$$

where T has units of degrees Celsius.

The shear stress tensor for each phase is given by [21]:

$$\tau_i = \mu_i\left(\frac{\nabla\mathbf{u}_i + (\nabla\mathbf{u}_i)^T}{2} - \frac{2}{3}(\nabla \cdot \mathbf{u}_i)\mathbf{I}\right), \tag{5}$$

where the effective viscosity of the solid μ_s is obtained from the expression for the slurry:

$$\mu_{mix} = \phi_s\mu_s + \phi_l\mu_l = \mu_l\left(1 + \frac{\phi_s}{\phi_{max}}\right)^{-[\eta]\phi_{max}}, \tag{6}$$

with μ_l the temperature-dependent dynamic viscosity of the liquid phase (in this case, water) by [22]:

$$\mu(T) = 2.414 \times 10^{-5} \times 10^{\frac{247.8}{T-140}}, \tag{7}$$

where T has units of Kelvin. It is noted that this mixture model can alternatively be extended using formulations suitable for dilute-granular regimes [23,24]. However, considering the mostly low concentrations in the present domain, the model represented by Equation (6) is a reasonable approximation of most of the mixture behavior.

The temperature field was calculated by an advection-diffusion equation:

$$\frac{\partial(\phi_l\rho_l T_l)}{\partial t} + \nabla \cdot (\phi_l\rho_l T_l\mathbf{u}_l) = \alpha\nabla^2(\phi_l\rho_l T_l). \tag{8}$$

In this equation, the thermal diffusivity of the liquid phase, α, has a fixed value of $1.43 \times 10^{-7}\,\mathrm{m^2\,s^{-1}}$, corresponding to ambient conditions (25 °C). Particles are dilute enough to assume that their thermal inertia is negligibly small when compared to water mass, and thus, the water phase temperature remains unaffected by a difference in particle temperature. On the other hand, it is assumed that particles are small enough to quickly reach the temperature of the liquid phase. This is seen noting that the heat diffuses within each particle with a time scale τ_p such as $\tau_p \sim d_s^2/\alpha_s$, where d_s and α_s are the particle diameter and thermal diffusivity, respectively. On the other hand, in a purely convective flow ($\phi = 0$), from dimensional arguments, a convective velocity scale can be estimated as $v_c \sim \sqrt{g \sin\theta \beta \Delta T b}$, where g, β, and ΔT are the acceleration due to gravity, water thermal expansion coefficient, and temperature jump, respectively. Comparing the time scale required for the thermally-diffusive flow to span the particle diameter with the time scale required to offset a particle of its own size due to convection (in the absence of a drag force), $\tau_c = d_p/v_c$, yields the dimensionless ratio $\tau_p/\tau_c = \frac{d_p}{\alpha_s}\sqrt{g\sin\theta\beta\Delta T b}$. Assuming $d_s \sim 10\,\mu\mathrm{m}$ (typical of fine particle content), $\Delta T \sim 10\,\mathrm{K}$, $\alpha_s \sim 10^{-6}\,\mathrm{m^2/s}$ [25], $\beta \sim 10^{-4}\,\mathrm{K^{-1}}$ [26], and b between 1 cm and 10 cm yields values of $\tau_p \sim 10^{-4}\,\mathrm{s}$ and the ratio τ_p/τ_c between about $1/3$ and $1/10$, which implies that in the slowest particle heating scenario, the time required for particle heating to reach the surrounding temperature is that required to displace one particle diameter. Therefore, it is assumed herein that $T_s(\mathbf{x}, t) = T_l(\mathbf{x}, t)$. Although this hypothesis is debatable at high particle concentrations (i.e., near the bottom boundary) due to the high mass of particles and a near-stagnant condition, the particle separation process occurs in dilute conditions near the downward facing wall. Although experimental evidence shows that there is a wide applicability of the mixture model even at high concentrations [18], the very weak flow conditions occurring near the bottom may have significant uncertainty, not only due to the thermal coupling between phases, but also due to the high viscosity values and particle contacts in the sediment zone, which are not modeled in detail herein.

The volumetric forces between phases, denoted by the last term of the right-hand side of Equation (3), are given by $\mathbf{f}_i = \mathbf{f}_D + \mathbf{f}_L + \mathbf{f}_{WL}$, where $\mathbf{f}_D$ is the drag force, $\mathbf{f}_L$ is the lift force, and $\mathbf{f}_{WL}$ is the wall lubrication force (see [19] and the references therein, except that in that work, there was a turbulent dispersion force that here is null because the flow is laminar). Here, $\mathbf{f}_i = \mathbf{f}_s = -\mathbf{f}_l$ is the force per unit volume the fluid applies on the solid. The computational implementation used in the present problem is, as in the previous reference, based on the OpenFOAM library.

3. Simulation Cases and Boundary Conditions

The Boycott effect implies the formation of a momentum boundary layer near the downward facing wall, creating a greater suspension-supernatant boundary velocity than if the tube was vertical [13]. In a monodisperse suspension of spheres, the separation efficiency in the absence of heating the downward facing wall is conditioned by the container height, H_0, along with ρ_s, ϕ_s, ρ_l, $\mu_{\mathrm{mix}}(\phi_s)$, g, and d_s. Table 1 shows the geometry of various cases considered to analyze the combined effect of heating and particle settling within the cell.

Kinematic boundary conditions at the walls are no slip, without particle flow through them. This implies a zero particle gradient boundary condition at the walls. In the particular case of the top (horizontal) boundary, it was necessary to define a thin layer without particles near the top, thus ensuring numerical stability at the early stages of the flow development. Initially, the temperature at the walls and in the bulk of the (also homogeneous) suspension was $T_0 = 20\,°\mathrm{C}$. Numerical experiments started from this condition, imposing a temperature jump at the downward facing wall, referred to as ΔT_{DFW} in Table 1, such that at this boundary, $T_{\mathrm{DFW}} = T_0 + \Delta T_{\mathrm{DFW}}$. The total number of cells considered for the present computation was 67,000 for a width equal to 5 cm, corresponding to a physical element size of 0.8 mm and 1 mm in the horizontal and vertical direction, respectively.

Table 1. Range of variables studied in the numerical simulations. The acronym DFW stands for downward facing wall. In all the cases, the initial temperature prior to the start of the heating of the downward facing wall was $T_0 = 20\,°C$. At the start of each experiment, the temperature at the downward facing wall was set as $T_{DFW} = T_0 + \Delta T_{DFW}$. Variables d_s, ϕ_0, θ, and W denote particle diameter (monosized), initial concentration (constant), cell inclination, and horizontal projection of cell spacing, respectively. H_0 is equivalent to $\cos(45)$, and the length of the cell is $1\,m$.

Variable	Cases Considered
H_0	$0.707\,m$
ΔT_{DFW}	20, 30, 40, and $50\,°C$
d_s	5, 10, and $50\,\mu m$
ϕ_0	2%, 5%, 8%, and 15%
θ	$45°$
W	$5\,cm$

4. Results and Discussion

Figure 2 shows an example of the flow and particle progression in the cell, for the case $\Delta T_{DFW} = 30\,°C$, $\phi_0 = 0.05$, and $d_s = 10\,\mu m$. The Boycott effect induces particle accumulation at the upward facing wall, which transport them towards the bottom. In this case, the particle settling process is enhanced by heating at the downward facing wall, which, besides, accelerating it, as will be discussed below, tends to create some degree of additional resuspension near the top, as seen by the low, but nonzero particle concentration near the top for $t = 800\,s$ and $1600\,s$ in Figure 2a. On the other hand, heating also introduces flow variations in the line parallel to the plates. This is observed especially for $t = 200\,s$ and $t = 1600\,s$ in Figure 2b. In the former case, near $Y = -0.3\,m$, there is an intense vorticity zone that tends to accelerate particle migration towards the upward facing wall. For the latter time, it is seen that vorticity is concentrated near the zone between $Y = -0.3\,m$ and $Y = -0.4\,m$, affecting resuspension. This complex and asymmetrical dynamics impedes relying solely on a simplified Boycott effect interpretation to predict particle segregation within the cell.

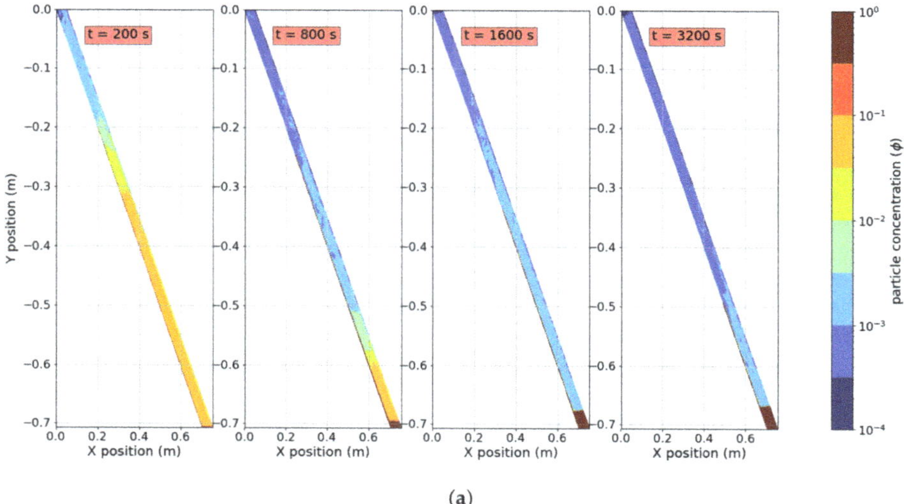

(a)

Figure 2. *Cont.*

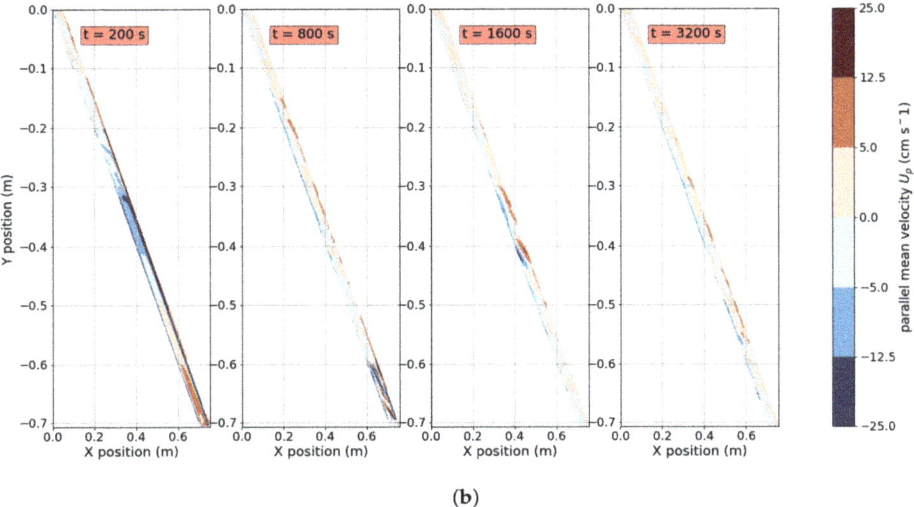

(b)

Figure 2. Typical flow and particle progression for the case $\Delta T_{\mathrm{DFW}} = 30\,°\mathrm{C}$, $\phi_0 = 0.05$, and $d_S = 10\,\mu\mathrm{m}$ and (from left to right) $t = 200\,$s, $800\,$s, $1600\,$s, and $3200\,$s. (a) Particle concentration (ϕ, with color scale in logarithmic scale) and (b) local parallel mean velocity, defined as $\langle \mathbf{u} \rangle \cdot \hat{p}$, with $\hat{p}$ a unit vector parallel to the upward facing wall.

Results have been analyzed in terms of the relative impact of the key variables shown in Table 1, in light of the distribution of particles within the cell. This will be seen directly from concentration maps or spatiotemporal diagrams in the present section. Additionally, data were interpreted in terms of the particle presence in specific areas of the cell, as depicted in Figure 3. Here, A1 denotes an initial concentration-dependent settling area at the bottom, A2 a fixed one at the top, and A3 a fixed one near the upward facing wall. In the former case, the height of A1 was assumed as that corresponding to the accumulation of particles at the bottom. To enable comparison of results between different experiments, the sediment accumulation length in the upright case, $Y_\infty = \frac{\phi_0 H_0 \cos\theta}{\phi_{\max}}$, was considered as a reference length scale. The purpose of the area A2 was to assess particle resuspension due to convection within the cell and was defined by the horizontal planes bound by $Y_z = 0.86 H_0$ and H_0.

The area A3 was defined as $Y_{\mathrm{sed}}\delta / \cos\theta$, with $\delta = 5\,$mm and $Y_{\mathrm{sed}}(t)$ the vertical projection of the length of the solid bed accumulated near the upward facing wall. In narrow tilted containers, Y_{sed} can have two connotations: one is to define the boundary between the supernatant region, where below such a boundary, there is a suspension roughly homogeneously distributed in the horizontal direction; the second possibility is that Y_{sed} denotes a narrow area near the upward facing wall (as we consider herein through A3), where particles are concentrated. Herbolzheimer and Acrivos [27] have identified, in the absence of heating, that these two modes of convection are defined by the particle Reynolds number, the geometry, and the dimensionless number $\Lambda = H_0^2 g(\rho_s - \rho_l)\phi_0 / w_0 \mu_{\mathrm{mix}}$ [13,28], where w_0 corresponds to the Stokes settling velocity of a particle, corrected by hindrance effects. As in the latter references, here, $\Lambda \gg 1$, implying that the convective process occurs in an enclosure where the vertical flow path is considerably higher than the characteristic particle size, a common assumption in lamellar settlers.

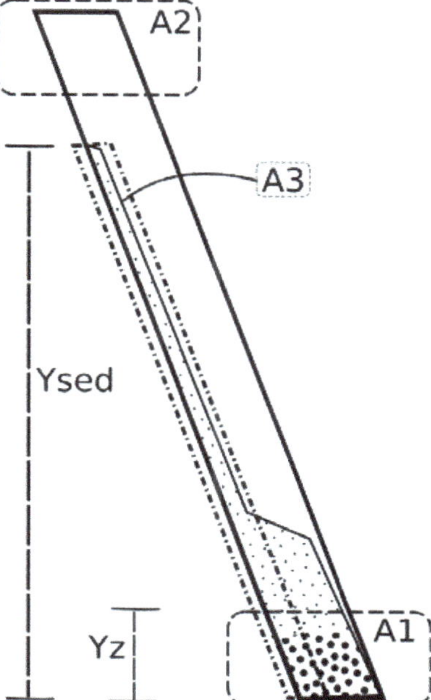

Figure 3. Zones of interest for the analysis.

4.1. Effect of Temperature

The effect of temperature was analyzed for a particle size where both the Boycott effect and natural convection are relevant. This is an intermediate situation between two opposites: one where particles are so coarse that only the Boycott effect (if any) exerts an influence on the separation process and a second instance where particles are so fine that the time scale for separation is too large, and therefore, the suspension behaves essentially as a homogeneous medium. It is thus convenient to use a fixed particle size to isolate the effect of temperature. In particular, a value of $d_s = 10\,\mu m$ is considered throughout this section. Additionally, to put the present problem into a suitable context for mineral particle separation, an analysis based on particle size is given in Section 4.2.

The effect of temperature on the settling dynamics can be seen looking at the average horizontal concentration as a function of the vertical coordinate. Figure 4 represents the spatiotemporal evolution of the suspension in the base case for $\Delta T_{DFW} = 0\,°C$, $20\,°C$, $30\,°C$, and $50\,°C$. The boundaries between colors represent average concentration isolines, which can be interpreted as the propagation velocity of constant concentration waves [8]. The dark zone near the bottom in the figure represents the sediment accumulation at the bottom of the cell, which has been previously characterized in terms of average values at A1. Particle resuspension near the top is denoted by the cyan-blue area near the top for times below about 20 min, where is it evident that this process is short term in comparison with the settling of the bulk of the suspension towards the bottom section. The orange zone near the left of each panel denotes the presence of a dense area close to the upward facing wall. Once the suspension migrated towards the bottom and left a specific height, the average concentration dropped to values on the order of 1% (yellow-green areas in the figure). Furthermore, a sharp transition between this (low) concentration and the virtually particle-free area corresponding to the supernatant (whose concentration has been expressed in previous figures herein in ppm) is denoted by the green-blue boundary.

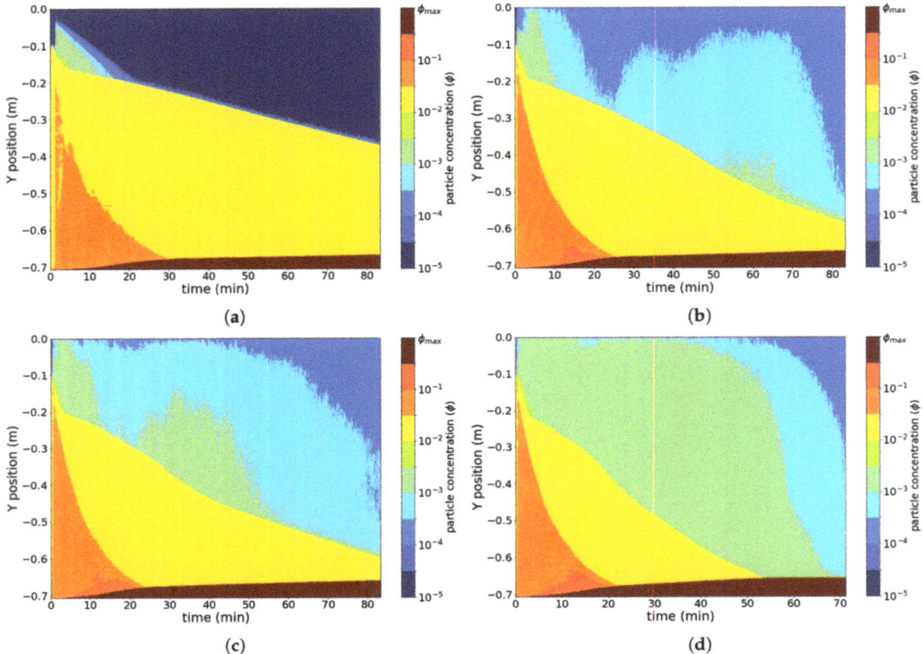

Figure 4. Spatiotemporal diagram of concentration for (**a**) $\Delta T_{\mathrm{DFW}} = 0$, (**b**) $\Delta T_{\mathrm{DFW}} = 20\,°\mathrm{C}$, (**c**) $\Delta T_{\mathrm{DFW}} = 30\,°\mathrm{C}$, and (**d**) $\Delta T_{\mathrm{DFW}} = 50\,°\mathrm{C}$. The x and y axes represent time and vertical position ($y = 0$ standing for the top of the cell). False color represents horizontally-averaged particle concentration.

The momentum boundary layer buildup at the downward facing wall when particle settling occurs enhanced when the latter was heated. This is not surprising as in the absence of particle settling, this process occurs as a direct consequence of natural convection, causing lighter fluid neighboring the downward facing wall (due to heating) to migrate towards the upper region. The impact of this heating enhancement on the filling process at the bottom is somewhat suggested in Figure 4a–d (brown zone), but is more evident from Figure 5, which shows the concurrent impact of the Boycott effect and heating of the downward facing wall for various values of ΔT_{DFW} and $d_s = 10\,\mu\mathrm{m}$, where it is evident, from the degree of particle accumulation at A1, that heating the downward facing wall enhanced the accumulation at the bottom of the enclosure.

Figure 5 reveals two different trends on the rate of particle accumulation, depending on whether the size of the control volume is $0.6Y_\infty$ or $0.9Y_\infty$. In both cases, there is a strong particle accumulation of the corresponding area; however, when considering a smaller control volume saturation when filling A1 occurs at relatively early values of time (close to $45\,\mathrm{min}$ for $\Delta T_{\mathrm{DFW}} = 0\,°\mathrm{C}$) and on the order of $23\,\mathrm{min}$ when heating at the downward facing wall is set (Figure 5a). When considering a larger control area of A1 the initial stage of sedimentation, due to the settling of the suspension, is followed by the re-accommodation of the sediment placed at the upward facing wall. This naturally occurs at a lower rate due to the high concentration gravitational flow at this boundary. However, the effect of heating becomes significant with $\Delta T_{\mathrm{DFW}} = 50\,°\mathrm{C}$, where the second slope is close to the first (Figure 5b). It is suggested that this occurs due to the additional shear, induced by natural convection due to heating, at the surface of the sediment zone above the upward facing wall. This also is shown in Table 2, where the end of the settling process can be between $\approx 4\%$ and 46% faster for $\Delta T_{\mathrm{DFW}} = 50\,°\mathrm{C}$ heating given both control areas.

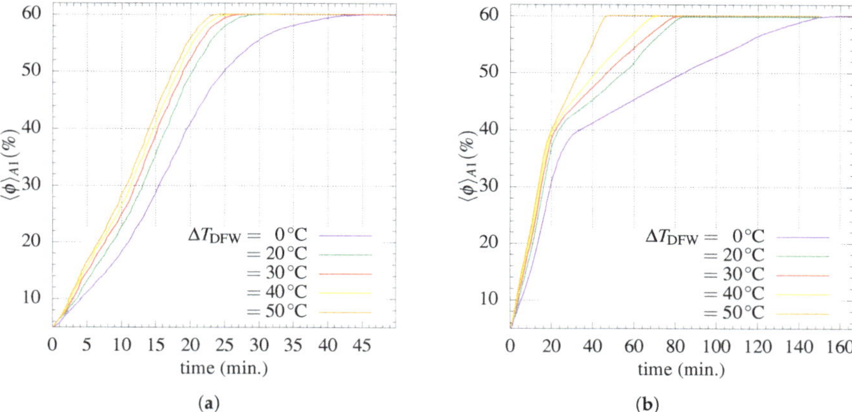

Figure 5. Effect of temperature on the mean particle volume concentration within $A1 = Y_z W$, for $d_s = 10\,\mu m$, $\theta = 45°$, $\phi_0 = 5\%$, and $W = 5\,cm$. The mass in A1 in these cases is equivalent to 60% of the total mass in the cell. (**a**) $Y_z = 0.6Y_\infty$ and (**b**) $Y_z = 0.9Y_\infty$.

Table 2. Dimensionless particle accumulation lapse in A1 for temperature difference ΔT_{DFW} at the downward facing wall ($t^*_{\Delta T_{DFW}}$), normalized by the particle accumulation time without heating ($t^*_{\Delta T_{DFW}=0}$), $\psi_{\Delta T_{DFW}} \equiv t^*_{\Delta T_{DFW}} / t^*_{\Delta T_{DFW}=0}$, where a and b are the case where $Y_z = 0.6Y_\infty$ and $Y_z = 0.9Y_\infty$, respectively. Here, $d_s = 10\,\mu m$, $\theta = 45°$, $\phi_0 = 5\%$, and $W = 5\,cm$.

ΔT_{DFW} (K)	$\psi_{\Delta T_{DFW},a}$	$\psi_{\Delta T_{DFW},b}$
20	0.62	0.54
30	0.58	0.53
40	0.53	0.43
50	0.51	0.29

The particle accumulation sequence was confirmed by the observation of both the settled particle height at the upward facing wall and the corresponding mass therein (Figure 6). Figure 6a shows the particle mass confined within A3, where two stages of particle accumulation are observed: a growing mass phase (Figure 6a), where the confined mass per unit width at $t = 0$ is $\phi_0 A3$ and increments as the combined thermal convection-Boycott effect pushes particles away the downward facing wall, towards the upward facing wall. This increasing mass near the upward facing wall reaches a limit, after which the part of the particle matter trapped in A3 is transferred horizontally to the part of A1 not included in A3. This mass transport process corresponds to a monotonic decrease of the dimensionless height of the dense thin layer that defines the length of A3, Y_{sed}, whose progression ($|dY_{sed}/dt|$) is faster when ΔT_{DFW} is higher. In particular, in light of the present simulations, compared to the base situation $\Delta T_{DFW} = 0$, the cases $\Delta T_{DFW} = 50\,°C$ and $\Delta T_{DFW} = 40\,°C$ caused decreased rates of Y_{sed} faster by about 50% and 18%, respectively. In this regard, the contribution of heating to the efficiency of the separation process is evident.

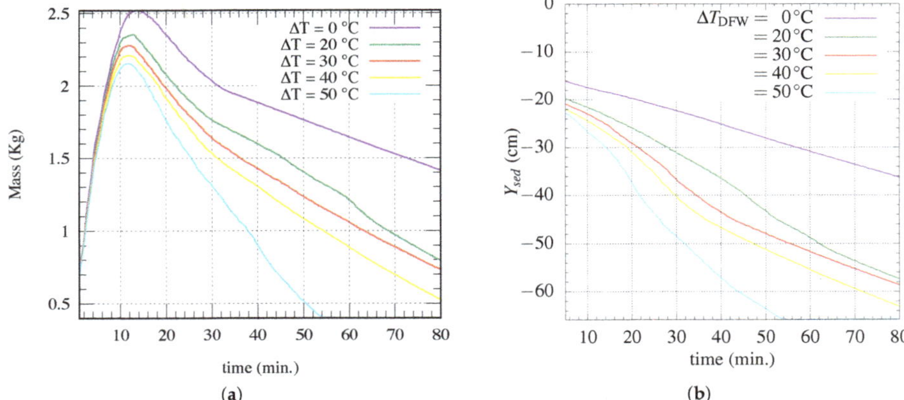

Figure 6. Particle accumulation at zone A3 for $d_s = 10\,\mu\text{m}$, $\theta = 45°$, $\phi_0 = 5\%$, and $W = 5\,\text{cm}$. (**a**) Particle mass and (**b**) the vertical position (Y_{sed}) of the particle sediment above the upward facing wall.

A second noticeable sign of the impact of heating at the downward facing wall is the departure from straight iso-concentration lines for $\Delta T_{\text{DFW}} > 0$. While in Figure 4a ($\Delta T_{\text{DFW}} = 0$), the mean suspension concentration neighboring the supernatant region (green-blue boundary) follows a straight line from early times, this is not so for $\Delta T_{\text{DFW}} > 0$. This early time behavior can be leading-order explained using the PNK theory, which states that the supernatant velocity is proportional to the settling velocity, a constant in the present problem. Besides the observation of the suspension acceleration effect of heating, it is also seen that this cannot be interpreted as an equivalent higher settling velocity as there is no longer a single higher (constant) slope, but the influence of a more complex mechanism that couples particle settling with flow velocity with stronger drag on particles. This effect of heating can be seen in the yellow/blue and yellow green intersections for times between 5 min and 40 min in Figure 4b,c and for times between 5 min and 20 min in Figure 4d, the latter corresponding to the strongest heating.

The present results show that, besides that heating promotes an overall faster settling process, it causes an increase of particle resuspension. This trend is apparent from the light blue/green zones in Figure 4b–d, denoting increasing (but yet very low) concentrations near the top for higher heating rates. Figure 7 shows the mean particle concentration at zone A2 (near the top). Although for the various instances of ΔT_{DFW} that were tested in simulations, a convergence was observed for simulation times in excess of 60 min, the transient behavior has been found to be significantly different. Higher temperature jumps at the downward facing wall tend to induce a stronger upward convective flow, pushing additional particles towards the top of the cell. This can also be seen in the spatiotemporal diagrams in Figure 4b–d, as a progressively lighter false color in the supernatant region in comparison with the base case with $\Delta T_{\text{DFW}} = 0$ (Figure 4a).

The convergence of mean concentration values in A2 shown in Figure 7 is explained by mass conservation. When particles reach the top of the enclosure, the same particle- and thermally-induced circulation that lifts the particles induces a downward motion that exerts a downward drag on the side near the upward facing wall, thus nudging particles away from the top.

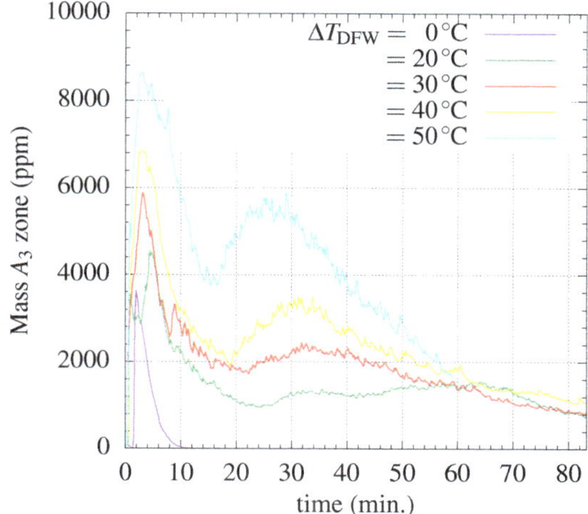

Figure 7. Mass resuspended at zone A2 for $d_s = 10\,\mu m$, $\theta = 45°$, $\phi_0 = 5\%$, and $W = 5\,cm$.

4.2. Effect of Particle Diameter

The PNK model states that efficiency, defined as the vertical speed of the supernatant-suspension interface, is proportional to the settling velocity, the latter being proportional to d_s^2. It is therefore no surprise that higher particle sizes are related to faster settling speeds and thus to considerably smaller overall particle deposition times.

Figure 8 shows the impact on heating via the opposing cases of $\Delta T_{\mathrm{DFW}} = 0\,°C$ (left column) and $30\,°C$ (right column), for particle sizes on the order of those found in copper sulfide comminution processes, i.e., $d_s = 10\,\mu m$, $50\,\mu m$ and $100\,\mu m$, in Rows 1–3, respectively. It is firstly noted that within this particle range, the higher the particle size, the more involved is the particle distribution. This can be attributed to the development of flow instabilities that depend on a monotonically increasing function of the settling velocity, whose primary effect is to accelerate the overall settling process. Herbolzheimer [29] analyzed the onset of strong mixing using linear stability analysis to long wave disturbances and found that, to first order disturbances, the system becomes unstable depending on a function of the cell geometry and the dimensionless number Λ, suggesting that the onset of stability occurs at a critical channel distance measured from the base that, in particular, depends on d_s^{-2}. The present combination of cell geometry, fluid, and particle characteristics shows that such instability occurs, in the presence of heating for some particle diameter between $10\,\mu m$ and $50\,\mu m$, even for $\Delta T_{\mathrm{DFW}} = 0$ (Figure 8c,d), but not for $d_s = 10\,\mu m$ (Figure 8a,b).

The spatiotemporal progressions in Figure 8 suggests that heating the downward facing wall causes the settling sequence to depart from being explained by the Boycott effect only. Although the whole process occurs considerably faster than in the smaller particle size, while in the case of $d_s = 10\,\mu m$ the particle input was smooth and occurred from the early stages of the settling process, as denoted by the contact line between the orange and the brown areas in Figure 8a,b, higher particle sizes are related to strong particle influxes to the bottom that proceed after a resuspension stage. This can be seen by the orange-red bands of Figure 8c–f, whose contact lines appear after a stage of smaller settling rate, denoted by the orange area near the beginning of the process. A comparison between Figure 8e,f denotes a slight retardation of the sediment bed formation due to heating, an aspect that can also be noticed by the presence of a second (reddish) high concentration characteristic above the main one, directed towards the bottom.

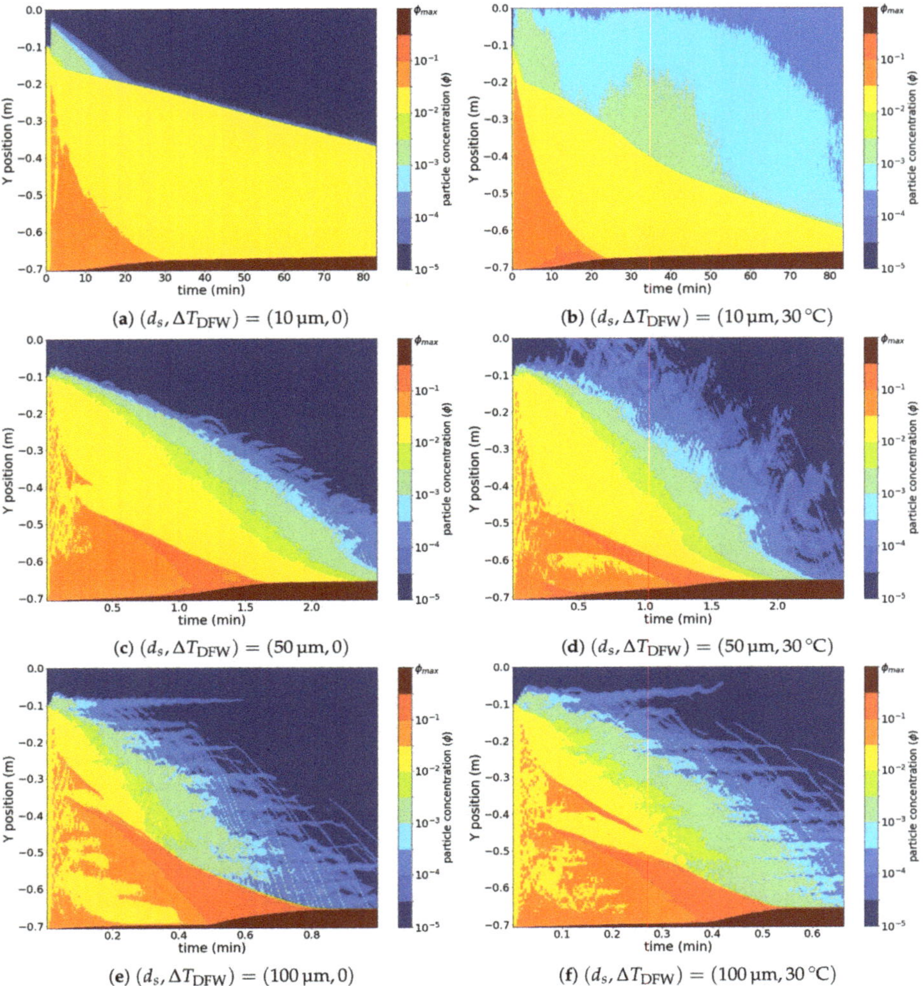

Figure 8. Spatiotemporal diagram of concentration for (**a**) $(d_s, \Delta T_{\mathrm{DFW}}) = (10\,\mu\mathrm{m}, 0\,^{\circ}\mathrm{C})$, (**b**) $(d_s, \Delta T_{\mathrm{DFW}}) = (10\,\mu\mathrm{m}, 30\,^{\circ}\mathrm{C})$, (**c**) $(d_s, \Delta T_{\mathrm{DFW}}) = (50\,\mu\mathrm{m}, 0\,^{\circ}\mathrm{C})$, (**d**) $(d_s, \Delta T_{\mathrm{DFW}}) = (50\,\mu\mathrm{m}, 30\,^{\circ}\mathrm{C})$, (**e**) $(d_s, \Delta T_{\mathrm{DFW}}) = (100\,\mu\mathrm{m}, 0\,^{\circ}\mathrm{C})$, and (**f**) $(d_s, \Delta T_{\mathrm{DFW}}) = (100\,\mu\mathrm{m}, 30\,^{\circ}\mathrm{C})$, with $\theta = 45^{\circ}$, $\phi_0 = 5\%$, and $W = 5\,\mathrm{cm}$. The x and y axes represent time and vertical position ($y = 0$ standing for the top of the cell). False color represents horizontally-averaged particle concentration.

Besides the observation of a considerable enhancement of sediment base formation (noticeable by the time scale of Figure 8c–e in contrast to Figure 8a,b), the effect of heating is to enhance, by higher values of the flow velocity near the downward facing wall, the kinetic energy. Figure 8e,f, corresponding to $d_s = 100\,\mu\mathrm{m}$, confirms this flow nonlinearity. Nonetheless, the quantitative implications on the sediment bed formation are not evident from the present set of simulations. Figure 9, showing the temporal progression of particle accumulation at A1 for $d_s = 50\,\mu\mathrm{m}$ and $d_s = 100\,\mu\mathrm{m}$, exposes the considerable difference on the settling time scale in comparison with $10\,\mu\mathrm{m}$ (Figure 5).

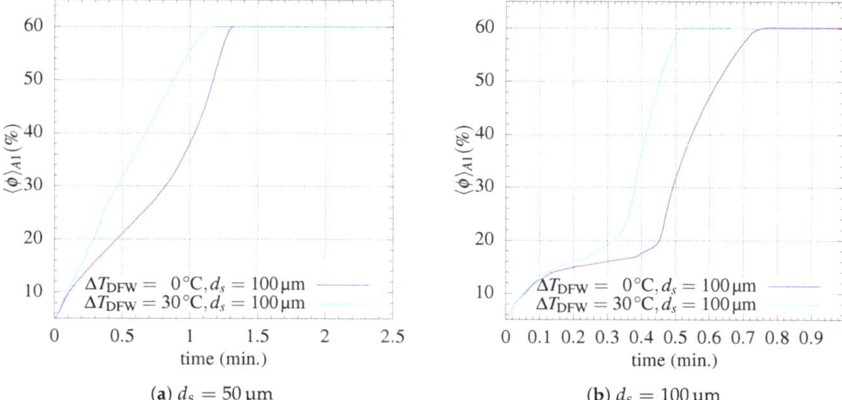

(a) $d_s = 50\,\mu\text{m}$ **(b)** $d_s = 100\,\mu\text{m}$

Figure 9. Average concentration at zone A1 as a function of particle diameter for **(a)** $d_s = 50\,\mu\text{m}$ and **(b)** $d_s = 100\,\mu\text{m}$.

When comparing the relative impact of the heating on the corresponding particle sizes for $\Delta T_{\text{DFW}} = 30\,°\text{C}$, there is no clear distinction on particle accumulation rates. Table 3 shows, for the case $\Delta T_{\text{DFW}} = 30\,°\text{C}$, that the dimensionless time scale of the end of the settling process at A1 was similar comparing $d_s = 10\,\mu\text{m}$ and $d_s = 50\,\mu\text{m}$, but much shorter for $d_s = 100\,\mu\text{m}$. Although this single result represents no proof of a trend due to the limitations of the numerical scheme used for computations—a mixture model, where close interaction between particles has only been regarded via a concentration-dependent viscosity and a solid phase pressure term—it is suggested that the effect of particle size on the separation efficiency should be analyzed in light of a time scale based both on the settling velocity and the convective process.

Table 3. Dimensionless particle accumulation lapse in A1 for temperature difference $\Delta T_{\text{DFW}} = 50\,°\text{C}$ at the downward facing wall for $d_s = 10\,\mu\text{m}$, $d_s = 50\,\mu\text{m}$, and $d_s = 100\,\mu\text{m}$ (t^*_{50,d_s}), normalized by the particle accumulation time without heating (t^*_{0,d_s}), $\psi_{50,d_s} \equiv t^*_{50,d_s}/t^*_{0,d_s}$. Here, $\theta = 45°$, $\phi_0 = 5\%$, and $W = 5\,\text{cm}$.

d_s (µm)	ψ_{50,d_s}
10	0.94
50	0.94
100	0.70

The impact of heating at the downward facing wall is also noticeable noting the concentration range of the supernatant layer. This is suggested from the spatiotemporal Figure 8b,d,f (heated), in comparison with their non-heated counterparts in Figure 8a,c,e, respectively.

While there was clear resuspension in the heated cell with 10 µm particles, there was not such a clear influence of heating on supernatant turbidity for $d_s = 50\,\mu\text{m}$ and 100 µm. This can also be seen from the computation of the particle accumulation at A2 in terms of the dimensionless time $\tau = tw_s/b$ (Figure 10), where the highest contrast between no heating and heating on the supernatant concentration was found for the smallest particle size tested. An explanation is based on the particle size being able to be obtained from two key aspects: first is the submerged weight of particles, equal to $(\rho_l - \rho_s)g\pi d_s^3/6\hat{k}$, exerting a greater force on particles if their size is larger; second is the drag of the liquid phase on particles, causing them to move upwards and also decreasing with particle diameter. In Stokes flow—applicable to the present particle range—the drag force on a single particle can be written as $\mathbf{F}_d = -3\pi\mu\mathbf{u}_r d_s$. Therefore, the work of drag required to move a single particle a distance z

is related to the change of mechanical energy as $\Delta E \sim \mathbf{F}_d \Delta z$. Note that the relative velocity, $\mathbf{u}_s$, can be related to the absolute velocity of the solid phase as $\mathbf{u}_r = \mathbf{u}_s/(1 - \phi)$. If u_s is a scale of the solid phase velocity and ϕ_0, the average particle concentration, is a scale of the concentration, then the single effect of drag is to modify the velocity of the solid phase as:

$$u_s\big|_{\text{drag}} \sim \frac{(1 - \phi_0)\Delta E}{\mu d_s \Delta z},$$ (9)

thus implying that increasing z and d_s within the cell causes a decrease of the drag force, proportional to u_s, thus reducing the opposing force to particle weight in the supernatant layer by a factor proportional to $1/d_s$.

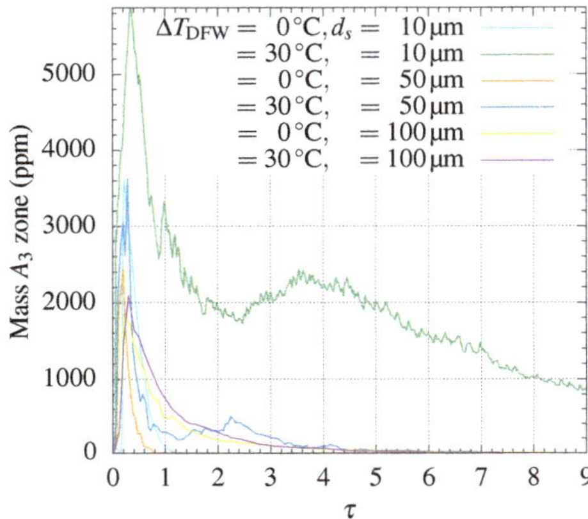

Figure 10. Average concentration at zone A2 for $d_s = 10\,\mu\text{m}$, $50\,\mu\text{m}$ and $100\,\mu\text{m}$, $\Delta T_{\text{DFW}} = 0$ and $50\,^{\circ}\text{C}$ as a function of the dimensionless time $\tau = t w_0/b$, with w_0 the Stokes velocity for each particle size.

4.3. Effect of Particle Concentration

Increasing mean particle concentration causes, at constant cell dimensions, a higher total particle volume, which implies an enhanced particle settling per unit time. In the absence of natural convection due to heating, following the PNK concept, then the position of the interface between the suspension and the supernatant layer ($H(t)$) can be written, in terms of the present set of variables, as [13]:

$$\frac{dH}{dt} = -w_0 \left(1 + \frac{H}{W} \tan\theta\right).$$ (10)

Assuming particle mass concentration in the sediment layer, the top of the sediment layer (with vertical position h_s, measured from the base of the cell) has the form:

$$\frac{dh_s}{dt} = w_0 \left(1 + \frac{H_0}{W} \tan\theta\right) F(\phi_0, t),$$ (11)

with:

$$F(\phi_0, t) = \frac{\phi_0 f(\phi_0)}{\phi_m - \phi_0} \exp\left[-\frac{w_0 f(\phi_0) \tan\theta}{W} t\right],$$ (12)

where at low volume concentrations (as in the present case), $f(\phi_0)$ is a weak function of the concentration ([30] has derived a similar expression with exponential dependence of the concentration). Assuming there is no flocculation and considering $f(\phi_0) \approx 1 - \phi_0^{1/3}$ [31], suitable for a low concentration, the particle concentration-dependent function for early values of time (neglecting the exponential term in Equation (12)) is given by $\frac{\phi_0(1-\phi_0^{1/3})}{\phi_m-\phi_0}$, which is a monotonically-increasing function with the concentration. For larger values of time, the effect of the exponential term is slight. Even considering times on the order of the total settling time, the exponential term cause no significant difference. Solving for the critical condition when $H(t_f) = h_s(t_f)$, based on the PNK model, a final time for settling is estimated, for low concentrations, as:

$$t_f = \frac{W}{w_0 \tan \theta} \ln \left(1 + \frac{H_0}{W} \tan \theta \right). \tag{13}$$

Figure 11 shows the time- and concentration-dependent term.

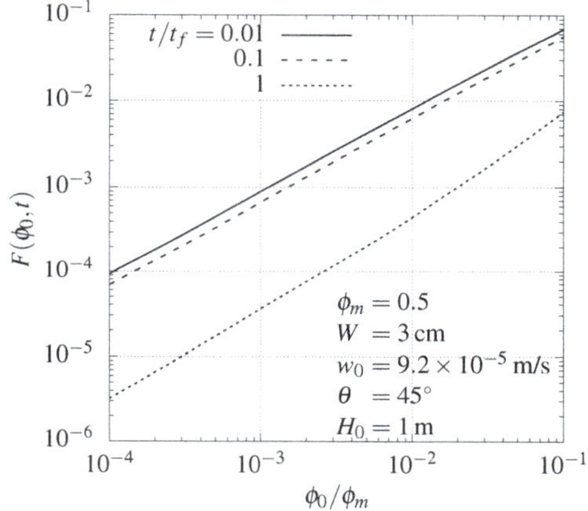

Figure 11. Concentration-dependent function of the rate of particle accumulation at the bottom in the absence of heating for three values of time, given by Equation (12). Time has been normalized by the scale of the particle settling process, given by Equation (13).

The same trend to increase the rate of particle accumulation with concentration was reproduced in the presence of heating at the downward facing wall. This can be seen in terms of the particle accumulation at zone A1, as shown in Figure 12. However, compared to the non-heating case, it was found, as expected, that the A1 filling process was faster in front of heating, as discussed above.

A related question is on the impact of increasing particle concentration on the overall efficiency of the separation process. An indication of this can be observed from the rate of particle accumulation at A1 in relation with the initial volume of particles or, equivalently, the dimensionless ratio $\langle \phi \rangle_{A1} / \phi_0$. This is depicted in Figure 13, where it is shown that, at equal values of ΔT_{DFW}, the concentration rise at A1 with time became slower at higher concentrations. This result reveals that particle settling at the bottom is a more efficient process for lower particle hindrance conditions. However, this implies that, at equal throughput, comparatively larger settler areas would be required. In this context, a final choice of particle concentration entering the cell should then be posed in terms of a trade-off analysis.

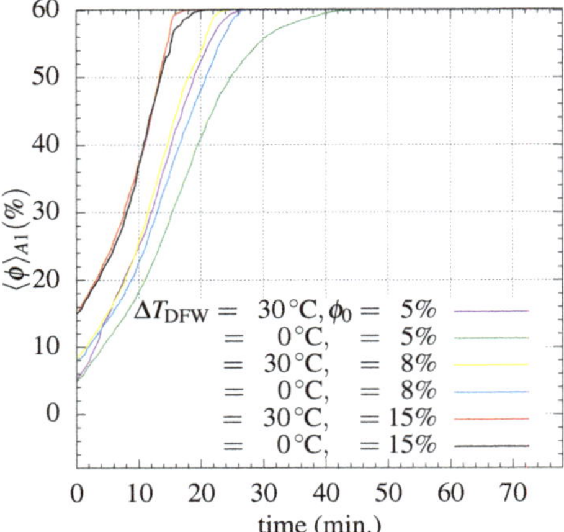

Figure 12. Effect of the initial particle concentration on the particle concentration in A1.

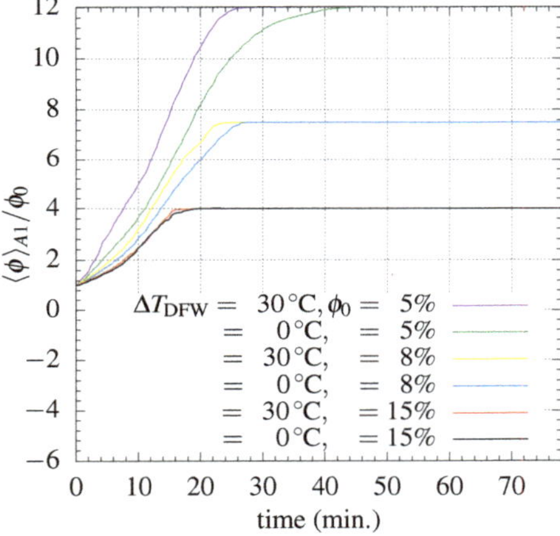

Figure 13. Dependence of the initial particle concentration on the normalized accumulation in A1, defined as $\langle\phi\rangle_{A1}/\phi_0$.

A significant impact of the initial concentration could be seen on the resuspended material near the top of the cell, in the supernatant layer. This is depicted in Figure 14 for the case $\Delta T_{\text{DFW}} = 30\,^\circ\text{C}$. Higher particle concentrations are bonded to smaller rates of resuspension, whose evidence, in Figure 14, was lower concentrations in zone A2. As in the previous section, this can also be interpreted in terms of the drag force between the solid and the liquid phase. The scale shown in Equation (9) suggests that the drag force, proportional to u_s, decreased with a factor proportional to $(1 - \phi_0)/\Delta z$.

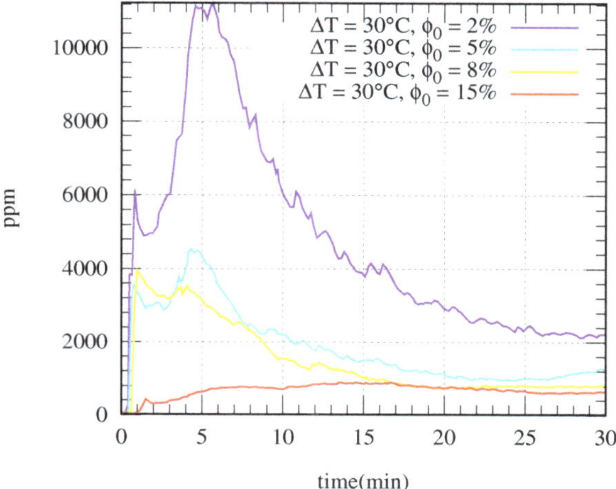

Figure 14. Effect of the initial particle concentration in the particle concentration in the clear water zone or A2.

5. Conclusions

The use of heat to enhance particle settling in inclined containers has been tested using high-resolution, two-dimensional numerical simulations in batch mode. Besides the finding that particle separation has a nonlinear effect with heating, the present study shows that both the particle size range and concentration corresponding to the highest enhancement of heating are within the particle size range of comminution processes and, in particular, copper ore tailings, thus rendering this approach viable for either primary or secondary solid separation stages in concentrator plants. Being solely gravity assisted and potentially energized by solar radiation, the use of heat-assisted lamellar settlers (HALS) could represent a viable option for a combined thickener-HALS unit process. Further development of this topic requires an extension to the continuous mode of operation, where a key aspect to consider is the interplay between enhanced settling and stronger mixing at the supernatant layer, both due to heating, the latter aspect promoting resuspension and therefore naturally suggesting an optimization problem on best operational conditions.

Author Contributions: Conceptualization, C.F.I. and C.R.; methodology, C.R. and C.F.I.; software, C.R.; validation, C.R. formal analysis, C.F.I. and C.R.; investigation, C.R. and C.F.I.; resources, C.F.I.; data curation, C.R.; writing, original draft preparation, C.F.I. and C.R.; writing, review and editing, F.A. and L.A.C.; visualization, C.R.; supervision, C.F.I.; project administration, C.F.I.; funding acquisition, C.F.I. and L.A.C.

Funding: This research was funded by the Chilean National Commission (CONICYT) for Science and Technology through Fondecyt Project Grant 1160971.

Acknowledgments: The authors acknowledge support from the Mining Engineering Department and the Advanced Mining Technology Center, Universidad de Chile.

Conflicts of Interest: The authors declare no conflict of interest.

References

1. Ihle, C.; Tamburrino, A. Variables affecting energy efficiency in turbulent ore concentrate pipeline transport. *Miner. Eng.* **2012**, *39*, 62–70. [CrossRef]
2. Ihle, C.F.; Kracht, W. The relevance of water recirculation in large scale mineral processing plants with a remote water supply. *J. Clean. Prod.* **2018**, *177*, 34–51. [CrossRef]

3. Herrera-León, S.; Lucay, F.A.; Cisternas, L.A.; Kraslawski, A. Applying a multi-objective optimization approach in designing water supply systems for mining industries. The case of Chile. *J. Clean. Prod.* **2019**, *210*, 994–1004. [CrossRef]

4. Wang, C.; Harbottle, D.; Liu, Q.; Xu, Z. Current state of fine mineral tailings treatment: A critical review on theory and practice. *Miner. Eng.* **2014**, *58*, 113–131. [CrossRef]

5. Tankosić, L.; Tančić, P.; Sredić, S.; Nedić, Z. Comparative Study of the Mineral Composition and Its Connection with Some Properties Important for the Sludge Flocculation Process-Examples from Omarska Mine. *Minerals* **2018**, *8*, 119. [CrossRef]

6. Ma, X.; Fan, Y.; Dong, X.; Chen, R.; Li, H.; Sun, D.; Yao, S. Impact of Clay Minerals on the Dewatering of Coal Slurry: An Experimental and Molecular-Simulation Study. *Minerals* **2018**, *8*, 400. [CrossRef]

7. Falconer, A. Gravity separation: Old technique/new methods. *Phys. Sep. Sci. Eng.* **2003**, *12*, 31–48. [CrossRef]

8. Concha, F. *Solid-Liquid Separation in the Mining Industry*; Springer: Berlin, Germany, 2014.

9. Chen, G.; Xiong, Q.; Morris, P.J.; Paterson, E.G.; Sergeev, A.; Wang, Y. OpenFOAM for computational fluid dynamics. *Not. AMS* **2014**, *61*, 354–363. [CrossRef]

10. Leung, W.F.; Probstein, R.F. Lamella and tube settlers. 1. Model and operation. *Ind. Eng. Chem. Process Des. Dev.* **1983**, *22*, 58–67. [CrossRef]

11. Boycott, A.E. Sedimentation of blood corpuscles. *Nature* **1920**, *104*, 532. [CrossRef]

12. Davis, R.H.; Zhang, X.; Agarwala, J.P. Particle classification for dilute suspensions using an inclined settler. *Ind. Eng. Chem. Res.* **1989**, *28*, 785–793. [CrossRef]

13. Acrivos, A.; Herbolzheimer, E. Enhanced sedimentation in settling tanks with inclined walls. *J. Fluid Mech.* **1979**, *92*, 435. [CrossRef]

14. Ortega, A.; Escobar, R.; Colle, S.; De Abreu, S.L. The state of solar energy resource assessment in Chile. *Renew. Energy* **2010**, *35*, 2514–2524. [CrossRef]

15. Solangi, K.H.; Islam, M.R.; Saidur, R.; Rahim, N.A.; Fayaz, H. A review on global solar energy policy. *Renew. Sustain. Energy Rev.* **2011**, *15*, 2149–2163. [CrossRef]

16. Kaviany, M. *Essentials of Heat Transfer: Principles, Materials, And Applications*; Cambridge University Press: Cambridge, UK, 2011.

17. Fujii, T.; Imura, H. Natural-convection heat transfer from a plate with arbitrary inclination. *Int. J. Heat Mass Transf.* **1972**, *15*, 755–767. [CrossRef]

18. Palma, S.; Ihle, C.F.; Tamburrino, A.; Dalziel, S.B. Particle organization after viscous sedimentation in tilted containers. *Phys. Fluids* **2016**, *28*. [CrossRef]

19. Reyes, C.; Ihle, C.F. Numerical simulation of cation exchange in fine-coarse seawater slurry pipeline flow. *Miner. Eng.* **2018**, *117*, 14–23. [CrossRef]

20. Maidment, D.R. *Handbook of Hydrology*; McGraw-HillL: New York, NY, USA, 1993; Volume 1.

21. Enwald, H.; Peirano, E.; Almstedt, A.E. Eulerian two-phase flow theory applied to fluidization. *Int. J. Multiph. Flow* **1996**, *22*, 21–66. [CrossRef]

22. Al-Shemmeri, T. *Engineering Fluid Mechanics*; Bookboon: London, UK, 2012.

23. Boyer, F.; Guazzelli, É.; Pouliquen, O. Unifying suspension and granular rheology. *Phys. Rev. Lett.* **2011**, *107*, 188301. [CrossRef] [PubMed]

24. Baumgarten, A.S.; Kamrin, K. A general fluid–sediment mixture model and constitutive theory validated in many flow regimes. *J. Fluid Mech.* **2019**, *861*, 721–764. [CrossRef]

25. Hanley, E.J.; Dewitt, D.P.; Roy, R.F. The thermal diffusivity of eight well-characterized rocks for the temperature range 300–1000 K. *Eng. Geol.* **1978**, *12*, 31–47. [CrossRef]

26. MacIntyre, S.; Romero, J.R.; Kling, G.W. Spatial-temporal variability in surface layer deepening and lateral advection in an embayment of Lake Victoria, East Africa. *Limnol. Oceanogr.* **2002**, *47*, 656–671. [CrossRef]

27. Herbolzheimer, E.; Acrivos, A. Enhanced sedimentation in narrow tilted channels. *J. Fluid Mech.* **1981**, *108*, 485. [CrossRef]

28. Hill, W.D.; Rothfus, R.; Li, K. Boundary-enhanced sedimentation due to settling convection. *Int. J. Multiph. Flow* **1977**, *3*, 561–583. [CrossRef]

29. Herbolzheimer, E. Stability of the flow during sedimentation in inclined channels. *Phys. Fluids* **1983**, *26*, 2043–2054. [CrossRef]

30. Dobashi, T.; Idonuma, A.; Toyama, Y.; Sakanishi, A. Effect of concentration on enhanced sedimentation rate of erythrocytes in an inclined vessel. *Biorheology* **1994**, *31*, 383–393. [CrossRef]
31. Davis, R.H.; Acrivos, A. Sedimentation of noncolloidal particles at low Reynolds numbers. *Annu. Rev. Fluid Mech.* **1985**, *17*, 91–118. [CrossRef]

Article

Optimization of the Heap Leaching Process through Changes in Modes of Operation and Discrete Event Simulation

Manuel Saldaña [1], Norman Toro [2,3,*], Jonathan Castillo [4], Pía Hernández [5] and Alessandro Navarra [6]

[1] Departamento de Ingeniería Industrial, Universidad Católica del Norte, Av. Angamos 610, Antofagasta 1270709, Chile

[2] Departamento de Ingeniería en Metalurgia y Minas, Universidad Católica del Norte, Av. Angamos 610, Antofagasta 1270709, Chile

[3] Departamento de Ingeniería Minera y Civil. Universidad Politécnica de Cartagena, Paseo Alfonso Xlll N°52, Cartagena 30203, Spain

[4] Departamento de Ingeniería en Metalurgia, Universidad de Atacama, Copiapó 1531772, Chile

[5] Departamento de Ingeniería Química y Procesos de Minerales, Universidad de Antofagasta, Avda. Angamos 601, Antofagasta 1240000, Chile

[6] Department of Mining and Materials Engineering, McGill University, 3610 University Street, Montreal, QC H3A 0C5, Canada

* Correspondence: ntoro@ucn.cl; Tel.: +56-552-651-021

Received: 19 May 2019; Accepted: 9 July 2019; Published: 10 July 2019

Abstract: The importance of mine planning is often underestimated. Nonetheless, it is essential in achieving high performance by identifying the potential value of mineral resources and providing an optimal, practical, and realistic strategy for extraction, which considers the greatest quantity of options, materials, and scenarios. Conventional mine planning is based on a mostly deterministic approach, ignoring part of the uncertainty presented in the input data, such as the mineralogical composition of the feed. This work develops a methodology to optimize the mineral recovery of the heap leaching phase by addressing the mineralogical variation of the feed, by alternating the mode of operation depending on the type of ore in the feed. The operational changes considered in the analysis include the leaching of oxide ores by adding only sulfuric acid (H_2SO_4) as reagent and adding chloride in the case of sulfide ores (secondary sulfides). The incorporation of uncertainty allows the creation of models that maximize the productivity, while confronting the geological uncertainty, as the extraction program progresses. The model seeks to increase the expected recovery from leaching, considering a set of equiprobable geological scenarios. The modeling and simulation of this productive phase is developed through a discrete event simulation (DES) framework. The results of the simulation indicate the potential to address the dynamics of feed variation through the implementation of alternating modes of operation.

Keywords: process optimization process; heap leaching; modes of operation; discrete event simulation

1. Introducion:

1.1. Overview

Conventional mine planning is traditionally applied in the industry through methodologies that consider an important part of the data to be deterministic. However, critical information used for mining calculations may exhibit statistical variations [1]. When a parameter is uncertain, the expected result is uncertain, since the calculations have considered a potentially unrepresentative value of the parameter, instead of another that could have the same or different probability of occurrence. Due to this, modern approaches consider the uncertainty and the risk associated with input parameters, which provide a wider vision of the possible losses and gains of the project [2]. The uncertainty of not knowing the real value of the metal content of interest to a certain process is indeed a real risk, so finding a way to organize resources or define alternative operational strategies is a very difficult calculation problem, mainly due to variables that are subject to geological uncertainty; there is generally a range of possible scenarios of mineral grade distribution, process capacities, and commodity market conditions, among others [3,4].

Modern approaches to mine production require simulation frameworks that can increase mineral recovery and are robust in mitigating feed variations [5]. This work presents a methodology for the evaluation of heap leaching, incorporating information of the mineralogical composition of the inputs; the approach is based on discrete event simulation (DES). In general, DES models are used to study systems and processes, in which state changes are computed only at discrete points in time (i.e., discrete events); the changes that occur between these events are not computed explicitly, but can be inferred *a posteriori*. The simulation of the heap leaching allows the planner to estimate the impact on the productivity of the implementation of different modes of operation [6] in response to variations in the mineralogical composition of ores.

1.2. Heap Leaching

After the comminution phase, the copper ores pass to the leaching stage, where the metals present in the mineralized rock are extracted through the application of water and leaching agents. This process is comparatively effective for low- to medium-grade copper oxide minerals (0.3–0.7%). Secondary copper sulfides and low-grade gold ores are also processed in this way [7–9], since it provides a low cost of capital compared to other methods, and since it does not require an intensive use of energy [10]. The agglomeration of the fines around the larger particles with water and concentrated sulfuric acid is known as "curing". This process improves the resistance of the material while having a good permeability of the mineral in the heap leaching, in order to reach adequate heap heights, improve copper recovery rates, and control processing times [11,12]. The acid solution is distributed by sprinklers or drippers, in which the copper (Cu^{2+}) dissolves in the leaching solution as it percolates the heap. The realization of tests at the laboratory level and in pilot plants determine the effectiveness of a heap. The amount of ore to be treated can vary considerably from hundreds to more than one million tons [12], depending on the mine.

Another emerging method is biohydrometallurgy, which plays an important role in the recovery of copper from copper sulfides with economic, environmental, and social benefits [13]. To date, many investigations on acid bioleaching of secondary sulfides [14,15] and primary sulfides [16–18] have been reported presenting good results.

Even in its role as a surplus generator, large-scale mining faces great challenges. These include an increase in costs due to various factors, such as the deterioration of grades and other factors associated with the aging of deposits and increased operating costs to be compatible with sustainable development demands. In typical operations, heap leaching processes operate in approximate times of three months for sulfide ores in chloride media, and also with lower ore grades [19].

2. Materials and Methods

There are several processes through which minerals can be leached, depending mainly on the physical and chemical considerations, such as the solubility of the metal, the kinetics of the solution, the consumption of the reagent, etc. [20]. Heap leaching is currently the most common leaching method in the Chilean mining industry.

2.1. Discrete Event Simulation

With a DES framework, an event is a random occurrence that occurs at a discrete point in time, and whose outcome depends on chance. An event is considered simple if it consists of a specific result or compound if it consists of two or more independent events [21].

Within a system of discrete events, one or more phenomena of interest change their value in discrete points in time [22]. Discrete event simulation considers the evolution of the system, but the states are modified only at discrete moments of time, and they are caused by the occurrence of some event. For this, the state of the system does not explicitly consider variations between two consecutive events. The event n occurs at time t_n, and event $n + 1$ will occur at time t_{n+1}, as the simulation clock jumps directly to the instant t_{n+1}. Upon advancing to t_{n+1}, the system statistics and state variables are updated, and this process repeated until a termination condition is met [23].

2.2. Mathematical Modeling of Heap Leaching

Around 20% of the world's copper production is obtained by heap leaching. This process has been modeled by many authors; however, the validation, verification and implementation of these models are difficult since there is uncertainty about the operating conditions and parameters of the leaching model [24–26].

The performance of heap leaching depends on many input variables (operational and design), which means its optimization is complex [27]. The materials are leached with various chemical solutions that extract valuable minerals. These chemical solutions are a weak sulfuric acid solution for copper oxide ores, and chloride media [28] for copper secondary sulfides. The valuable minerals are irrigated with a chemical solution that dissolves the valuable metal of the ore, as the resulting pregnant leaching solution (PLS) passes through the ore, and is recuperated at the base of the heap. The valuable material is then extracted from the PLS, and the chemical solution is recycled back into the heap. The most common methods for recovery of valuable minerals are solvent extraction and electro-winning processes [12].

The following is an analytical model for heap leaching developed by Mellado et al. [29–31], using a system of first order equations:

$$\frac{\partial y}{\partial \tau} = -k_\tau y^{n_\tau} \tag{1}$$

where "y" is a dynamic quantity, such as the concentration or recovery R_t, k_τ are kinetic constants associated with the characteristics of the heap and grade of the mineral respectively, and n_τ is the order of the reaction. The subscript τ represents a time scale that depends on the phenomenon to be modeled. To solve Equation (1), an initial condition is required. Mellado et al. introduced a delay (i.e., a time ω where R_t begins to change ($R_t(\omega) = 0$)); the general solution for $n_\tau = 1$ is given by (see Mellado et al. [29] for the general solution):

$$R_\tau = R_\tau^\infty \left(1 - e^{-k_\tau(\tau-\omega)}\right) \tag{2}$$

Dixon and Hendrix [32,33] considered that the leaching phenomenon occurs at different scales of size and time, and that different phenomena participate in the leaching process. On the other hand, Mellado et al. [31] incorporated the different scales in an analytical model of the leaching process, introducing the parameters K_θ and K_τ, related to size and time, respectively, as can be seen in Equation (3):

$$R(t) = \frac{\alpha}{Z^\gamma + \beta} \left[1 - \lambda e^{K_\theta(t-\omega^*)} - (1-\lambda)e^{K_\tau(t-\omega^*)}\right] \tag{3}$$

Mellado et al. develop the parameters K_θ, K_τ, and ω^* in Equations (4)–(6) respectively:

$$K_\theta = k_\theta \frac{\mu_s}{\varepsilon_b Z} \tag{4}$$

$$K_\tau = k_\tau \frac{D_{Ae}}{\varepsilon_o r^2} \tag{5}$$

$$\omega^* = \frac{\varepsilon_b Z}{\mu_s} \omega \tag{6}$$

where α, β, and γ are mathematical constants of fit, Z is the height of the heap, λ is a factor of kinetic weight, k_θ and k_τ are kinetic constants, μ_s is the surface velocity of the leaching flow in the heap, ε_b is the volumetric fraction of the bulk solution in the heap, ω is the delay of the reaction, D_{Ae} is the effective diffusivity of the solute within the pores of the particles, ε_o is the porosity of the particles, and r is the radius of the particles.

The goodness-of-fit statistics used to study the model adjusted to observations (operational data supplied from an industrial heap leaching operation at a copper mine in Antofagasta, Chile) are: The mean absolute deviation (MAD, Equation (7)), a statistic that measures the dispersion of forecast error; the mean square error (MSE, Equation (8)), measure of error dispersion that penalizes the periods or values where the error module is higher than the average value; and the absolute average percentage error (MAPE, Equation (9)), a statistic that gives the deviation in percentage terms, calculating the averages of the absolute values between the real value [21].

$$\text{MAD} = \frac{\sum |\text{Real} - \text{Forecast}|}{n} \tag{7}$$

$$\text{MSE} = \frac{\sum (\text{Real} - \text{Forecast})^2}{n} \tag{8}$$

$$\text{MAPE} = \frac{1}{n} \sum \frac{|\text{Real} - \text{Forecast}|}{|\text{Real}|} \tag{9}$$

2.3. Adjustment of the Analytical Model for the Recovery of Copper from Copper Oxides

Adjusting the analytical model by means of a linear optimization model that minimizes the error measurements of the adjustment to operational data, considering the theoretical restrictions of the analytical model, results in the following equation:

$$R(t) = 0.9993\left(1 - 0.4e^{-0.0844(t-2.3684)} - 0.6e^{-0.0055(t-2.3684)}\right) \tag{10}$$

Figure 1 shows the adjusted models from operational data and analytical model respectively for the leaching process operating only with sulfuric acid as a leaching agent, while the goodness-of-fit statistics are shown in Table 1.

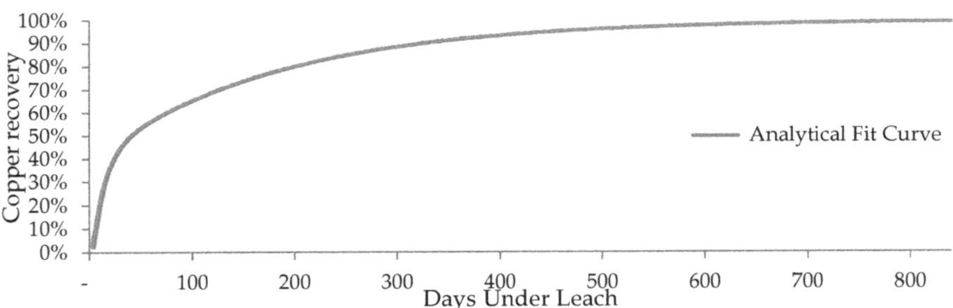

Figure 1. Operational fit curve versus analytical fit curve for copper recovery from oxide ores.

Table 1. Statistics of analytical models of leaching of copper oxides adding sulfuric acid.

Curve/Statistic	MAD	MSE	MAPE
R(t) (Oxides)	1.008×10^{-2}	1.222×10^{-4}	1.28×10^{-2}

The interpretation of the error statistics indicates the degree to which the generated model explains the system to be modeled, from which it is possible to conclude that the difference between the real and predicted values is negligible, which means that the analytical model explains the operational values.

2.4. Adjustment of Analytical Model for Copper Recovery from Secondary Copper Sulfides

The analytical model for copper recovery as a function of time for sulfide minerals (secondary sulfides) is modeled by Equation (11).

$$R(t) = 0.8841\left(1 - 0.2e^{-0.0072(t-1.91)} - 0.8e^{-0.0771(t-1.91)}\right) \tag{11}$$

The adjusted curve of Figure 2 and the error measures of the adjusted model presented in Equation (3) have the goodness-of-fit statistics and low error statistics shown in Table 2, indicating that the analytical model fits the sample data of the operation.

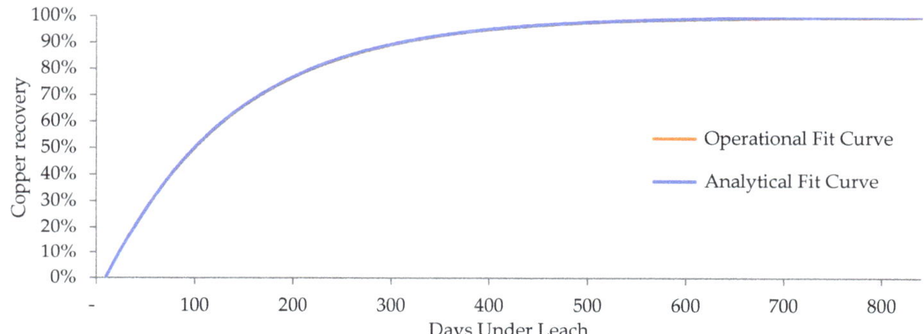

Figure 2. Operational fit curve versus analytical fit curve for copper recovery from sulfide ores.

Table 2. Statistics of analytical models of leaching of secondary copper sulfides adding sulfuric acid.

Curve/Statistic	MAD	MSE	MAPE
R(t) (Oxides)	6.63×10^{-4}	5.068×10^{-7}	8.93×10^{-4}

2.5. Adjustment of Analytical Models for Copper Recovery from Secondary Copper Sulfide Ores Adding Chlorides

Adjusting the curves for the leaching of copper sulfide minerals for two levels of chloride concentration (20 and 50 g/L) as shown in Figure 3, produces the following equations:

$$R(t)[\text{Chloride 20 g/L}] = 0.9159\left(1 - 0.3e^{-0.0168t-2.3684} - 0.7e^{-0.0057t-2.3684}\right) \tag{12}$$

$$R(t)[\text{Chloride 50 g/L}] = 0.9291\left(1 - 0.3e^{-0.0633t-2.3684} - 0.7e^{-0.0071t-2.3684}\right) \tag{13}$$

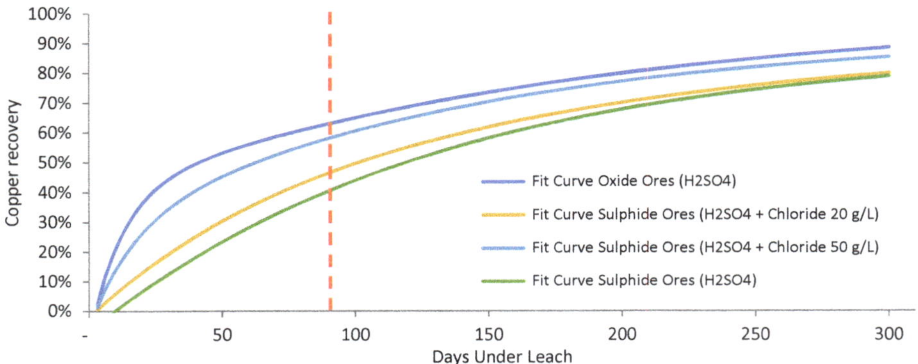

Figure 3. Copper recovery from oxide and sulfide ores using H_2SO_4 and chlorides as an additive.

The goodness-of-fit statistics for the leaching of copper sulfide minerals for two levels of chloride concentration (20 and 50 g/L) are shown in Table 3. Low error statistics indicate that the generated analytical model fits the sample data

Table 3. Statistics of analytical models of leaching adding chlorides.

Curve/Statistic	MAD	MSE	MAPE
R(t) (Chloride 20 g/L)	1.68×10^{-4}	4.59×10^{-7}	5.40×10^{-4}
R(t) (Chloride 50 g/L)	9.17×10^{-5}	5.23×10^{-7}	5.89×10^{-4}

The expected recovery of copper for the different configurations in 90 days of leaching is presented in Table 4.

Table 4. Recovery for each configuration in a 90-day leaching time.

Configuration	Recovery (%)
Leaching of secondary copper sulfides with sulfuric acid	40.5
Leaching of secondary copper sulfides adding chlorides (20 g/L)	46.5
Leaching of secondary copper sulfides adding chlorides (50 g/L)	58.1
Leaching of copper oxides with sulfuric acid	64.6

2.6. Modeling and Simulation of Heap Leaching Using a DES Framework

Once the process workflow of heap leaching has been characterized, it is possible to model the heap leach stage sequentially with the Arena simulation software. The update of copper recovery over time is simulated by parametrizing the analytical models retrieved from the literature, and incorporating them into the Arena simulation [23].

The schematic of the simulation model is presented in Figure 4, next to the subprocess responsible for the update in discrete time. The update of the recovery state is carried out whenever a production campaign is in development, while the use of the operational parameters is updated in the module "Assignment of attributes to the piles", and the recovery of ore is obtained from the analytical models derived from Equations (1) and (2), (these equations depend on the leaching time and operating conditions of the site).

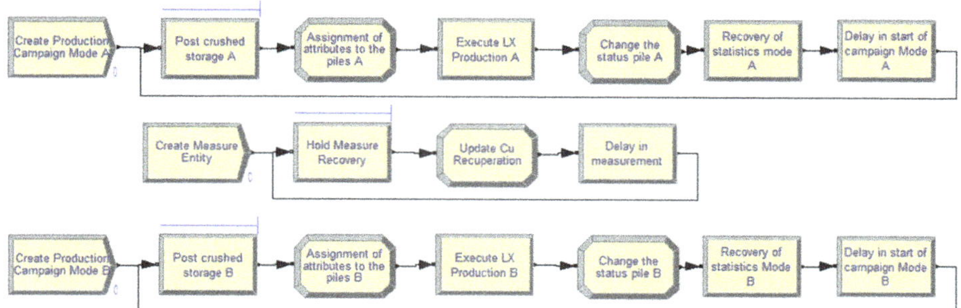

Figure 4. High-level diagram of heap leaching in Arena Simulation Software.

The heap leaching process is modeled by production campaigns, whose start is determined by the availability of inventories of the crushing phase, the development of the campaign corresponding to the production of each heap, the limited production capacity due to available physical space, and the downstream storage capacity. For each heap, the expected recovery of ore is measured according to the adjusted analytical models and the production in tons considering the variations in ore grades of the feed.

The storage of crushed material works under the logic of inventory theory [3], where the comminution product is kept waiting until the end of the leaching campaign. The module "Post crushed storage" stores the ore that will enter the leaching process when its respective mode of operation is activated. Each mode of operation is determined by the type of ore to be leached, and the decision to apply a given mode depends on the maximum and minimum stock levels established for each type of mineral. The current context considers two modes of operation:

- Mode A: Leaching of copper oxides.
- Mode B: Leaching of copper sulfide minerals (secondary sulfides).

The assignment of attributes to the heaps, such as the grade of each type of copper ore, is obtained from ore data from the *Empresa Nacional de Minería* (ENAMI), which is a Chilean state-owned enterprise. These attributes are taken as input variables for the analytical models used to estimate the expected recovery of ore under operational conditions. After a simulated leaching campaign, recovery results are saved. A comparative analysis of simulated leaching operations, with and without an additional mode, allows us to quantify the benefit of implementing the additional mode.

3. Discussion of Results

3.1. Simulated Scenarios

With the objective of evaluating the variation in the leaching productivity through the incorporation of analytical models that integrate mineralogical characteristics under conditions of uncertainty, the following scenarios are defined:

- Scenario 1 (standard operation): Leaching of copper oxides and secondary copper sulfides adding sulfuric acid only. The leaching of secondary sulfides with sulfuric acid slows down the process of extracting ore from the rock, increasing the time required until the marginal extraction of ore is negligible [12,34].
- Scenario 2 (proposed operation): Leaching of oxides with sulfuric acid and leaching of secondary sulfides with chloride. The leaching of secondary sulfides by adding chloride accelerates the recovery of copper from sulfide minerals, decreasing the leaching time [34–37].

Scenario 1:

From the graphical analysis of copper recovery for each production campaign (see Figure 5), a decrease in the expected recovery of copper ore can be observed in sulfide mineral leaching campaigns using sulfuric acid as reagent (without incorporating additives), due to the slower dissolution kinetics of the secondary copper sulfides.

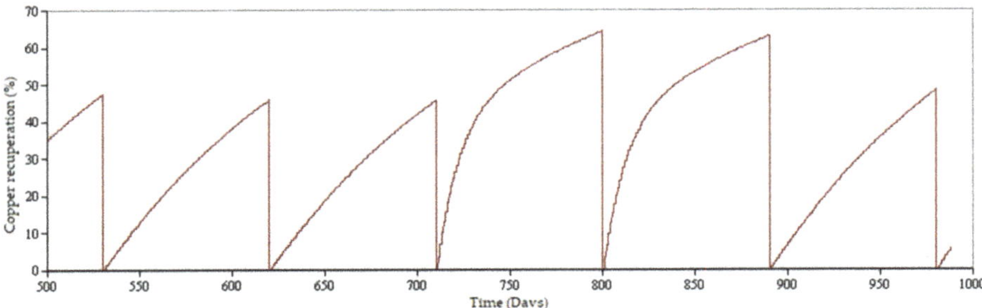

Figure 5. Copper recovery of base case maintaining a single mode of operation

Then the average copper recovery is approximately 65% in the case of oxide ores, and 40% in the case of sulfide ores. Of the total production time, 61% of the time was for processing oxide ores, and 39% for sulfide ores, hence an average recovery of approximately 55%.

Scenario 2:

A similar analysis for scenario 2 reveals that 61% of the time was spent on oxide ores, for which only sulfuric acid was used as reagent, while 39% was on sulfide minerals, using sulfuric acid and chlorides as additives. The average recovery of ore is maintained at 65% for operational mode A and increases to 58% for mode B (improvements in extraction derived from the addition of chlorides), working at a chloride concentration of 50 g/L. The resulting average recovery is approximately 62%.

The benefit of having alternate modes of operation is illustrated in Figures 6 and 7, showing that the expected copper recovery from sulfide ores is greater when varying the mode of operation, being independent of the characteristics of the feed and considering that the leaching time remains constant. (Leaching time is kept constant due to the increase in opportunity costs of maintaining a longer time of a leaching heap whose recovery rate decreases over time).

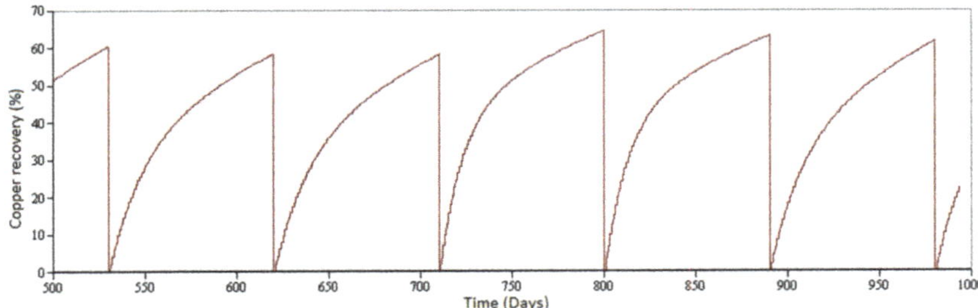

Figure 6. Copper recovery for the proposed scenario (two operation modes).

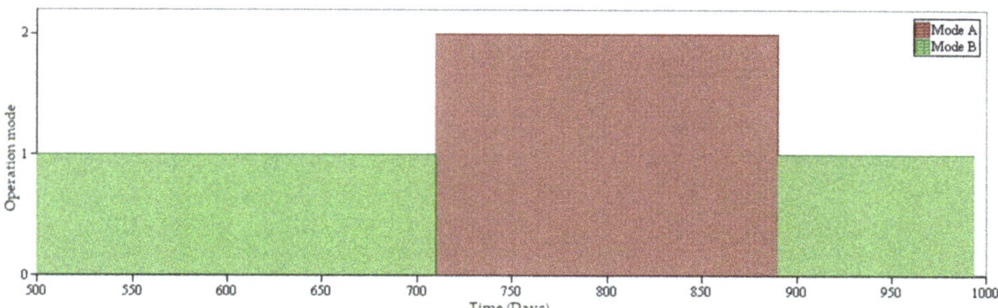

Figure 7. Mode of operation for the proposed scenario.

Considering two modes of operation, the objective is to optimize production by alternating the modes of operation as a function of the feed, avoiding instances of lower recovery of copper with mode A (leaching of oxide ores with sulfuric acid) whenever mode B (leaching of sulfides by adding chloride) will be more appropriate; this is the case when there are sufficient sulfides that a detrimental passivation layer will form in the presence of sulfuric acid. This passivation causes a decrease in recoveries when leaching secondary sulfides with sulfuric acid, that can be mitigated with longer exposure time to the leaching agent, but this means an increase in production costs, considering the increases in the consumption of acid and the opportunity costs of the use of the leaching equipment; the chloride counteracts this phenomenon.

3.2. Comparison of Samples

In order to compare the productivity of the leaching phase under the scenarios considered, a hypothesis test is carried out [38], for which the null hypothesis is defined as:

$$H_0{:}\mu_2 = \mu_1$$

where μ_2 represents the average production in thousands of tons of the leaching phase considering changes in the modes of operation, and μ_1 represents the average value of production considering a single mode of production. The alternate hypothesis is given by:

$$H_a{:}\mu_2 > \mu_1$$

Developing the hypothesis test in the statistical analysis software Minitab 18 [39], and considering a sample size of 100 simulations, it can be concluded that the size of the production average of the proposed situation is greater than the current situation, as shown in Figure 8. It is further concluded that the hypothesis test is significant, since the *p*-value is less than the level of significance, as shown in Figure 9.

Individual Samples		
Statistics	**Actually**	**Proposal**
Sample size	100	100
Mean	3852.1	4399.0
90% CI	(3800, 3904)	(4340.1, 4457.9)
Standard deviation	313.11	354.61

Figure 8. Statistics test of 2 samples (CI: Confidence interval of 90%).

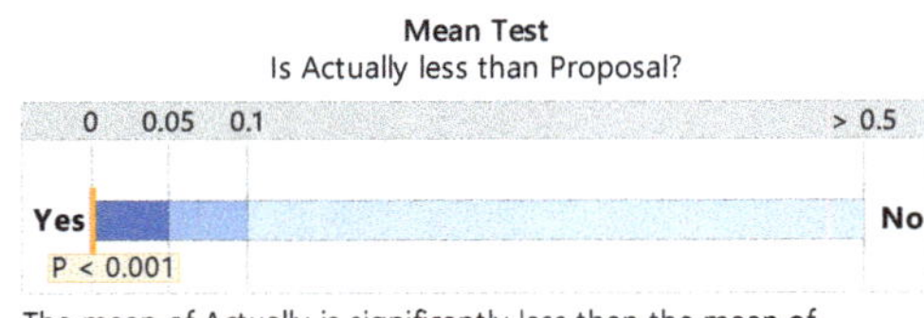

Figure 9. *p*-value of the hypothesis test.

The confidence interval quantifies the uncertainty associated with the estimation of the difference in the means from the data of the samples, so it is possible to have more than 90% certainty that the true difference is between −625.08 and −468.71, and a 95% assurance that it is less than −468.71, as shown in Figures 10 and 11.

Figure 10. Differences of the samples.

Figure 11. A 90% confidence interval for the difference.

Finally, comparing the data distributions for both samples (Figure 12), the difference of the mean values of the samples can be observed graphically. Although the distributions have some overlap, the means are several error bars away.

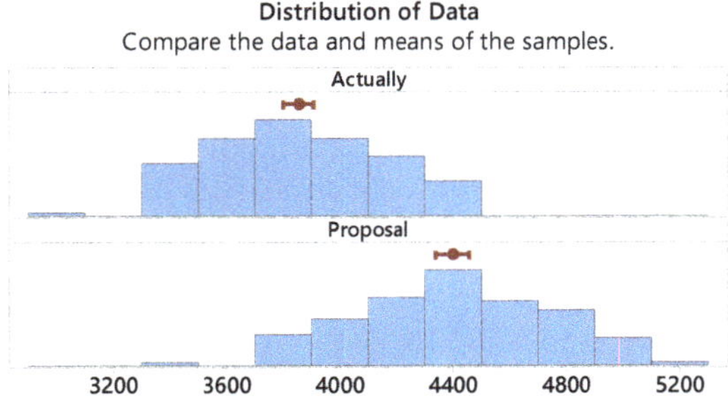

Figure 12. Distribution of production for both samples.

In summary, when copper recovery is carried out by means of a single mode of operation (simulation based on analytical models extracted from the literature, which does not consider variations in reagent concentrations), there may not be a systemic response to the changing mineralogical characteristics of the feed, resulting in lost production.

4. Conclusions

4.1. Conclusions

Mineral deposits tend to be heterogeneous, which forces the processing parameters to evolve over time. In this document, a simulation of the production sequence of the heap leaching was considered, simplifying the feeding to only two types of ore. However, the framework could be extended to a greater number of mineral types that could come from a range of geological domains in a mine, as long as the ore to be processed is of interest and it is technically and economically possible to process it through the hydrometallurgical route.

The use of alternating modes of operation has the potential to improve the strategic mine plan, making the value chain more flexible by making a better use of assets and improving mineral recovery, addressing the varying mineralogical characteristics of the feed. The hypothesis test indicates the average production increase to incorporate the dynamics of operating modes in heap leaching, in this case increasing the expected recovery of copper, from 55% to 62%.

The quantification of the improvements by addressing uncertainty in the processing of minerals through alternating modes of operation, the incorporation of analytical models for the unit processes and the sequential simulation through a discrete event simulation framework constitute an opportunity to effectively model and plan leaching operations, from a system-wide perspective. The approach can assist in local and ultimately global mine optimizations for cash flows and asset utilization.

4.2. Future Work

To further advance the operation research of leaching processes, the following avenues are being considered:

1. Include other modes of operation and analytical models that incorporate more operational variables to the process, together with parameters that have a significant impact on recovery.
2. Study the impact on an industrial scale of operating the leaching process with alternating modes of operation, including the analysis operating and capital costs.

Author Contributions: M.S. and N.T. contributed in the methodology, conceptualization and simulation; J.C. and P.H. investigation and resources and A.N. contributed with supervision and validation.

Funding: This research received no external funding.

References

1. Navarra, A.; Alvarez, M.; Rojas, K.; Menzies, A.; Pax, R.; Waters, K. Concentrator operational modes in response to geological variation. *Miner. Eng.* **2019**, *134*, 356–364. [CrossRef]
2. Hustrulid, W.; Kuchta, M.; Martin, R. *Open Pit Mine Planning and Design*; CRC Press: Boca Raton, FL, USA, 2013; ISBN 9781466575127.
3. Dimitrakopoulos, R. Strategic mine planning under uncertainty. *J. Min. Sci.* **2011**, *47*, 138–150. [CrossRef]
4. Rahmanpour, M.; Osanloo, M. Determination of value at risk for long-term production planning in open pit mines in the presence of price uncertainty. *J. S. Afr. Inst. Min. Metall.* **2016**, *116*, 8–11. [CrossRef]
5. Upadhyay, S.P.; Askari-Nasab, H. Simulation and optimization approach for uncertainty-based short-term planning in open pit mines. *Int. J. Min. Sci. Technol.* **2018**, *28*, 153–166. [CrossRef]
6. Drebenstedt, C.; Singhal, R. *Mine Planning and Equipment Selection*; Brookfield: Rotterdam, The Netherlands, 2014; ISBN 978-3-319-02677-0.
7. Harris, G.B.; White, C.W.; Demopoulos, G.P.; Ballantyne, B. Recovery of copper fromamassive polymetallic sulphide by high concentration chloride leaching. *Can. Metall. Q.* **2008**, *47*, 347–356. [CrossRef]
8. Robertson, S.W.; Van Staden, P.J.; Seyedbagheri, A. Advances in high-temperature heap leaching of refractory copper sulphide ores. *J. S. Afr. Inst. Min. Metall.* **2012**, *112*, 1045–1050.

9. Petersen, J. Heap leaching as a key technology for recovery of values from low-grade ores—A brief overview. *Hydrometallurgy* **2016**, *165*, 206–212. [CrossRef]

10. COCHILCO. *Caracterización de los Costos de la Gran Minería del Cobre*; COCHILCO: Santiago, Chile, 2015.

11. Lu, J.; Dreisinger, D.; West-Sells, P. Acid curing and agglomeration for heap leaching. *Hydrometallurgy* **2017**, *167*, 30–35. [CrossRef]

12. Schlesinger, M.; King, M.; Sole, K.; Davenport, W. *Extractive Metallurgy of Copper*, 5th ed.; Elsevier: Amsterdam, The Netherlands, 2011; ISBN 9780080967899.

13. Liu, H.; Xia, J.; Nie, Z.; Ma, C.; Zheng, L.; Hong, C.; Zhao, Y.; Wen, W. Bioleaching of chalcopyrite by Acidianus manzaensis under different constant pH. *Miner. Eng.* **2016**, *98*, 80–89. [CrossRef]

14. Ruan, R.; Zou, G.; Zhong, S.; Wu, Z.; Chan, B.; Wang, D. Why Zijinshan copper bioheapleaching plant works efficiently at low microbial activity-Study on leaching kinetics of copper sulfides and its implications. *Miner. Eng.* **2013**, *48*, 36–43. [CrossRef]

15. Lee, J.; Acar, S.; Doerr, D.L.; Brierley, J.A. Comparative bioleaching and mineralogy of composited sulfide ores containing enargite, covellite and chalcocite by mesophilic and thermophilic microorganisms. *Hydrometallurgy* **2011**, *105*, 213–221. [CrossRef]

16. Zhang, R.; Sun, C.; Kou, J.; Zhao, H.; Wei, D.; Xing, Y. Enhancing the leaching of chalcopyrite using acidithiobacillus ferrooxidans under the induction of surfactant Triton X-100. *Minerals* **2018**, *9*, 11. [CrossRef]

17. Ma, L.; Wang, X.; Liu, X.; Wang, S.; Wang, H. Intensified bioleaching of chalcopyrite by communities with enriched ferrous or sulfur oxidizers. *Bioresour. Technol.* **2018**, *268*, 415–423. [CrossRef] [PubMed]

18. Ma, L.; Wu, J.; Liu, X.; Tan, L.; Wang, X. The detoxification potential of ferric ions for bioleaching of the chalcopyrite associated with fluoride-bearing gangue mineral. *Appl. Microbiol. Biotechnol.* **2019**, *103*, 2403–2412. [CrossRef] [PubMed]

19. CESCO. *La Minería Como Plataforma Para el Desarrollo: Hacia Una Relación Integral y Sustentable de la Industria Minera en Chile*; CESCO: Santiago, Chile, 2013.

20. Beiza, L. *Lixiviación de Mineral y Concentrado de Calcopirita en Medios Clorurados*; Universidad Católica del Norte: Antofagasta, Chile, 2012.

21. Devore, J. *Probability & Statistics for Engineering and the Sciences*, 8th ed.; Cengage Learning: Boston, MA, USA, 2010; ISBN 0-538-73352-7.

22. Neeraj, R.R.; Nithin, R.P.; Niranjhan, P.; Sumesh, A.; Thenarasu, M. Modelling and simulation of discrete manufacturing industry. *Mater. Today Proc.* **2018**, *5*, 24971–24983. [CrossRef]

23. Kelton, W.D. *Simulation with Arena*; McGraw-Hill Education: New York, NY, USA, 2015; ISBN 978-0-07-340131-7.

24. Van Staden, P.J.; Kolesnikov, A.V.; Petersen, J. Comparative assessment of heap leach production data—1. A procedure for deriving the batch leach curve. *Miner. Eng.* **2017**, *101*, 47–57. [CrossRef]

25. Ordóñez, J.; Condori, A.; Moreno, L.; Cisternas, L. Heap leaching of caliche ore. modeling of a multicomponent system with particle size distribution. *Minerals* **2017**, *7*, 180.

26. Morrison, R.D.; Shi, F.; Whyte, R. Modelling of incremental rock breakage by impact—For use in DEM models. *Miner. Eng.* **2007**, *20*, 303–309. [CrossRef]

27. Leiva, C.; Flores, V.; Salgado, F.; Poblete, D.; Acuña, C. Applying softcomputing for copper recovery in leaching process. *Sci. Prog.* **2017**, *2017*, 6. [CrossRef]

28. Miki, H.; Nicol, M.; Velásquez-Yévenes, L. The kinetics of dissolution of synthetic covellite, chalcocite and digenite in dilute chloride solutions at ambient temperatures. *Hydrometallurgy* **2011**, *105*, 321–327. [CrossRef]

29. Mellado, M.E.; Cisternas, L.A.; Gálvez, E.D. An analytical model approach to heap leaching. *Hydrometallurgy* **2009**, *95*, 33–38. [CrossRef]

30. Mellado, M.E.; Gálvez, E.D.; Cisternas, L.A. Stochastic analysis of heap leaching process via analytical models. *Miner. Eng.* **2012**, *33*, 93–98. [CrossRef]

31. Mellado, M.; Cisternas, L.; Lucay, F.; Gálvez, E.; Sepúlveda, F. A posteriori analysis of analytical models for heap leaching using uncertainty and global sensitivity analyses. *Minerals* **2018**, *8*, 44. [CrossRef]

32. Dixon, D.G.; Hendrix, J.L. A mathematical model for heap leaching of one or more solid reactants from porous ore pellets. *Metall. Trans. B* **1993**, *24*, 1087–1102. [CrossRef]

33. Dixon, D.G.; Hendrix, J.L. A general model for leaching of one or more solid reactants from porous ore particles. *Metall. Trans. B* **1993**, *24*, 157–169. [CrossRef]

34. Helle, S.; Jerez, O.; Kelm, U.; Pincheira, M.; Varela, B. The influence of rock characteristics on acid leach extraction and re-extraction of Cu-oxide and sulfide minerals. *Miner. Eng.* **2010**, *23*, 45–50. [CrossRef]

35. Jones, D.A.; Paul, A.J.P. Acid leaching behavior of sulfide and oxide minerals determined by electrochemical polarization measurements. *Miner. Eng.* **1995**, *8*, 511–521. [CrossRef]

36. Cheng, C.Y.; Lawson, F. The kinetics of leaching chalcocite in acidic oxygenated sulphate-chloride solutions. *Hydrometallurgy* **1991**, *27*, 249–268. [CrossRef]

37. Ruiz, M.C.; Honores, S.; Padilla, R. Leaching kinetics of digenite concentrate in oxygenated chloride media at ambient pressure. *Metall. Mater. Trans. B Process Metall. Mater. Process. Sci.* **1998**, *29*, 961–969. [CrossRef]

38. Douglas, C. *Montgomery: Design and Analysis of Experiments*, 8th ed.; John Wiley & Sons: New York, NY, USA, 2012; ISBN 978-1-118-14692-7.

39. Mathews, P.G. *Design of Experiments with MINITAB*; William, A., Ed.; ASQ Quality Press: Milwaukee, WI, USA, 2005; ISBN 0873896378.

Article

Design of Flotation Circuits Using Tabu-Search Algorithms: Multispecies, Equipment Design, and Profitability Parameters

Freddy A. Lucay [1],*, Edelmira D. Gálvez [2] and Luis A. Cisternas [3]

[1] Escuela de Ingeniería Química, Pontificia Universidad Católica de Valparaíso, Valparaíso 2340000, Chile
[2] Departamento de Ingeniería Metalurgica y Minas, Universidad Católica del Norte,
    Antofagasta 1240000, Chile; egalvez@ucn.cl
[3] Departamento de Ingeniería Química y Procesos de Minerales, Universidad de Antofagasta,
    Antofagasta 1240000, Chile; luis.cisternas@uantof.cl
* Correspondence: freddy.lucay@pucv.cl; Tel.: +56-322272-2611

Received: 31 January 2019; Accepted: 9 March 2019; Published: 15 March 2019

Abstract: The design of a flotation circuit based on optimization techniques requires a superstructure for representing a set of alternatives, a mathematical model for modeling the alternatives, and an optimization technique for solving the problem. The optimization techniques are classified into exact and approximate methods. The first has been widely used. However, the probability of finding an optimal solution decreases when the problem size increases. Genetic algorithms have been the approximate method used for designing flotation circuits when the studied problems were small. The Tabu-search algorithm (TSA) is an approximate method used for solving combinatorial optimization problems. This algorithm is an adaptive procedure that has the ability to employ many other methods. The TSA uses short-term memory to prevent the algorithm from being trapped in cycles. The TSA has many practical advantages but has not been used for designing flotation circuits. We propose using the TSA for solving the flotation circuit design problem. The TSA implemented in this work applies diversification and intensification strategies: diversification is used for exploring new regions, and intensification for exploring regions close to a good solution. Four cases were analyzed to demonstrate the applicability of the algorithm: different objective function, different mathematical models, and a benchmarking between TSA and Baron solver. The results indicate that the developed algorithm presents the ability to converge to a solution optimal or near optimal for a complex combination of requirements and constraints, whereas other methods do not. TSA and the Baron solver provide similar designs, but TSA is faster. We conclude that the developed TSA could be useful in the design of full-scale concentration circuits.

Keywords: design; flotation circuits; Tabu-search algorithm; multispecies

1. Introduction

Froth flotation is a process used in mining, based on the different surface properties of ore, for separating the valuable mineral from gangue. This process is performed in an aerated tank where the ore is mixed with water and reagents to render the valuable mineral hydrophobic [1]. Due to the complexity of the process in practice, multiple interconnected stages are used to form a flotation circuit.

The design of flotation circuits using optimization techniques has been widely studied in the literature [2,3]. The design requires three elements: (1) defining superstructures for representing a set of alternatives for design; (2) a mathematical model for modeling the different alternatives of the design, here considered goals and constraints, and determining the objectives to be optimized; and (3) an optimization technique for solving the problem.

The optimization techniques can be broadly classified into exact and approximate methods [4]. Exact methods obtain optimal solutions and guarantee their optimality. These methods use the analytical properties of the problem for generating a sequence of points converging to a global optimal solution [5]. This category includes methods such as the branch and bound algorithm, branch and cut algorithm, dynamic programming, Bayesian search algorithms, and successive approximation methods. Approximate methods are aimed at providing a good quality solution in a reasonable amount of time, but finding a global optimal solution is not guaranteed. Approximate methods can be classified into approximation algorithms and heuristic methods. The latter may be divided into two families: specific heuristics and metaheuristics. Specific heuristics are tailored and designed for solving a specific problem. Metaheuristics are general-purpose algorithms that can be applied to solve almost any optimization problem [4].

The term metaheuristics was introduced in 1986 by Glover [6], and is defined as an iterative generation process guiding a subordinate heuristic by combining intelligently different concepts for exploring and exploiting the search space; learning strategies are used to structure information in order to efficiently find the near-optimal solution [7]. Some of the proposed metaheuristics algorithms in the literature are: differential evolution [8], genetic algorithm (GA) [9], memetics algorithm [10], artificial immune system [11], simulated annealing [12], ant colony optimisation [13], particle swarm optimisation [14], and Tabu-search [15].

The Tabu-search algorithm (TSA) is a local search methodology used for solving combinatorial optimization problems [6]. The TSA is an adaptive procedure with the ability to apply other methods, such as linear and nonlinear programming algorithms [16]. The TSA uses the information gathered during the iterations to create a more efficient search process. TSA uses short-term memory to prevent previously visited solutions from being accepted. This memory prevents the algorithm from being trapped in cycles.

The literature shows that TSA has been used and compared to other optimization techniques in different disciplines. Han et al. [17] used TSA for training neural networks for wind power prediction. They reported that, compared with the backpropagation algorithm, TSA can improve prediction precision as well as convergence rate. Ting et al. [18] hybridized TSA and GA; this strategy was called the Tabu genetic algorithm (TGA). TGA integrates TSA into GA's selection. GA and TSA structures are not modified in these approaches. The classic traveling salesman problem was used for validating the proposed algorithm. Along this line, Soto et al. [19] hybridized TSA with multiple neighborhood searches for addressing multi-depot open vehicle routing. The proposed hybrid system provided good results, which was attributed to the successful exploration of neighborhoods. This allowed the search to achieve a good balance between intensification and diversification. Lin and Miller [20] applied TSA to chemical engineering problems. When TSA was compared with simulated annealing, TSA provided superior performance; this was attributed to the use of short-term memory, which enables it to escape from local optima [20]. Konak et al. [21] demonstrated that TSA is more efficient than GA because TSA does not require objective function gradient information. Kis [22] solved job-shop scheduling problems using TSA and GA. Kis reported that TSA was superior to GA both in terms of solution quality and computation time. Pan et al. [23] proposed a particle swarm optimization (PSO) algorithm for a no-wait flow shop scheduling problem. They compared PSO and TSA. The results indicated that the TSA and their hybrids generate better results than PSO. Mandami and Camarda [24] proposed a multi-objective optimization technique for plant design using TSA. They used a bounding technique for increasing the efficacy of the algorithm. They evaluated the effectiveness of the algorithm using a nonlinear model of 10 variables. The authors concluded that it is feasible to generate a Pareto-optimality curve for plant design problems using multi-objectives.

As stated by Cisternas et al. [2], the exact methods only guarantee finding the global solution when the design problem is small. This observation was corroborated by the works of Mehrotra and Kapur [25], Reuter et al. [26,27], Schena et al. [28,29], Mendez et al. [3], Maldonado et al. [30], Cisternas et al. [31,32], Calisaya et al. [33], and Acosta-Flores et al. [34]. Similar to exact methods, the probability of finding the global solution using approximate methods decreases as the problem size increases [5]. In this context, according to Acosta-Flores et al. [34] and Cisternas et al. [2], the genetic algorithms are the only metaheuristic algorithms that have been used for designing flotation circuits [1,35–38]. However, GA has been used only for small problems because its convergence is slow.

The methodologies proposed in the literature considered different assumptions for simplifying the design problem. For example, many of these methodologies considered a maximum of six flotation banks and a maximum of eight cells; however, these studies usually considered only two species in the fed ore to the flotation circuit. The exceptions are the works of Calisaya et al. [33] and Acosta-Flores et al. [34]. These authors examined several species in the fed ore, but their works were based on the fact that there are few structures that are optimal for a given problem [39]. These few structures were identified before applying the optimization search, which reduces the size of the optimization problem. Therefore, the studied problem is complex and difficult to solve when obtaining a global solution.

In this study, we propose using the TSA for designing flotation circuits. The developed algorithm incorporates diversification and intensification, the first of which is used for exploring new regions, and the second, for exploring regions close to a good solution. The algorithm was implemented for designing circuits that process several species. The superstructure implemented involves five flotation banks and each bank could use 3–15 cells, and the objective functions are economical. Four case studies were developed for illustrating the applicability of the algorithm.

2. Background

2.1. Superstructure

The proposed superstructures in the literature usually correspond to an equipment superstructure, which allows the streams of concentrate and tail of one equipment to be sent to any other equipment. The superstructures are important because they define the alternatives and the size of the problem [2]. According to Sepúlveda et al. [40], the circuits generally use between three and five flotation stages. Then, the equipment superstructure (Figure 1) used in this work considers the following stages: rougher stage (R), cleaner stage (C1), re-cleaner stage (C2), scavenger stage (S1), and re-scavenger stage (S2).

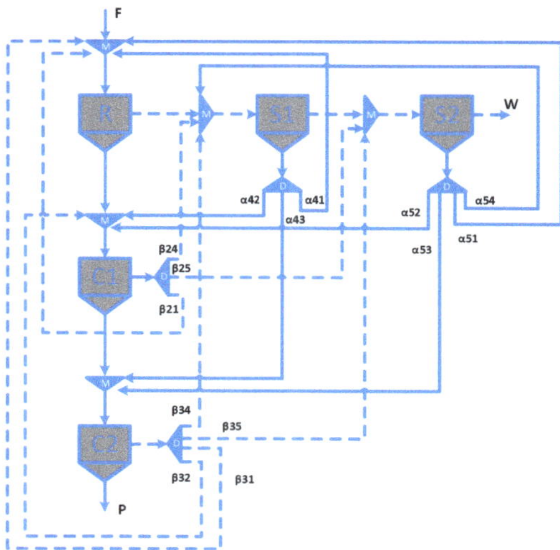

Figure 1. Equipment superstructure. R: rougher stage, C1: cleaner stage, C2: re-cleaner stage, S1: scavenger stage, S2: re-scavenger stage, M: stream mixer, D: stream splitter, $\alpha_{ij} \in \{0,1\}$ decision variables indicating the destination of the concentrate stream from stage i, $\beta_{ij} \in \{0,1\}$ decision variables indicating the destination of the tail stream from stage i.

Many of the proposed superstructures in the literature considered nonsensical alternatives and/or presented degeneracy, i.e., alternatives that are equivalent [2]. All the design alternatives shown in Figure 1 have sense and do not present degeneracy. In Figure 1, the triangles with label M represent mixers of the feed that arrive at stage i, where $i \in \{1,2,3,4,5\} = I$. Note that the numbers 1, 2, 3, 4, and 5 are related to $R, C_1, C_2, S_1,$ and S_2, respectively. The triangles with label D represent splitters that allow sending the concentrate and tail streams from stage i to the other stages of the superstructure.

2.2. Mathematical Model

The model includes mass balances in flotation stages, splitters, and mixers, and goals, constraints, and objective function are considered. The mass balance in flotation stages is determined with:

$$C_{ik} = R_{ik} \cdot F_{ik}, \ C_i = \sum_{k=1}^{m} C_{ik} \tag{1}$$

$$T_{ik} = (1 - R_{ik}) \cdot F_{ik}, \ T_i = \sum_{k=1}^{m} T_{ik} \tag{2}$$

where F_{ik} is the mass flow of the species k fed to the flotation stage i, C_{ik} is the mass flow of the species k in the concentrate stream C_i of the flotation stage i, T_{ik} is the mass flow of the species k in the tail stream T_i of the flotation stage i, and R_{ik} is the recovery of the species k in the flotation stage i, where $k = 1, 2, \ldots, m, i \in I$, and m is the number of species. The flotation model used for representing the recovery in the flotation stages was proposed by Yianatos and Henríquez [41]:

$$R_{ik} = R_{max,i,k} \cdot \left(1 - \frac{1 - (1 + k_{max,i,k} \cdot \tau_i)^{1-N_i}}{(N_i - 1) \cdot k_{max,i,k} \cdot \tau_i} \right) \tag{3}$$

where R_{ik} is the recovery of the species k in the flotation stage i, $k_{max,i,k}$ is the maximum rate constant of the species k in the flotation stage i, τ_i is the cell residence time in the flotation stage i, N_i is the number of flotation cells used in the flotation stage i, and $R_{max,i,k}$ is the maximum recovery at the infinite time of the species k in the flotation stage i.

In practice, the flotation circuits do not use stream branching because a large number of pumps and junction boxes are necessary for carrying out the concentration process [37]. Then, the mass balances in the splitters of the concentrate streams are expressed as:

$$C_{ik} = \sum_{j \in I} \alpha_{ij} \cdot C_{ijk}, \ \sum_{j \in I} \alpha_{ij} = 1, \ i \in I \tag{4}$$

where $\alpha_{ij} \in \{0,1\}$ are decision variables indicating the destination of the concentrate stream (Figure 1), C_{ijk} is the mass flow of species k in the concentrate stream from stage i to stage j, and C_{ik} is the mass flow of species k in the concentrate stream of stage i. Similarly, for the tail streams:

$$T_{ik} = \sum_{j \in I} \beta_{ij} \cdot T_{ijk}, \ \sum_{j \in I} \beta_{ij} = 1, \ i \in I \tag{5}$$

where $\beta_{ij} \in \{0,1\}$ are decision variables indicating the destination of the tail stream (Figure 1), T_{ijk} is the mass flow of species k in the tail stream from stage i to stage j, and T_{ik} is the mass flow of species k in the tail stream of stage i. The mass balances in mixers are expressed as:

$$F_{jk} = \begin{cases} M_{1k} + \sum_{i \in I} \alpha_{ij} \cdot C_{ijk} + \sum_{i \in I} \beta_{ij} \cdot T_{ijk}, & j = 1 \\[2ex] \sum_{i \in I} \alpha_{ij} \cdot C_{ijk} + \sum_{i \in I} \beta_{ij} \cdot T_{ijk}, & j \neq 1 \end{cases} \tag{6}$$

where M_{1k} is the mass flow of the species k fed to the flotation circuit. Note that mass balance for each species k processed in the circuit can be rewritten as a matrix system. These systems are solved numerically for C_{ik} and T_{ik} using a linear programming algorithm. The TSA has the ability to make use of this type of algorithm.

The market for copper concentrate establishes a minimum grade (g_{LO}); then, the following equation is included in the mathematical model:

$$gradeCu = \frac{\sum_k C_{5Pk} \cdot g_{k,cu}}{\sum_k C_{5Pk}} \geq g_{LO} \tag{7}$$

where $g_{k,cu}$ is the copper grade of the species k, and $gradeCu$ is the copper grade in the final concentrate of flotation circuit. In this work, the value of g_{LO} is 0.25.

Several objective functions have been used in the literature. Mehrotra and Kapur [25], Green [42], and Pirouzan et al. [38] implemented technical expressions for designing flotation circuits. The technical functions are difficult to define. For example, if the recovery is maximized, it is necessary to restrict the concentration grade to a value that is not known a priori. This difficulty was overcome by some authors using a multi-objective function; however, it is difficult to define the relative weight of each objective. Schena et al. [29] and Cisternas et al. [32,39,43] implemented economic expressions for designing circuits. These expressions highlight the maximization of revenues, maximization of profits, and the maximization of the net present worth. These last two expressions require estimating both the equipment costs and the operating costs. Cisternas et al. [43] found that the objective function has a significant effect on both the solution and circuit structure obtained. In this work, we used the maximization of revenues and maximization of the net present worth as the objective functions.

The revenue can be calculated using different models depending on the type of product and its market. In the case of copper concentrate, the net-smelter-return formula can be used [32,43]:

$$Revenue = \sum_k CF_k \left[p \left(\sum_k CF_k \cdot g_{k,cu} - \mu \right) (q - Rfc) - Trc \right] H \tag{8}$$

where p is the fraction of metal paid, μ is the grade deduction, Trc is the treatment charge, q is the metal price, Rfc is the refinery charge, CF_k is the mass flow of the species k in the final concentrate, and $g_{k,cu}$ is the copper grade of the species k.

For determining net present worth, the capital costs and total costs of the process must first be estimated. The capital cost considers the fixed capital and working capital. The first is estimated with the following equation:

$$I_F = FL \cdot \sum_{i \in I} I_{F,i} \cdot N_i \tag{9}$$

where [44]:

$$I_F(i) = 105.7 + 10.72 \cdot V_i - 149.1 V_i^2 \tag{10}$$

with

$$V_i = \frac{F_i \cdot \tau_i \cdot E_g}{\rho_p} \tag{11}$$

where V_i is the cell volume (m^3) of flotation stage i, F_i is the feed stream to stage i, E_g is the gas factor, ρ_p is the pulp density, $I_{F,i}$ is the fixed capital cost to stage i, I_F is the fixed capital cost of circuit, and FL is the Lang factor [45]. Equation (10) is valid for volume between 5 m^3 and 200 m^3. The working capital costs are estimated with the following equation:

$$I_w = FL_w \sum_{i \in I} I_{F,i} \cdot N_i \tag{12}$$

where FL_w is the Lang factor for working capital and is assumed to be 0.9. The total costs of the process are estimated with the following equation:

$$Total\ costs = \sum_{i \in I} C_{op,i} + MCM \sum_{k \in K} F_k \tag{13}$$

with

$$C_{op,i} = H \cdot N_i \cdot V_i \cdot P_k \cdot E_g \tag{14}$$

where $C_{op,i}$ is the operating cost of flotation stage i, P_k is the kilowatt-hours cost, E_g is the gas factor, and MCM are the costs of mine-crushing-grinding per ton of fed ore to the flotation plant. The profits generated by the project are estimated with:

$$Profits(P_B) = Revenue - total\ costs - D \tag{15}$$

The annual cash flows are estimated using the following expression:

$$F_c = P_A + D \tag{16}$$

with

$$P_A = (1 - r_t)P_B \tag{17}$$

$$D = \frac{I_F}{n} \tag{18}$$

where P_B are the profits before taxes, r_t is the tax rate, D is the annual depreciation, and n is the life time of the project (the value of n used here is 15). The net present worth is calculated using:

$$W_{NP} = -I_{cap} + \sum_{i=1}^{n} \frac{F_{c,i}}{(1 + r_d)^i} \tag{19}$$

where W_{NP} is the net present worth, $I_{cap} = I_F + I_w$, and r_d is the discount rate. It is assumed that cash flows are equal in all years of the project, so Equation (19) is simplified:

$$W_{NP} = -I_{cap} + F_C \frac{(1 + r_d)^n - 1}{r_d(1 + r_d)^n} - \gamma \tag{20}$$

with

$$\gamma = \sum_i v_i \tag{21}$$

where γ denotes the penalty parameter [46]. For each violation of the constraints of the mathematical model, a value $v_i > 0$ is considered, defined by the user. This penalty parameter must also be considered when the objective function is the maximization of revenues.

2.3. Optimization Technique: Tabu-Search Algorithm

TSA is a local search methodology that was proposed by Glover and Laguna in 1998 [15]. TSA is a strategy for solving combinatorial optimization problems ranging from graph theory to mixed integer programming problems. It is an adaptive procedure with the ability to making use of other methods, such as linear programming algorithms, which help to overcome the limitations of local optimality [16]. Gogna and Tayal [47] stated that the TSA uses the information gathered during the iterations to produce a more efficient search process. Here, the search space is simply the space of all possible solutions that can be considered during the search. For example, in vehicle routing problems, the search space considers both binary and continuous variables [48]. The TSA accepts non-improving solutions to the global solution to move out of local optima. The distinguishing feature of the TSA is the use of memory structures.

The main memory structure used by the TSA is the Tabu list (TL) short-term memory, which has a record of previously visited solutions. This key idea can be linked to artificial intelligence concepts [48]. The TL should be carefully formulated for an effective search while minimizing the computation time and the memory requirements. Medium- and long-term memories can be used for improving the intensification and diversification of the TSA [4].

Usually, the TSA starts with an initial solution, randomly selected, which is entered to TL. Then the algorithm uses a local search procedure or neighborhoods to move iteratively from a potential solution x to an improved solution x', also called the best neighbor, in the neighborhood of x ($N(x)$). Local search procedures often become stuck in local optima. In order to avoid these pitfalls and to explore regions of search space that would be left unexplored by other local search procedures, TSA carefully explores the neighborhood of each solution as the search progresses [47]. The solutions admitted to the new neighborhood are determined through the use of memory structures. The best neighbor (x_{best}) of $N(x)$ is accepted as a global solution if $x_{best} \notin TL$ and if it maximizes the objective function. If the best neighbor $x_{best} \in TL$, then the next best neighbor of $N(x)$ is the new postulant to enter into TL. This procedure is repeated until x_{best} is entered into T. Next, the new neighborhood of the best neighbor is generated, and the procedure described earlier is repeated until some stopping criterion is satisfied [49].

The search space of the design problem studied in this work considers binary, discrete, and continuous variables, which are related to circuit structure, the number of cells in flotation stages, and the operating conditions in flotation stages, respectively. The developed TSA implements short-term and long-term memories: the TL and the frequency matrix (FM), respectively. The latter was proposed by De los Cobos [50], which allows the exploration of new regions of the search space.

In our case, the algorithm starts with an initial solution ($x = ((\alpha_{ij}, \beta_{ij})_{i,j}, (N_t, \tau_t)_t)$), which is entered into TL. Here, the variables $(\alpha_{ij}, \beta_{ij})_{i,j}$ are associated with the circuit structure, and $(N_t, \tau_t)_t$ are associated with number of equipment and operating conditions of circuit. Then, the neighborhood $N(x)$ of the initial solution x is generated to create natural permutations of the structural variables, i.e., a set of structures is generated. Subsequently, the operating conditions and equipment number are assigned to each structure through a uniform distribution function. The uniform distribution function is defined using the operating conditions and the equipment number of the initial solution. This method of generating neighborhoods is based on the structure being more influential on the objective function than the operating conditions and equipment number [43]. Subsequently, the equipment size, copper grade of the concentrate, and profitability parameters of each flotation circuit, neighbor of $N(x)$, are determined via mass balances. If the constraints of the mathematical model are violated for some neighbor of $N(x)$, then its objective function (f) is penalized (γ). Then, the best neighbor (x_{best}) of $N(x)$ is determined, i.e., the neighbor maximizing the objective function. The best solution (x_{best}) is accepted as a global solution if $x_{best} \notin TL$ and $f\left(x_{global}\right) < f(x_{best})$. If $x_{best} \in TL$, the next best neighbor of $N(x)$ is taken as x_{best}. This step is repeated until x_{best} is entered to TL. The structural variables of x_{best} are entered into FM. Subsequently, the neighborhood of x_{best} is generated and the search procedure described earlier is repeated $Iteration_{max}$ times. Notably, TL has a determined number of rows (n_r), i.e., once TL is full, the update of its information is carried out at each iteration of the algorithm. Next, we explain how diversification and intensification strategies are included in the search procedure. Each time that x_{global} does not improve before D_{max} iterations of algorithm, then the diversification in the search procedure is incorporated. The diversification allows the generation of a new best neighbor, whose structural variables are obtained from the gathered information in FM, and operating conditions and equipment number are assigned through a uniform distribution function. Note that FM records all the structural variables of x_{best} from the first iteration of algorithm. This information allows us to determine the structures more often visited (high frequency in FM) by the algorithm. A structure not visited, or rarely visited (low frequency in FM), is assigned to the new best neighbor. The intensification is incorporated in the search procedure each time passing I_{max} iterations of the algorithm. The intensification allows the exploration of regions close to a good solution. In this work, this is carried out using the gathered information in TL. The new best neighbor is aleatorily selected among the better neighbors recorded in TL. The full procedure is shown in Figure 2.

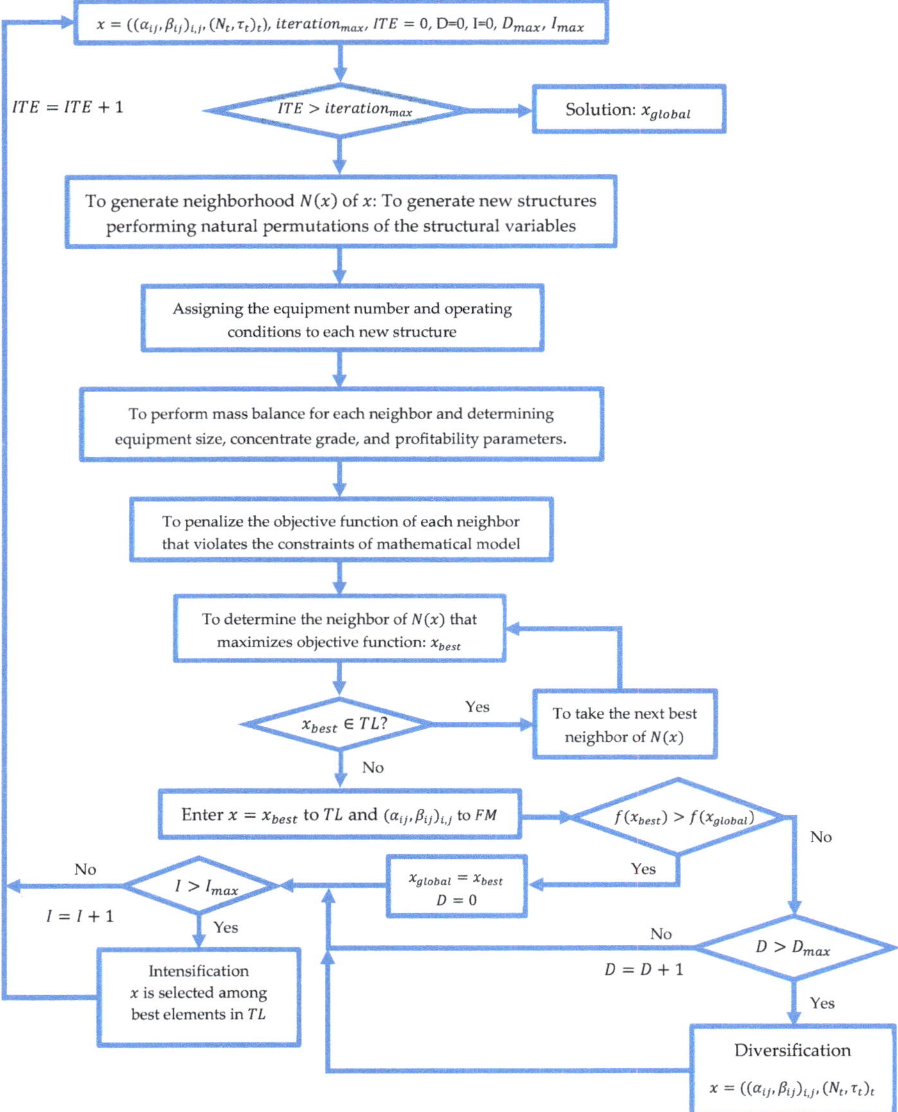

Figure 2. Block diagram of Tabu-search algorithm. ITE: iteration of algorithm, TL: Tabu list, FM: frequency matrix, D: iteration related to diversification, I: iteration related to intensification, $\left(\alpha_{ij},\ \beta_{ij}\right)_{i,j}$: circuit structure, $(N_t,\ \tau_t)_t$: number of equipment and operating conditions of circuit, $N(x)$: neighborhood of the solution x, x_{best}: best neighbor of $N(x)$, f: objective function, D_{max} : number of iteration related to the implementation of diversification, I_{max} : number of iteration related to the implementation of intensification.

3. Applications

Four case studies were developed for illustrating the proposed algorithm. The first and second cases involve designing a copper ore concentrator plant considering the maximization of revenues and maximization of the net present worth as the objective function, respectively. The third case analyses a benchmarking between the Tabu-search algorithm and the Baron solver. Finally, the fourth case provides the comparison between our approach and the methodology proposed by Acosta-Flores et al. [34].

The feed is composed of seven species: $k = 1$ (chalcopyrite fast), $k = 2$ (chalcopyrite slow), $k = 3$ (chalcocite fast), $k = 4$ (chalcocite slow), $k = 5$ (pyrite), $k = 6$ (silica), and $k = 7$ (gangue). The mass flows of the fed species are shown in Table 1. The values of the constants in Equation (3) are shown in Table 2. The values for the constants in Equation (8) are: $p = 0.975$, $\mu = 0.015$, $Trc = 300$ US\$/ton, $q = 4000$ US\$/ton, $Rfc = 200$ US\$/ton, and $H = 7200$ h/year. The TSA could assign between 3 and 15 cells, and between three and five minutes of residence time to flotation stages.

Table 1. Species in feed to process.

Species	Copper Grade wt %	Feed (t/h)
Chalcopyrite fast (Cpf)	0.35	15
Chalcopyrite slow (Cpy)	0.25	8
Chalcocite fast (Cf)	0.1	5
Chalcocite slow (Cs)	0.07	3
Pyrite (P)	0.0	4
Silica (S)	0.0	200
Gangue (G)	0.0	300

Table 2. Values of $k_{max,i,k}$ and $R_{max,i,k}$ for each stage and species in Equation (3).

	$k_{max,i,k}$							$R_{max,i,k}$						
Stage\Species	Cpf	Cpy	Cf	Cs	P	S	G	Cpg	Cpy	Cf	Cs	P	S	F
R	1.85	1.50	1.00	0.70	0.80	0.60	0.30	0.90	0.85	0.85	0.75	0.80	0.60	0.20
C1	1.30	1.00	0.80	0.40	0.70	0.30	0.20	0.75	0.70	0.70	0.60	0.60	0.50	0.15
C2	1.30	1.00	0.80	0.40	0.70	0.30	0.20	0.70	0.65	0.65	0.50	0.60	0.50	0.15
S1	1.85	1.50	1.00	0.70	0.80	0.60	0.30	0.90	0.85	0.85	0.75	0.80	0.60	0.20
S2	1.85	1.50	1.00	0.70	0.80	0.60	0.30	0.90	0.85	0.85	0.75	0.80	0.60	0.20

3.1. Maximization of Revenues

The equipment superstructure, the mathematical model, included maximization of revenues as the objective function, and the TSA were used for solving the design problem. The developed algorithm was capable of solving the design problem despite the number of species processed, and the requirements and constraints established in the design procedure. The obtained structure is shown in Figure 3, and the revenue, the net present worth, profit before taxes, total capital investment, and total costs of the circuit were USD \$130,960,079/year, USD \$595,475,455, USD \$91,716,855/year, USD \$53,265,087, and USD \$36,358,032/year, respectively. The τ_R, τ_{C1}, τ_{C2}, τ_{S1}, τ_{S2}, N_R, N_{C1}, N_{C2}, N_{S1}, N_{S2}, V_R, V_{C1}, V_{C2}, V_{S1}, and V_{S2} were 5 min, 3 min, 3 min, 5 min, 5 min, 15 cells, 3 cells, 3 cells, 15 cells, 15 cells, 197.60 m^3, 22.63 m^3, 9.98 m^3, 167.59 m^3, and 155.91 m^3, respectively. The final concentrate of the circuit was 14.962 ton/h of chalcopyrite fast, 7.916 ton/h of chalcopyrite slow, 4.938 ton/h of chalcocite fast, 2.633 ton/h of chalcocite slow, 0.008 ton/h of pyrite, 0.017 ton/h of silica, and 0.001 ton/hr of gangue, and its copper grade was 25.70%.

The number of cells and residence time in rougher, scavenger, and re-scavenger are the maxima available. These results are logical because the metallurgical aim of these stages is increasing the recovery of the valuable ore. Also, the objective function does not consider either the equipment costs or the operating costs. The number of cells and residence time in the cleaner and recleaner stages are the minima available. Again, these results are logical because the aim of these stages is increasing the copper grade in the concentrate. These results were obtained by previously evaluating different combinations of the algorithm parameters. In this case, the TSA used 170 neighbors in each neighborhood, 2000 iterations, diversification was applied each time the global solution did not improve after 30 iterations, the intensification was applied each time passing 50 iterations of the algorithm, and the number of rows of the TL was 50. The runtime of the TSA was 601.63 s.

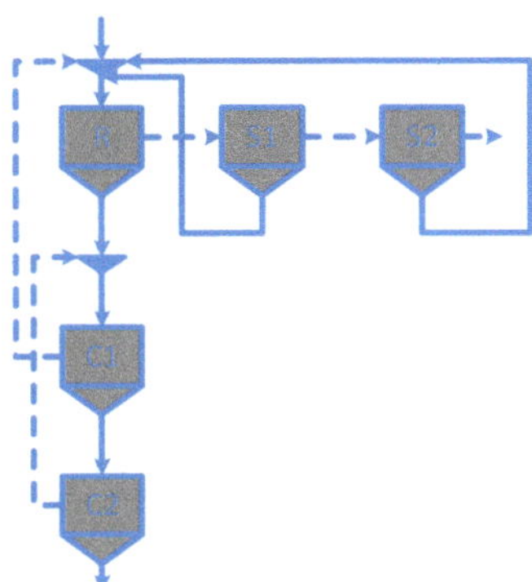

Figure 3. Structure obtained by maximizing the revenues.

3.2. Maximization of the Net Present Worth

The equipment superstructure and mathematical model included maximization of the net present worth as the objective function, and the TSA was used for solving the design problem. Again, the developed algorithm was capable of solving the design problem despite the number of species processed, and the requirements and constraints established in the design procedure. The obtained structure is shown in Figure 4, and the net present worth, revenue, profit before taxes, total capital investment, and total costs of the circuit were USD \$724,828,320, USD \$129,830,340/year, USD \$107,977,400/year, USD \$ 8,386,189, and USD \$21,398,690/year, respectively. The values of τ_R, τ_{C1}, τ_{C2}, τ_{S1}, τ_{S2}, N_R, N_{C1}, N_{C2}, N_{S1}, N_{S2}, V_R, V_{C1}, V_{C2}, V_{S1}, and V_{S2} were 3 min, 3 min, 3 min, 3 min, 3 min, 3 cells, 4 cells, 3 cells, 3 cells, 3 cells, 106.84 m^3, 19.84 m^3, 9.60 m^3, 95.72 m^3, and 91.41 m^3, respectively. The final concentrate of circuit contained 14.846 ton/h of chalcopyrite fast, 7.757 ton/h of chalcopyrite slow, 4.739 ton/h of chalcocite fast, 2.164 ton/h of chalcocite slow, 0.009 ton/h of pyrite, 0.019 ton/h of silica, and 0.001 ton/h of gangue, with a copper grade of 26.07%.

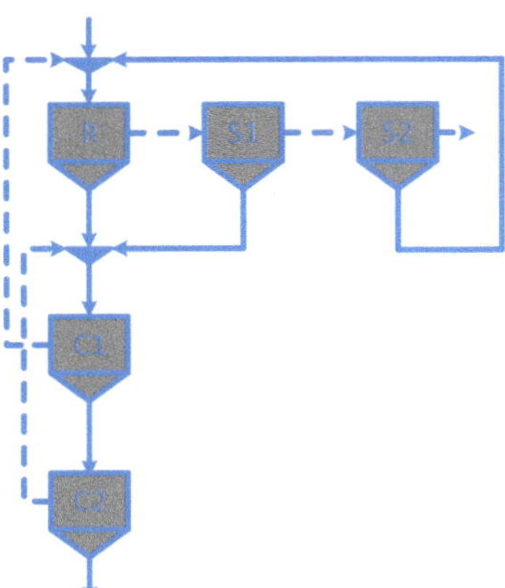

Figure 4. Structure obtained by maximizing the net present worth.

The number of cells and residence time used in the circuit and shown in Figure 4 are lower compared with the obtained in the case outlined in Section 3.1. These results are logical because the net present worth considers both the equipment costs and the operating costs. The obtained designs in the cases in Sections 3.1 and 3.2 are different, corroborating the results reported by Cisternas et al. [43], i.e., the objective function affects the design of concentration plants. The TSA used 170 neighbors at each neighborhood, 2000 iterations, the diversification was applied each time that the global solution was not improved after 30 iterations, the intensification was applied each time passing 50 iterations of the algorithm, and the number of rows of the TL was 50. The runtime of the algorithm was 527.61 s.

3.3. Benchmarking between the Tabu-Search Algorithm and the Baron Solver

Many authors have used the Baron solver to solve design problems. As such, we completed benchmarking between the TSA and the Baron solver. The mathematical model for the Baron solver appears in Appendix A. Note that the Baron solver is based on the branch and cut algorithm, i.e., belongs to the family of exact methods.

Initially, we completed benchmarking using the maximization of the revenues as the objective function; the results are shown in Table 3 and Figures 3–5. Then, we performed benchmarking using the maximization of the net present worth as the objective function, and the results are shown in Table 4 and Figures 3–5. Many authors only used the revenues for designing the circuit, i.e., they included neither equipment design nor operational costs in the mathematical model [1,31,32,34,51]. Thus, we performed benchmarking using a simplified mathematical model, and the results are shown in Table 5 and Figures 3–5. Note that the superstructure of Figure 1 represents a total of 144 flotation circuit configurations and that only 11 configurations were selected in all the analyzed cases. These structures correspond to 4 for the case maximization of the revenues as the objective function, 4 for the case the maximization of the net present worth as the objective, and 3 when a simplified mathematical model is used. The structures of Figures 3 and 4 were the most frequently selected circuits.

Table 3 shows the benchmarking between the TSA and the Baron solver when the maximization of revenues was the objective function. This table depicts five cases. The first considered the species Cpf, Cps, S, and G; the second considered the species Cpf, Cps, P, S, and G; the third considered the species Cpf, Cps, Cf, P, S, and G; the fourth considered the species Cpf, Cps, Cf, Cs, P, S, and G; and the fifth considered the species Cpf, Cps, Cf, Cs, Bof (bornite fast), P, S, and G. In each case, the following output variables were considered: the revenues, net present worth, profit before taxes, total capital investment, total annual cost, equipment size, operating conditions, number of equipment, copper grade in concentrate, circuit structure, runtime of algorithm, number of iterations of TSA, neighborhood size, number of iterations of each diversification, number of iterations of each intensification, and number of rows of *TL*. Table 3 shows that TSA is capable to converge independently of the number of species processed in the flotation circuit. The results provided by the TSA and the Baron solver were similar in all analysed cases, except in the fifth case. In this last case, both optimization techniques provided similar revenue; however, the other output variables were different. This result could be related to the simultaneous imposition of all goals and constraints of the mathematical model. Perhaps a relaxation of constraints could help with the exploration of regions, offering a better quality solution. The runtime of TLA in the first, second, third, fourth, and fifth cases was 21.3, 183.4, 287.7, 344.7, and 386.2 s, respectively. The runtime for the Baron solver in the first, second, third, fourth, and fifth cases was 233.0, 423.0, 9026.2, 14,640.0, and 10,257.0 s, respectively. The TSA provided results faster than the Baron solver, but only provides a good quality solution.

Table 3. Benchmarking between the Tabu-search algorithm and the Baron Solver (revenues, Bof = bornite fast).

Case	1		2		3		4		5	
Algorithm	Tabu	Baron	Tabu	Baron	Tabu	Baron	Tabu	Baron	Tabu search	Baron
Species	Cpf, Cps, S, G		Cpf, Cps, P, S, G		Cpf, Cps, Cf, P, S, G		Cpf, Cps, Cf, Cs, P, S, G		Cpf, Cps, Cf, Cs, Bof, P, S, G	
Revenue, USD/year	132,318,860	132,323,459	132,852,410	132,854,057	130,981,546	130,981,549	130,960,079	130,958,748	129,185,242	130,335,529
Net present worth, USD	612,961,580	612,702,132	618,316,980	617,392,845	598,070,573	598,081,544	595,475,455	599,377,620	671,113,149	692,374,613
Profit before taxes, USD/year	94,235,451	94,208,659	94,972,404	94,849,038	92,078,958	92,080,433	91,716,855	92,213,521	101,068,398	103,931,519
Total capital investment, USD	52,137,292	52,317,036	51,261,610	51,471,333	52,917,663	52,915,338	53,265,087	52,005,813	24,608,590	20,166,465
Total annual cost, USD/year	35,259,307	35,280,960	35,103,337	35,216,989	36,036,215	36,034,869	36,358,032	35,928,246	26,783,878	25,311,660
V_R, m^3	182.822	182.54	181.825	185.596	195.500	195.46	197.704	197.60	119.350	111.037
V_{C1}, m^3	21.222	20.31	22.805	22.872	22.442	22.44	22.712	22.63	25.841	20.083
V_{C2}, m^3	7.761	7.15	8.973	8.800	9.964	9.96	9.984	9.98	14.138	10.287
V_{S1}, m^3	163.211	163.17	163.697	163.700	165.782	165.78	167.565	167.59	159.679	101.602
V_{S2}, m^3	153.893	153.88	153.971	153.972	154.464	154.46	155.908	155.91	151.817	155.590
τ_R, min	5.000	5.000	4.900	5.000	5.000	5.000	5.000	5.000	3.300	3.000
τ_{C1}, min	3.130	3.000	3.000	3.000	3.000	3.000	3.000	3.000	4.300	3.000
τ_{C2}, min	3.320	3.000	3.070	3.000	3.000	3.000	3.000	3.000	4.400	3.000
τ_{S1}, min	5.000	5.000	5.000	5.000	5.000	5.000	5.000	5.000	4.400	3.000
τ_{S2}, min	5.000	5.000	5.000	5.000	5.000	5.000	5.000	5.000	4.800	4.920
N_R	15.000	15.000	14.000	14.000	15.000	15.000	15.000	14.000	3.000	3.000
N_{C1}	4.000	5.000	5.000	5.000	3.000	3.000	3.000	3.000	3.000	3.000
N_{C2}	3.000	3.000	3.000	3.000	3.000	3.000	3.000	3.000	3.000	3.000
N_{S1}	15.000	15.000	15.000	15.000	15.000	15.000	15.000	15.000	9.000	3.000
N_{S2}	15.000	15.000	15.000	15.000	15.000	15.000	15.000	15.000	10.000	13.000
Grade Cu	0.3123	0.312	0.274	0.274	0.257	0.257	0.257	0.257	0.250	0.250
Circuit structure	Figure 4	Figure 4	Figure 4	Figure 4	Figure 3	Figure 3	Figure 3	Figure 3	Figure 5d	Figure 5c
Time, s	46.610	233.000	176.580	423.030	287.670	9026.190	601.630	14,640.00	433.930	10,257.020
Iterations of algorithm	1000	-	1000	-	1500	-	2000	-	3000	-
Neighborhood size	40	-	130	-	150	-	170	-	60	-
Iterations of diversification	50	-	20	-	30	-	30	-	20	-
Iteration of intensification	10	-	40	-	40	-	50	-	50	-
No. rows of Tabu list	100	-	50	-	50	-	50	-	50	-

Table 4. Benchmarking between the Tabu-search algorithm and the Baron solver (net present worth; Bof = bornite fast, Bos = bornite slow).

Case	1		2		3		4		5		6	
Algorithm	Tabu	Baron	Tabu	Baron	Tabu	Baron	Tabu	Baron	Tabu	Baron	Tabu	Baron
Species	Cpf, Cps, S, G		Cpf, Cps, P, S, G		Cpf, Cps, Cf, P, S, G		Cpf, Cps, Cf, Cs, P, S, G		Cpf, Cps, Cf, Cs, Bof, P, S, G		Cpf, Cps, Cf, Cs, Bof, Bos, P, S, G	
Net present worth, USD	735,935,440	735,938,754	737,697,720	737,697,912	726,139,400	726,139,411	724,828,320	724,828,404	723,842,180	723,842,865	719,710,450	719,712,493
Revenue, USD/year	130,918,800	130,920,451	131,435,920	131,433,456	129,853,889	129,853,889	129,830,340	129,830,357	129,839,150	129,840,240	129,548,420	129,551,233
Profit before taxes, USD/year	109,609,160	109,606,145	109,882,230	109,881,842	108,168,025	108,168,032	107,977,400	107,977,414	107,833,120	107,833,404	107,267,450	107,268,177
Total capital inv., USD	8,162,377	8,108,261	8,345,130	8,338,775	8,329,184	8,329,184	8,386,189	8,386,191	8,415,244	8,418,023	9,135,362	9,141,936
Total annual cost, USD/year	20,867,504	20,875,109	21,101,667	21,099,930	21,234,694	21,234,694	21,398,690	21,398,690	21,550,197	21,550,859	21,786,145	21,787,868
V_R, m^3	101.244	101.240	103.058	103.070	105.819	105.820	106.847	106.850	111.845	111.807	111.390	111.325
V_{C1}, m^3	15.985	17.190	18.905	18.700	19.595	19.600	19.847	19.840	25.510	25.520	21.292	21.620
V_{C2}, m^3	6.619	8.390	8.112	8.110	9.578	9.580	9.600	9.600	10.266	10.412	10.252	10.260
V_{S1}, m^3	93.280	93.280	93.795	93.800	94.817	94.820	95.722	95.720	97.236	97.225	99.287	99.265
V_{S2}, m^3	90.073	90.070	90.219	90.220	90.615	90.620	91.412	91.410	91.901	91.898	92.295	92.322
τ_R, min	3.000	3.000	3.000	3.000	3.000	3.000	3.000	3.000	3.000	3.000	3.000	3.000
τ_{C1}, min	3.000	3.200	3.080	3.040	3.000	3.000	3.000	3.000	3.960	3.973	3.200	3.256
τ_{C2}, min	3.000	3.730	3.000	3.000	3.000	3.000	3.000	3.000	3.000	3.051	3.000	3.000
τ_{S1}, min	3.000	3.000	3.000	3.000	3.000	3.000	3.000	3.000	3.000	3.000	3.000	3.000
τ_{S2}, min	3.000	3.000	3.000	3.000	3.000	3.000	3.000	3.000	3.000	3.000	3.000	3.001
N_R	3.000	3.000	3.000	3.000	3.000	3.000	3.000	3.000	3.000	3.000	3.000	3.000
N_{C1}	4.000	4.000	4.000	4.000	4.000	4.000	4.000	4.000	3.000	3.000	3.000	3.000
N_{C2}	4.000	3.000	4.000	4.000	3.000	3.000	3.000	3.000	3.000	3.000	3.000	3.000
N_{S1}	3.000	3.000	3.000	3.000	3.000	3.000	3.000	3.000	3.000	3.000	4.000	4.000
N_{S2}	3.000	3.000	3.000	3.000	3.000	3.000	3.000	3.000	3.000	3.000	3.000	3.000
Grade Cu	0.3125	0.3130	0.276	0.276	0.2608	0.261	0.2607	0.261	0.250	0.250	0.250	0.250
Circuit structure	Figure 5a	Figure 5a	Figure 5a	Figure 5a	Figure 4	Figure 4	Figure 4	Figure 4	Figure 3	Figure 3	Figure 5c	Figure 5c
Time, s	106.260	94.200	424.720	294.920	464.990	533.400	527.610	355.200	710.040	1082.640	875.980	4299.860
No. rows Tabu list	50	-	50	-	50	-	50	-	50	-	50	-
No. iterations of algorithm	2000	-	2000	-	2000	-	2000	-	2000	-	2000	-
Iteration of diversification	30	-	30	-	30	-	30	-	30	-	30	-
Iteration of intensification	50	-	40	-	40	-	50	-	50	-	50	-
Neighborhood size	70	-	130	-	150	-	170	-	190	-	210	-

Table 5. Benchmarking between the Tabu algorithm and the Baron solver (revenues; reduced mathematical model; Pf = pyrite fast, Ps = pyrite slow).

Case	1		2		3		4		5		6	
Algorithm	**Tabu**	**Baron**	**Tabu**	**Baron**	**Tabu**	**Baron**	**Tabu**	**Baron**	**Tabu**	**Baron**	**Tabu**	**Baron**
Species	**Cpf, S, G**		**Cpf, Cps, S, G**		**Cpf, Cps, Pf, S, G**		**Cpf, Cps, Pf, Ps, S, G**		**Cpf, Cps, Cf, Pf, Ps, S, G**		**Cpf, Cps, Cf, Cs, Pf, Ps, S, G**	
Revenue, USD/year	26,455,569	26,455,571	38,755,316	did not converge after 5 days	38,568,418	did not converge after 5 days	38,567,734	did not converge after 5 days	36,714,037	did not converge after 5 days	There is no solution	There is no solution
τ_R, min	5.000	5.000	5.000	-	5.000	-	5000	-	3.260	-	-	-
τ_{C1}, min	3.000	3.000	3.090	-	3.000	-	3000	-	3.000	-	-	-
τ_{C2}, min	3.000	3.000	3.000	-	3.000	-	3000	-	3.000	-	-	-
τ_{S1}, min	5.000	5.000	5.000	-	5.000	-	5000	-	5.000	-	-	-
τ_{S2}, min	5.000	5.000	5.000	-	5.000	-	5000	-	3.800	-	-	-
N_R	15.000	15.000	15.000	-	15.000	-	15,000	-	5.000	-	-	-
N_{C1}	5.000	5.000	3.000	-	4.000	-	4000	-	3.000	-	-	-
N_{C2}	3.000	3.000	3.000	-	3.000	-	3000	-	3.000	-	-	-
N_{S1}	15.000	15.000	15.000	-	15.000	-	15,000	-	6.000	-	-	-
N_{S2}	15.000	15.000	15.000	-	15.000	-	15,000	-	6.000	-	-	-
Grade Cu	0.345	0.345	0.304	-	0.303	-	0.303	-	0.250	-	-	-
Circuit structure	Figure 3	Figure 3	Figure 3	-	Figure 4	-	Figure 4	-	Figure 5i	-	-	-
Time, s	54.900	1961.830	120.040	-	165.340	-	216.390	-	472.720	-	-	-
No. rows Tabu list	100	-	50	-	50	-	100	-	50	-	-	-
No. iterations of algorithm	1000	-	1000	-	1000	-	1000	-	1000	-	-	-
Iteration of diversification	30	-	20	-	20	-	20	-	20	-	-	-
Iteration of intensification	100	-	40	-	40	-	40	-	40	-	-	-
Neighborhood size	90	-	110	-	130	-	150	-	170	-	-	-

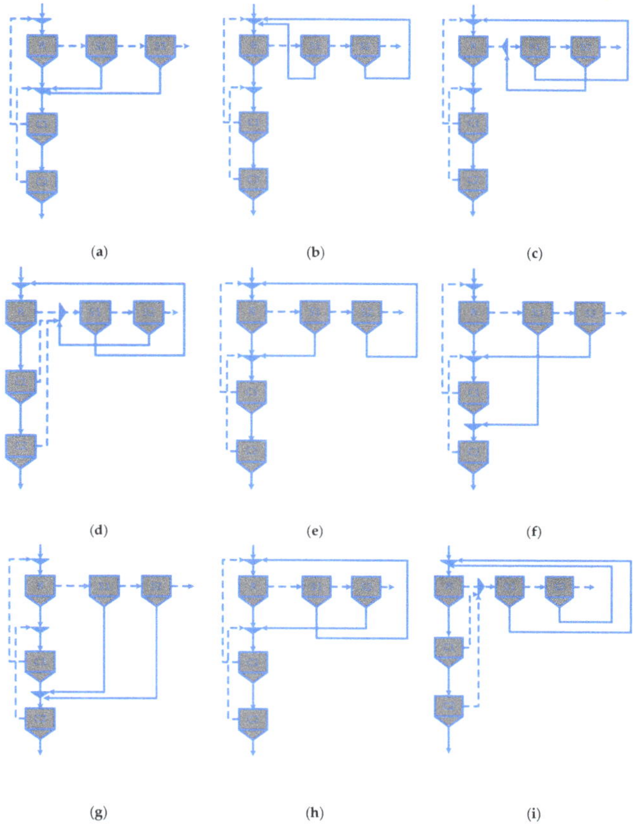

Figure 5. Structures of circuits obtained by maximization of net present worth (**a,c**) or revenues (**c,d,i**) in the case 3.3. Structures (**a,b,e–h**) are obtained by maximization revenues in the case 3.4.

Table 4 shows the benchmarking between TSA and the Baron solver when the maximization of the net present worth was the objective function. This table demonstrates six cases. The first considered the species Cpf, Cps, S, and G; the second considered the species Cpf, Cps, P, S, and G; the third considered the species Cpf, Cps, Cf, P, S, and G; the fourth considered the species Cpf, Cps, Cf, Cs, P, S, and G; the fifth considered the species Cpf, Cps, Cf, Cs, Bof, P, S, and G; and the sixth considered the species Cpf, Cps, Cf, Cs, Bof, Bos (bornite slow), P, S, and G. In each case, the following output variables were considered: net present worth, revenues, profit before taxes, total capital investment, total annual cost, equipment size, operating conditions, number of equipment, copper grade in concentrate, circuit structure, runtime of algorithm, number of iterations of TSA, neighborhood size, number of iterations of each diversification, number of iterations of each intensification, and number of rows of TL. Table 4 shows that TSA is capable to converge independently of the number of species processed in the flotation circuit and objective function used. The results provided by both optimization techniques were similar in all analysed cases. The runtime of TLA in the first, second, third, fourth, fifth, and sixth cases was 106.260 secs, 424.72, 464.99, 527.61, 710.04, and 875.98 s, respectively. The runtime for the Baron solver in the first, second, third, fourth, fifth, and sixth cases was 94.2, 294.92, 533.40, 355.20, 1082.64, and 4299.86 s, respectively. In the first, second, and fourth cases, the Baron solver provided results before the TSA. In the third, fifth, and sixth cases, the TSA provided results before the Baron solver.

Table 5 shows the benchmarking between the TSA and the Baron solver when the mathematical model is simplified. This table demonstrates six cases. The first considered the species Cpf, S and G; the second considered the species Cpf, Cps, S, and G; the third considered the species Cpf, Cps, Pf (pyrite fast), S, and G; the fourth considered the species Cpf, Cps, Pf, Ps (pyrite slow), S, and G; the fifth considered the species Cpf, Cps, Cf, Pf, Ps, S, and G; and the sixth considered the species Cpf, Cps, Cf, Cs, Pf, Ps, S, and G. In each case, are the following output variables were considered: revenue, operating conditions, number of equipment, copper grade in concentrate, circuit structure, runtime of algorithm, number of iterations of TSA, neighborhood size, number of iterations of each diversification, number of iterations of each intensification, and number of rows of TL. The runtime of TLA in the first, second, third, fourth, and fifth cases was 162.36, 344.06, 368.83, 480.36, and 1098.57 s, respectively. The runtime for the Baron solver in the first case was 1961.8 s, and in the second to fifth cases, the Baron solver did not converge after five days. Both optimization techniques do not provide a solution for the sixth case because the minimal copper grade constraint in the final concentrate cannot be satisfied. Note, the results provided by the TSA could be used for reducing the runtime of the Baron solver. When we delimited the search space of the Baron solver in the fifth case in Table 5, based on the TSA results, runtime of the Baron solver was 101.22 s, and the output variables of mathematical model were similar to those provided by the TSA.

3.4. Comparison with Another Approach

Section 3.3 showed that the Baron solver could converge after a long time period depending on the mathematical model and number of species used in the design procedure. Maybe, due to these results, Acosta-Flores et al. [34] proposed a design methodology based on two phases. In the first phase, they identified a set of optimal structures using discrete values for the flotation stages, then, in the second phase, they determined the optimal design of each structure obtained in the previous phase. Note that they used a superstructure of six flotation stages. However, the optimal structures only used five stages (R, C1, C2, S1, and S2). These authors modeled the flotation stages using the expressions proposed by Yianatos et al. [41]. When all parameters used by Acosta-Flores et al. [34] were included in the proposed methodology, the design shown in Table 6 was obtained. Considering that versions of the Baron solver in the General Algebraic Modeling System (GAMS) environment improve over time, we determined the final design of the structures proposed by Acosta-Flores et al. [34] using our version of GAMS. The results are shown in Table 6.

Table 6. Comparison between approaches.

Algorithm	Tabu Search	Baron			
Revenue, USD $1000/year	49,792	49,306	49,783	49,543	49,792
τ_R, min	6.000	6.000	6.000	6.000	6.000
τ_{C1}, min	3.020	6.000	2.349	6.000	2.978
τ_{C2}, min	0.500	0.500	0.500	0.500	0.500
τ_{S1}, min	6.000	2.424	6.000	1.816	6.000
τ_{S2}, min	6.000	2.424	6.000	6.000	6.000
N_R	15.000	15.000	15.000	15,000	15.000
N_{C1}	8.000	8.000	8.000	8.000	8.000
N_{C2}	5.000	2.000	6.000	2.000	5.000
N_{S1}	15.000	15.000	15.000	15.000	15.000
N_{S2}	15.000	15.000	15.000	10.000	15.000
Grade Cu	0.222	0.220	0.222	0.222	0.222
Circuit structure	Figure 4	Figure 5g (circuit 1) *	Figure 5a (circuit 2) *	Figure 5f (circuit 3) *	Figure 4 (circuit 4) *
Time, s	798.06	605,789.65	628,906.07	630,906.7	80,193.67

* number of circuits in Acosta-Flores et al. [34].

Table 6 shows that TSA is capable of converging despite changing the parameters in the methodology proposed. The best design provided by both approaches was similar. However, our method required fewer computational resources. The TSA used 320 neighbors at each neighborhood, 1500 iterations, the diversification was applied each time that the global solution was not improved after 20 iterations; the intensification was applied each time passing 40 iterations of the algorithm, and the number of rows of the TL was 70. Note that the TSA provided secondary designs (Table 7). However, neither approach guarantees finding a global solution.

Table 7. Best design and secondary designs provided by the Tabu-search algorithm.

Algorithm	Best Design	Secondary Designs		
Revenue, USD/year	49,792,2192	49,698,998	49,670,988	49,575,119
τ_R, min	6.000	4.190	4.780	3.200
τ_{C1}, min	3.020	5.160	5.680	5.700
τ_{C2}, min	0.500	5.390	4.260	6.000
τ_{S1}, min	6.000	5.850	5.640	5.950
τ_{S2}, min	6.000	6.000	5.200	5.210
N_R	15.000	15.000	15.000	9.000
N_{C1}	8.000	4.000	3.000	2.000
N_{C2}	5.000	2.000	2.000	3.000
N_{S1}	15.000	15.000	15.000	7.000
N_{S2}	15.000	15.000	15.000	11.000
Grade Cu	0.222	0.222	0.222	0.222
Circuit structure	Figure 4	Figure 5b	Figure 5h	Figure 5e

In general, Tables 3–6 show that the TSA converges faster than the Baron solver, and, the TSA always provided a solution in a reasonable amount of time when the mathematical model was well-defined. Notably, the TSA parameters must be adjusted. Evidently, this procedure took time. For example, Lin et al. [52] proposed determining the neighborhood size using the number of free variables of a mathematical model. A similar strategy was applied in this work; however, in general, the TSA did not provide good results. This behavior could be related to the considered multispecies in the design procedure. Then, the neighborhood size was tuned by trial and error.

Cisternas et al. [2] and Lin et al. [5] mentioned that both exact and approximate methods have a low probability of finding the global solution in a reasonable amount of time for large. However, our results contradict these observations. Tables 3 and 4 show that both optimization techniques converge and provided similar designs despite the design procedure considering several species, operating costs, capital costs, size of equipment, number of equipment, and metallurgical parameters, among other variables. In the case of the TSA, these results could be explained by a minimally dynamic and noncomplex superstructure, which did reduce the number of alternatives to 144. In the case of the Baron solver, these results could be explained by the delimitation of the search space due to large number of equations used by the mathematical model, which helped the solver to converge in a reasonable amount of time. This finding is corroborated by the results shown in Table 5. In this case, the mathematical model included neither equipment design nor operational cost, and generally, the solver did not converge.

The results indicate that the diversification and neighborhood have a strong influence on the results of TSA. When the diversification was not incorporated into the search procedure, the results of the TSA were distinct compared to the results provided by the Baron solver. Although a large neighborhood did not help with obtaining good solutions, better solutions were obtained using small neighborhoods and many iterations of the TSA, i.e., achieving a gradual improvement in the best neighbor.

The TSA developed in this work is flexible, i.e., would allow including into the design procedure grinding models, cell models, or the geological uncertainty (non-linear algorithm). The latter is related to the variation of copper grade that occurs during real operation of the circuit. Initially, the consideration of geological uncertainty or degree of liberation of valuable minerals (grinding) drives optimization under uncertainty. There are several approaches for addressing this type of problem such as stochastic optimization [53]. A possible strategy for solving the stochastic optimization problem is scenario trees based on stochastic parameters of the model. The solution of each scenario would be determined via TSA. A summary of the advantages and disadvantages of the optimization techniques used in this work is shown in Table 8.

Table 8. Comparison between the Tabu-search algorithm and the Baron solver.

Algorithm	Tabu Search	Baron Solver
Advantages	The convergence is fast. The algorithm always provides a solution when the mathematical model is well defined. The algorithm is flexible, i.e., it allows the use of other methods, such as linear programming algorithms. The algorithm provides good quality solutions. The algorithm provides secondary designs.	The solver provides a global optimal design when converged. The solver does not need to adjust parameters for providing a solution. The obtained designs do not change if the solver is run again.
Disadvantages	Some algorithm parameters, such as neighborhood size and number of iterations of the algorithm, among others, must be adjusted for finding a good quality solution. Penalty parameters must be used for satisfying the constraints of the mathematical model. The obtained designs could change if the algorithm is run again. The algorithm must incorporate diversification and intensification for finding a good quality solution.	Depending on the mathematical model and the number of species, the convergence is slow or the algorithm does not converge. The variables of the model must be bounded for guaranteeing the finding of global optimal design, i.e., experience in circuit design is required. The solver provides a single design. The obtained designs depend on the version of the solver.

4. Discussion and Conclusions

An algorithm based on Tabu-search was developed and used for designing flotation circuits for several species. The algorithm incorporates diversification for exploring new regions and intensification for exploring regions close to a good solution. The circuits designed using the Tabu-search algorithm were logical and allowed incorporating the influence of the objective function into the design of concentration plants. A comparison between the Tabu-search algorithm and the Baron solver in the GAMS environment was performed. In general, the solutions provided by both optimization techniques were similar. The Tabu-search algorithm always quickly provided a solution when the mathematical model was well-defined, whereas the Baron solver, in some cases, converged slowly or did not converge at all. Finally, we compared our approach and the methodology proposed by Acosta-Flores et al. [34]. The results indicated the best design provided by both proposals was similar. Both approaches provided secondary designs, but our proposal required fewer computational resources. Thus, the developed algorithm can converge to an optimal solution or near optimal for a complex combination of requirements and constraints in a design problem, whereas other methods do not. The developed algorithm represents progress in the design of flotation circuits, which could be useful in the design of full-scale mineral concentration circuits.

In future work, we will seek to include geological uncertainty, operating uncertainty, and grinding in the design procedure. Furthermore, we will analyze the hybridization between TSA and genetic algorithms.

Author Contributions: F.A.L. developed the ideas and objectives of the work. F.A.L. developed the algorithms, simulations and prepared the original draft manuscript. L.A.C. and E.D.G. were responsible for supervision and project administration. E.D.G. and L.A.C. reviewed the final manuscript. L.A.C. analyzed the results.

Acknowledgments: The financial support from CONICYT (Fondecyt 1171341 and 1180826) is gratefully acknowledged.

Conflicts of Interest: The authors declare no conflict of interest.

Nomenclature

$C1$	Cleaner stage
$C2$	Re-cleaner stage
C_i	Concentrate stream of the flotation stage i
C_{ik}	Mass flow of species k in concentrate C_i
C_{ijk}	Mass flow of species k in the concentrate stream from stage j to stage i
$C_{op,i}$	Operating cost of flotation stage i
D	Annual depreciation
E_g	Gas factor
F_C	Annual cash flows
FL	Lang factor
FL_w	Lang factor for working capital
FM	Frequency matrix
g	Grade
H	Number of hours per year of plant operation
I_{cap}	Capital cost
I_F	Fixed capital cost
I_w	Working capital cost
F_{ik}	Mass flow of species k in feed streams of stage i
$k_{max,i,k}$	Maximum rate constant of the species k in flotation stage i
M_{1k}	Mass flow of species k fed to the flotation circuit
N_i	Number of flotation cell in stage i
$N(x)$	Neighborhood of x
n	Life time of the project
n_r	Number of rows of Tabu list
P	Final concentrate
P_B	Profits before taxes
P_k	Kilowatt-hours cost
p	Fraction of metal paid
R	Rougher stage
R_{ik}	Recovery of stage i for species k
$R_{max,i,k}$	Maximum recovery at infinite time of stage i for species k
Rfc	Refinery charge
r_t	Tax rate
$S1$	Scavenger stage
$S2$	Re-scavenger stage
T_i	Tail stream of the flotation stage i
T_{ik}	Mass flow of the species k in tail T_i
T_{ijk}	Mass flow of species k in the tail stream from stage j to stage i
TL	Tabu list
Trc	Treatment charge
V_i	Cell volumen in stage i
W	Final tail
W_{NP}	Net present worth
x_{best}	Best neighbor of $N(x)$
Greek symbols	
α_{ij}	Decision variables
β_{ij}	Decision variables
γ	Penalty parameter
ρ_p	Pulp density
μ	Grade deduction
τ_i	Cell residence time in stage i

Appendix A

Mathematical model in the GAMS environment.
Note that GAMS implements sets for defining the mathematical model. The main sets are:

$$M1 = \{r/r \text{ is } F, R, C1, C2, S1, S2, W, P\}$$

$$M2 = \{r/r \text{ is } R, C1, C2, S1, S2\}$$

$$L = \{(r,s)/(r,s) \text{ is a stream from stage } r \text{ to stage } s, \ r,s \in M1\}$$

$$K_1 = \{k_1/k_1 \text{ is a species}\}$$

$$LC = \{(r,s)/(r,s) \text{ is a concentrate stream}, \ (r,s) \in M1\}$$

$$LT = \{(r,s)/(r,s) \text{ is a tail stream}, \ (r,s) \in M1\}$$

For the mass balance in splitters, the following equations are considered:

$$CC(r,k1) = \sum_s F_c(r,s,k_1), \ (r,s) \in LC, \ k_1 \in K_1, \ r \in M2$$

$$CT(r,k1) = \sum_s F_w(r,s,k_1), \ (r,s) \in LT, \ k_1 \in K_1, \ r \in M2$$

where $F_c(r,s,k_1)$ is the concentrate flows of species k_1 in the stream from r to s, $F_w(r,s,k_1)$ is tail flow of species k_1 from r to s, $CC(r, k_1)$ and $CT(r, k_1)$ are the mass flow of species k_1 in the concentrate and tail streams of stage r, respectively. Stream branching is not allowed, so the following equations are included:

$$\sum_s F_c(r,s,k_1) \le F_c^{UP} y_{rs}^c, \ \sum_s y_{rs}^c = 1, \ (r,s) \in LC$$

$$\sum_s F_w(r,s,k_1) \le F_w^{UP} y_{rs}^w, \ \sum_s y_{rs}^w = 1, \ (r,s) \in LT$$

where y_{rs}^c and y_{rs}^w are binary variables indicating the choice of the destination of the concentrate and tail streams, respectively; F_c^{UP} and F_w^{UP} correspond to the lower and upper bounds of mass flow of species k_1 in the concentrate and tail, respectively. For the mass balance in the mixer, the following equations are considered:

$$FS(s,k_1) = \sum_{is.t. \ (r,s) \in LC} F_c(r,s,k_1) + \sum_{is.t. \ (r,s) \in LT} F_w(r,s,k_1), \ k_1 \in K_1, \ s \in M2$$

where $FS(s,k_1)$ is the mass flow of species k_1 in feed streams to stage s. The final concentrate of the flotation circuits is calculated using:

$$CF(k_1) = \sum_r F_c(r,P,k_1), \ (r,P) \in LC, \ k_1 \in K_1$$

where $CF(k_1)$ is the mass flow of species k_1 in final concentrate. The mass balance in the flotation stages is determined using:

$$CC(r,k_1) = T(r,k_1) \cdot FS(s,k_1), \ k_1 \in K_1, \ r \in M2$$

$$CT(r,k_1) = (1 - T(r,k_1)) \cdot FS(s,k_1), \ k_1 \in K_1, \ r \in M2$$

where $T(r,k_1)$ is the flotation model proposed by Yianatos and Henríquez [41]. The objective function is defined using the Equations (8)–(20), without considering the penalty parameter.

References

1. Hu, W.; Hadler, K.; Neethling, S.J.; Cilliers, J.J. Determining flotation circuit layout using genetic algorithms with pulp and froth models. *Chem. Eng. Sci.* **2013**, *102*, 32–41. [CrossRef]
2. Cisternas, L.A.; Lucay, F.A.; Acosta-Flores, R.; Gálvez, E.D. A quasi-review of conceptual flotation design methods based on computational optimization. *Miner. Eng.* **2018**, *117*, 24–33. [CrossRef]
3. Mendez, D.A.; Gálvez, E.D.; Cisternas, L.A. State of the art in the conceptual design of flotation circuits. *Int. J. Miner. Process.* **2009**, *90*, 1–15. [CrossRef]
4. Talbi, E.G. *Metaheuristics: From Design to Implementation*; John Wiley & Sons: Hoboken, NJ, USA, 2009; ISBN 9780470278581.

5. Lin, M.H.; Tsai, J.F.; Yu, C.S. A review of deterministic optimization methods in engineering and management. *Math. Probl. Eng.* **2012**, *2012*, 756023. [CrossRef]

6. Glover, F. Future paths for integer programming and links to artificial intelligence. *Comput. Oper. Res.* **1986**, *13*, 533–549. [CrossRef]

7. Osman, I.H.; Laporte, G. Metaheuristics: A bibliography. *Ann. Oper. Res.* **1996**, *63*, 511–623. [CrossRef]

8. Price, K.; Storn, R.M.; Lampinen, J.A. *Differential Evolution: A Practical Approach to Global Optimization (Natural Computing Series)*; Springer: Berlin/Heidelberg, Germany, 2005.

9. Holland, J.H. Adaptation in Natural and Artificial Systems. MIT Press: Cambridge, MA, USA, 1975.

10. Moscato, P. On Evolution, Search, Optimization, Genetic Algorithms and Martial Arts: Towards Memetic Algorithms. Caltech Concurrent Computation Program; C3P Report. 1989. Available online: http://citeseer.ist.psu.edu/viewdoc/download?doi=10.1.1.27.9474&rep=rep1&type=pdf (accessed on 31 January 2019).

11. Farmer, J.D.; Packard, N.H.; Perelson, A.S. The immune system, adaptation, and machine learning. *Phys. D Nonlinear Phenom.* **1986**, *22*, 187–204. [CrossRef]

12. Kirkpatrick, S.; Gelatt, C.D.; Vecchi, M.P. Optimization by simulated annealing. *Science* **1983**, *220*, 671–680. [CrossRef] [PubMed]

13. Dorigo, M. Optimization, Learning and Natural Algorithms. Ph.D. Thesis, Dipartimento di Elettronica, Politecnico di Milano, Milan, Italy, 1992. [CrossRef]

14. Kennedy, J.; Eberhart, R. Particle Swarm Optimization. In Proceedings of the IEEE International Conference on Neural Networks, Perth, WA, Australia, 27 November–1 December 1995; Volume IV, pp. 1942–1948.

15. Glover, F.; Laguna, M. Tabu Search. In *Handbook of Combinatorial Optimization*; Du, D.Z., Pardalos, P.M., Eds.; Springer: Boston, MA, USA, 1998; pp. 2093–2229. ISBN 978-1-4613-0303-9.

16. Glover, F. Tabu Search-Part I. *ORSA J. Comput.* **1989**, *1*, 190–206. [CrossRef]

17. Han, S.; Li, J.; Liu, Y. Tabu search algorithm optimized ANN model for wind power prediction with NWP. *Energy Procedia* **2011**, *12*, 733–740. [CrossRef]

18. Ting, C.K.; Li, S.T.; Lee, C. On the harmonious mating strategy through tabu search. *Inf. Sci.* **2003**, *156*, 189–214. [CrossRef]

19. Soto, M.; Sevaux, M.; Rossi, A.; Reinholz, A. Multiple neighborhood search, tabu search and ejection chains for the multi-depot open vehicle routing problem. *Comput. Ind. Eng.* **2017**, *107*, 211–222. [CrossRef]

20. Lin, B.; Miller, D.C. Application of tabu search to model indentification. In Proceedings of the AIChE Annual Meeting, Los Angeles, CA, USA, 12–17 November 2000.

21. Kulturel-Konak, S.; Smith, A.E.; Coit, D.W. Efficiently solving the redundancy allocation problem using tabu search. *IIE Trans. (Inst. Ind. Eng.)* **2003**, *35*, 515–526. [CrossRef]

22. Kis, T. Job-shop scheduling with processing alternatives. *Eur. J. Oper. Res.* **2003**, *151*, 307–332. [CrossRef]

23. Pan, Q.K.; Fatih Tasgetiren, M.; Liang, Y.C. A discrete particle swarm optimization algorithm for the no-wait flowshop scheduling problem. *Comput. Oper. Res.* **2008**, *35*, 2807–2839. [CrossRef]

24. Mandani, F.; Camarda, K. Multi-objective Optimization for Plant Design via Tabu Search. *Comput. Aided Chem. Eng.* **2018**, *43*, 543–548.

25. Mehrotra, S.P.; Kapur, P.C. Optimal-Suboptimal Synthesis and Design of Flotation Circuits. *Sep. Sci.* **1974**, *9*, 167–184. [CrossRef]

26. Reuter, M.A.; van Deventer, J.S.J.; Green, J.C.A.; Sinclair, M. Optimal design of mineral separation circuits by use of linear programming. *Chem. Eng. Sci.* **1988**, *43*, 1039–1049. [CrossRef]

27. Reuter, M.A.; Van Deventer, J.S.J. The use of linear programming in the optimal design of flotation circuits incorporating regrind mills. *Int. J. Miner. Process.* **1990**, *28*, 15–43. [CrossRef]

28. Schena, G.; Villeneuve, J.; Noël, Y. A method for a financially efficient design of cell-based flotation circuits. *Int. J. Miner. Process.* **1996**, *46*, 1–20. [CrossRef]

29. Schena, G.D.; Zanin, M.; Chiarandini, A. Procedures for the automatic design of flotation networks. *Int. J. Miner. Process.* **1997**, *52*, 137–160. [CrossRef]

30. Maldonado, M.; Araya, R.; Finch, J. Optimizing flotation bank performance by recovery profiling. *Miner. Eng.* **2011**, *24*, 939–943. [CrossRef]

31. Cisternas, L.A.; Méndez, D.A.; Gálvez, E.D.; Jorquera, R.E. A MILP model for design of flotation circuits with bank/column and regrind/no regrind selection. *Int. J. Miner. Process.* **2006**, *79*, 253–263. [CrossRef]

32. Cisternas, L.A.; Gálvez, E.D.; Zavala, M.F.; Magna, J. A MILP model for the design of mineral flotation circuits. *Int. J. Miner. Process.* **2004**, *74*, 121–131. [CrossRef]

33. Calisaya, D.A.; López-Valdivieso, A.; de la Cruz, M.H.; Gálvez, E.E.; Cisternas, L.A. A strategy for the identification of optimal flotation circuits. *Miner. Eng.* **2016**, *96–97*, 157–167. [CrossRef]

34. Acosta-Flores, R.; Lucay, F.A.; Cisternas, L.A.; Galvez, E.D. Two-phase optimization methodology for the design of mineral flotation plants, including multispecies and bank or cell models. *Miner. Metall. Process.* **2018**, *35*, 24–34. [CrossRef]

35. Guria, C.; Verma, M.; Mehrotra, S.P.; Gupta, S.K. Multi-objective optimal synthesis and design of froth flotation circuits for mineral processing, using the jumping gene adaptation of genetic algorithm. *Ind. Eng. Chem. Res.* **2005**, *44*, 2621–2633. [CrossRef]

36. Guria, C.; Verma, M.; Gupta, S.K.; Mehrotra, S.P. Simultaneous optimization of the performance of flotation circuits and their simplification using the jumping gene adaptations of genetic algorithm. *Int. J. Miner. Process.* **2005**, *77*, 165–185. [CrossRef]

37. Ghobadi, P.; Yahyaei, M.; Banisi, S. Optimization of the performance of flotation circuits using a genetic algorithm oriented by process-based rules. *Int. J. Miner. Process.* **2011**, *98*, 174–181. [CrossRef]

38. Pirouzan, D.; Yahyaei, M.; Banisi, S. Pareto based optimization of flotation cells configuration using an oriented genetic algorithm. *Int. J. Miner. Process.* **2014**, *126*, 107–116. [CrossRef]

39. Cisternas, L.A.; Jamett, N.; Gálvez, E.D. Approximate recovery values for each stage are sufficient to select the concentration circuit structures. *Miner. Eng.* **2015**, *83*, 175–184. [CrossRef]

40. Sepúlveda, F.D.; Lucay, F.; González, J.F.; Cisternas, L.A.; Gálvez, E.D. A methodology for the conceptual design of flotation circuits by combining group contribution, local/global sensitivity analysis, and reverse simulation. *Int. J. Miner. Process.* **2017**, *164*, 56–66. [CrossRef]

41. Yianatos, J.B.; Henríquez, F.D. Short-cut method for flotation rates modelling of industrial flotation banks. *Miner. Eng.* **2006**, *19*, 1336–1340. [CrossRef]

42. Green, J.C.A. The optimisation of flotation networks. *Int. J. Miner. Process.* **1984**, *13*, 83–103. [CrossRef]

43. Cisternas, L.A.; Lucay, F.; Gálvez, E.D. Effect of the objective function in the design of concentration plants. *Miner. Eng.* **2014**, *63*, 16–24. [CrossRef]

44. Ruhmer, W. *Handbook on the Estimation of Metallurgical Process Cost*, 2nd ed.; Mintek: Randburg, South Africa, 1991.

45. Peters, M.S.; Timmerhaus, K.D.; West, R.E. *Plant Design and Economics for Chemical Engineers*, 4th ed.; McGraw-Hill Publishing Company: New York, NY, USA, 1991; ISBN 0070496137.

46. Masuda, K.; Kurihara, K.; Aiyoshi, E. A penalty approach to handle inequality constraints in particle swarm optimization. In Proceedings of the IEEE International Conference on Systems, Man and Cybernetics, Istanbul, Turkey, 10–13 October 2010.

47. Gogna, A.; Tayal, A. Metaheuristics: Review and application. *J. Exp. Theor. Artif. Intell.* **2013**, *25*, 503–526. [CrossRef]

48. Gendreau, M.; Potvin, J.-Y. Variable Nneighborhood search (chapter). In *Handbook of Metaheuristics*; Springer: Berlin, Germany, 2010. [CrossRef]

49. López, E.; Salas, Ó.; Murillo, Á. A deterministic algorithm using tabu search. *Revista de Matemática Teoría y Aplicaciones* **2014**, *21*, 127–144. [CrossRef]

50. De los cobos, S.G.; Goddard, J.; Gutiérrez, M.; Martínez, A. *Búsqueda y Exploración Estocástica*, 1st ed.; Universidad Autonoma Metropolitana: Mexico City, Mexico, 2010; ISBN 9786074772395.

51. Jamett, N.; Cisternas, L.A.; Vielma, J.P. Solution strategies to the stochastic design of mineral flotation plants. *Chem. Eng. Sci.* **2015**, *134*, 850–860. [CrossRef]

52. Lin, B.; Chavali, S.; Camarda, K.; Miller, D.C. Using Tabu search to solve MINLP problems for PSE. *Comput. Aided Chem. Eng.* **2003**, *15*, 541–546. [CrossRef]

53. Fouskakis, D.; Draper, D. Stochastic optimization: A review. *Int. Stat. Rev.* **2002**, *70*, 315–349. [CrossRef]

Article

Optimizing Flotation Circuit Recovery by Effective Stage Arrangements: A Case Study

Vahid Radmehr, Sied Ziaedin Shafaei *, Mohammad Noaparast and Hadi Abdollahi

School of Mining, College of Engineering, University of Tehran, Tehran 16846-13114, Iran;
v.radmehr@ut.ac.ir (V.R.); noparast@ut.ac.ir (M.N.); H_abdollahi@ut.ac.ir (H.A.)
* Correspondence: zshafaie@ut.ac.ir; Tel.: +98-021-820-84238

Received: 16 June 2018; Accepted: 18 September 2018; Published: 20 September 2018

Abstract: Recovery is one of the most important metallurgical parameters in designing and evaluating flotation circuits. The present study used the recovery arrangement for two and three stage circuits to evaluate the effect of stage recovery on the overall circuit recovery and flotation circuit configuration. The results showed that mainly the highest recovery value should be assigned to the rougher stage in order to achieve the maximum overall circuit recovery. Countercurrent rougher-cleaner and rougher-scavenger circuits, in which recycling streams step back one stage at a time, follow a general rule for the assignment of recovery. Finally, a flotation plant containing six flotation banks was examined as a case study. A program for calculating total circuit recovery, for all possible combinations of recovery was developed in MATLAB software. 720 recovery combinations were evaluated. The results showed that optimal recovery allocation in stages could be effective in achieving overall circuit recovery. It was shown that the use of a large number of stages in some of the flotation circuits leads to the loss of equipment and additional costs. The proposed approach can be employed as an effective tool for designing and optimizing various flotation circuits and their operational parameters.

Keywords: froth flotation; recovery arrangement; circuit configuration; process design

---

1. Introduction

Froth flotation is a process used for selective separation of hydrophobic materials from hydrophilic. Over the years, various criteria have been introduced for evaluating flotation circuits. Technically, grade and recovery are the most common indicators for evaluation of a flotation circuit performance. If a flotation circuit meets the acceptable grade for a concentrate, achieving the highest recovery of the circuit plays an important role in the desirability of the flotation plant [1]. In such a situation, knowing the effect of stage recovery on the overall circuit recovery and identifying the most effective stage in the circuit is essential. The recovery of a cell or bank depends on the operational and design parameters. Some of the main factors affecting the efficiency of the flotation circuit are the pulp chemical conditions (chemical reagent concentration, pH, aeration), physical conditions in the cell (mixing, supply and distribution of air, pulp level), and arrangement (size of the cells in each bank, the connection between banks) [2]. Previous studies in the field of simulation and optimization of flotation circuits have investigated some of these factors and their effect on the flotation process [3,4]. Most of the existing flotation models are empirical [5], and they are not capable of predicting circuit behavior based on operating conditions and circuit configuration. In other words, accurate first-order models are difficult to use in flotation circuit design using optimization [6]. Even the use of a first order kinetic model can be a challenge when many separation stages or abundant species are considered. The computational cost for this type of optimization is very high. First-order models with a limited number of species have been widely used in the design of flotation circuits that are based on optimization [7]. In these studies, identical flotation rate constant for the specific species in each cell is assumed. This assumption may be very simplistic, but is commonly used in industrial operations [8].

Despite extensive researches had been done on the effect of design and operational parameters on the efficiency of flotation circuits, very few publications can be found in the literature that discusses the role of stage recovery on the overall circuit recovery and the effect of circuit configuration on the stage recovery [6,7,9,10]. Some previous studies have evaluated the circuit configuration with circuit analysis method assuming the same recovery in various flotation stages [7,11,12]. Some industrial studies have also used the identical stage recovery in separation circuits, along with separation sharpness as an index for assessing the separation circuit configuration [13–15]. Assuming identical recovery in all stages was criticized because of its simplicity and inadequate ability in distinguishing the circuits with the same separation sharpness [16]. The present study is going to compare and evaluate the configuration of flotation circuits, assuming that an identical recovery in stages is an acceptable assumption and the results are valid for non-identical stage recovery. Recently, the effect of stage recovery on the overall circuit recovery has been investigated from two aspects of sensitivity and uncertainty analyses [6,7,9,17,18]. These studies have shown that the flotation circuit configuration is not very sensitive to stage recovery [9,19]. The overall output suggests that choosing the best flotation circuit is not sensitive to the stage recovery, and as a result, only approximate values of the recovery can be used to select the optimal flotation circuit [6,9,20].

This paper using simple numerical examples shows that the assumption of circuit configuration insensitivity to the stage recovery is not correct in all cases. The aim of this paper is to indicate that stage recovery is an essential parameter in the design of flotation circuits and in some cases is strongly influenced by the structure of the circuit. The remainder of the paper is organized into four sections. After the introduction, application of recovery as a transfer function (Section 2) is described, Section 3 analyzed recovery arrangement in simple circuits and its application is shown by an industrial example (Section 4). Section 5 summarizes the results.

2. Methodology

2.1. Transfer Function

Transfer function is a mathematical expression that links a conserved quantity or component, j, entering a stage to the output of that stage in a steady state condition [7,21]. With respect to Figure 1, for the jth component of the input to a separation stage (F_j), the conservation relation is written as follows:

$$C_j = R_j F_j$$
$$T_j = (1 - R_j) F_j \tag{1}$$

R_j = Recovery of the component j
C_j = mass flowrate of the component j into the concentrate
T_j = mass flowrate of component j into the tail

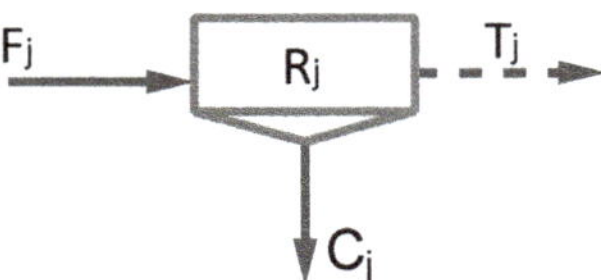

Figure 1. The mass conservation component j in a single unit.

The flotation process uses a variety of empirical, probabilistic and kinetic transfer functions to evaluate the mass ratio of output to input [22]. The empirical transfer function is obtained from the curve fitting to the recovery data in a steady state [23]. These curves do not attempt to establish logical relations or calculate the mass transfer function. The shape of these functions depends on the properties of the ore and on the structure of the circuit. Probabilistic transfer functions suggested that flotation is a sequence of events that must occur before the particle is collected [24]. As a result, probabilistic transfer functions are based on theoretical considerations of flotation sub-process occurrences. The probability transfer function correlates the particle recovery in flotation with the probability of floatation success, P_x, for a specific size:

$$P_x = P_c P_a F \tag{2}$$

P_c = probability of collision
P_a = probability of attachment
F = froth stability factor

If the flotation probability function in a single cell is considered to be R_j, then the recovery of bank of cells with the same flotation probabilities R_j is equal to [25]:

$$R = R_\infty (1 - (1 - R_j)^n) \tag{3}$$

In Equation (3), R depends on factors, such as the probability of particle coating with the collector; the probability of the particle entering the bubble-particle collision region in the cell; the probability of a bubble-particle collision in the bubble-particle collision zone; the probability of attaching the bubble-particle after collision with each other; and, the probability of bubble-particle attachment report to the concentrate. In the case of flotation cells in plants for different stages of the flotation circuit, it is better to use cells of the same size and only different in numbers (e.g., rougher, cleaner, scavenger) [2]. This, in addition to providing the residence time, also provides the possibility of creating the same R. Kinetic transfer functions provide mineral recovery on a time basis. Usually, the first order kinetic is used to obtain the transfer function of a single cell [4]. When considering the assumption of perfect mixing for flotation cells, the recovery of a j species with a constant rate k_j is related to the average residence time τ_j:

$$R_j = \frac{k_j \tau_j}{1 + k_j \tau_j} \tag{4}$$

In which:

R_j = the steady state transfer function of the mineral species j
k_j = constant flotation rate of the mineral species j, min^{-1}
τ_j = residence time for species j in the flotation cell, min
n = number of flotation cells in the bank

By placing the following equation in the probability transfer function,

$$1 - R_j = \frac{1}{1 + k_j \tau_j} \tag{5}$$

The kinetic transfer function is obtained by calculating the recovery of species j in a flotation bank:

$$R = 1 - (1 + k_j \tau_j)^{-n} \tag{6}$$

After obtaining the optimal arrangement of recovery at different stages of a flotation circuit, one can use a variety of transfer functions to calculate the optimal design and operational parameters.

2.2. Circuit Function Calculation

In mineral processing engineering, the circuit function can be directly used in optimizing the separation circuit. Establishing functional relationships between operational parameters, such as pulp density and the transfer function of a single unit, can easily be accomplished through empirical operations. Obtaining the recovery function for mineral separation circuits can lead to the simplification of complex flowsheets. When considering the recovery as a transfer function for each stage/bank (R_a and R_b), it is possible to calculate the transfer function of the circuit while using the manual method [26]. In Figure 2, a closed rougher-scavenger circuit, analytical solution can be obtained while using algebraic expressions.

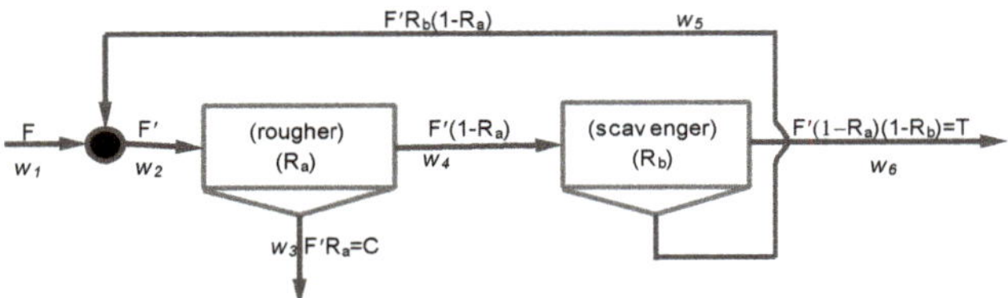

Figure 2. Example of rougher-scavenger circuit for circuit function calculation.

Manual calculations:

$$F' = F + F'R_b(1 - R_a) \tag{7}$$

$$F = F' - F'R_b(1 - R_a) \tag{8}$$

$$\frac{C}{F} = \frac{F'R_a}{F' - F'R_b(1 - R_a)} = \frac{R_a}{1 - R_b(1 - R_a)} \tag{9}$$

Here, a method for finding circuit function with any number of stages by writing the equations between the inputs and outputs of each bank/mixing point based on stage recovery is developed. The input and output equations for Figure 2 and the corresponding matrix are expressed, as follows, in which the last column is related to the input stream:

$$
\begin{array}{l}
w_2 = w_1 + w_5 \\
w_3 = R_a w_2 \\
w_4 = (1 - R_a)w_2 \\
w_5 = R_b w_4 \\
w_6 = (1 - R_b)w_4
\end{array}
\left|
\begin{array}{cccccc|c}
w_2 & w_3 & w_4 & w_5 & w_6 & w_1 \\
1 & 0 & 0 & -1 & 0 & -1 \\
-R_a & 1 & 0 & 0 & 0 & 0 \\
-(1 - RR_a) & 0 & 1 & 0 & 0 & 0 \\
0 & 0 & -R_b & 1 & 0 & 0 \\
0 & 0 & -(1 - R_b) & 0 & 1 & 0
\end{array}
\right|
\tag{10}
$$

These equations can be considered as linear systems AX = b, in which A is an m × n matrix. By solving the above equations, the transfer function is obtained for each stream:

$w_2/w_1 = 1/k,$

$w_3/w_1 = R_a/k,$

$w_4/w_1 = -(R_a - 1)/k,$

$w_5/w_1 = -(R_b \times (R_a - 1))/k,$

$w_6/w_1 = ((R_a - 1) \times (R_b - 1))/k,$

in which:

$$k = (R_a \times R_b - R_b + 1) \tag{11}$$

w is the name of the stream. If w is considered as the mass of a given mineral in a particular stream, by determining the transfer function of each of the streams, it is possible to do mass balance for the whole circuit. In this study, all of the calculations were carried out in MATLAB.

3. Results and Discussion

Recovery in Simple Circuits

Flotation circuit uses various stages of the rougher, scavenger, cleaner and cleaner-scavenger units. The overall circuit recovery can be obtained through the recovery of these units. In order to illustrate the effect of stage recovery on the overall circuit recovery, typical circuits of two and three stages have been investigated. Figure 3 schematically shows the typical or basic types of configuration for two and three stage flotation circuits. Assuming 0.4, 0.5, and 0.6 as available recovery values for a particular mineral species, all stage recovery combinations to obtain overall circuit recovery were evaluated. Here, analysing the circuit recovery is for one valuable component, but it can be developed to multi-components that increase the potential circuit stage recovery combinations. In this case, powerful search algorithms, like genetic algorithm, can be used to handle the optimization problem [27].

The signs of F, C, and T show the mass of particles with a specific property in feed, concentrate and tailings. R, C, and S, respectively, represent the rougher, cleaner, and scavenger stages. For all configurations, the numbering of the stages is started from the left to have the calculated recovery function in a specific order. Table 1 shows all possible recovery combinations in stages, six possible states, and the total recovery achieved with that circuit. Due to the structure of some circuits, their transfer function is in such a way that it does not matter how the recovery arrangement is in the stages. In these circuits, the overall circuit recovery will be the same in any way that the recovery is distributed in stages (S_2, S_3, S_6, and S_{12} circuits). By increasing the amount of recovery in each of the stages of these circuits, the total recovery also increased. The S_6 and S_{12} circuits have the highest and lowest overall recovery in the 33 circuits under investigation (0.88 and 0.26, respectively). By examining the circuits of Figure 3 and Table 1, it can be seen that all R-C-C type circuits show lower recovery than the single rougher stage. By adding a cleaner-scavenger unit to these circuits and creating R-C-CS circuits, the circuit recovery increases, but even in this situation, the overall circuit recovery is less than that of a single rougher unit.

Table 1. Recovery of the circuit with respect to different recovery in the stages (bold numbers are the highest recovery in the circuit).

Circuit	Circuit Recovery					
	Recovery in Each Stage (R_1, R_2, R_3)					
	(0.40, 0.50, 0.60)	(0.40, 0.60, 0.50)	(0.50, 0.40, 0.60)	(0.50, 0.60, 0.40)	(0.60, 0.40, 0.50)	(0.60, 0.50, 0.40)
S_1	0.40	0.40	0.50	0.50	0.60	0.60
S_2	0.20	0.24	0.20	**0.30**	0.24	**0.30**
S_3	0.70	0.76	0.70	**0.80**	0.76	**0.80**
S_4	0.25	0.29	0.29	0.38	0.38	**0.43**
S_5	0.63	0.71	0.57	**0.75**	0.63	0.71
S_6	0.88	0.88	0.88	0.88	0.88	0.88
S_7	**0.75**	0.68	0.74	0.58	0.66	0.57
S_8	**0.87**	**0.87**	0.86	0.86	0.85	0.85
S_9	0.85	0.83	**0.86**	0.81	0.85	0.83
S_{10}	**0.80**	0.78	0.78	0.73	0.73	0.70
S_{11}	**0.83**	0.81	**0.83**	0.77	0.81	0.77
S_{12}	0.12	0.12	0.12	0.12	0.12	0.12
S_{13}	0.15	0.17	0.14	**0.19**	0.15	0.17
S_{14}	0.13	0.14	0.13	**0.15**	0.14	**0.15**
S_{15}	0.15	0.14	0.17	0.15	**0.19**	0.17
S_{16}	0.20	0.22	0.22	0.27	0.27	**0.30**
S_{17}	0.16	0.17	0.19	0.19	**0.23**	**0.23**
S_{18}	0.620	**0.68**	0.58	**0.68**	0.58	0.62
S_{19}	0.59	**0.65**	0.49	**0.65**	0.49	0.57
S_{20}	0.58	0.67	0.52	**0.68**	0.54	0.62
S_{21}	0.64	**0.70**	0.63	**0.71**	0.66	0.69
S_{22}	0.20	0.24	0.20	**0.30**	0.24	**0.30**
S_{23}	0.35	0.35	0.43	0.43	**0.51**	**0.51**
S_{24}	0.32	0.33	0.39	0.42	0.46	**0.48**
S_{25}	0.29	0.30	0.31	0.36	0.34	**0.38**

Table 1. *Cont.*

Circuit	Circuit Recovery					
	Recovery in Each Stage (R_1, R_2, R_3)					
	(0.40, 0.50, 0.60)	(0.40, 0.60, 0.50)	(0.50, 0.40, 0.60)	(0.50, 0.60, 0.40)	(0.60, 0.40, 0.50)	(0.60, 0.50, 0.40)
S_{26}	**0.42**	0.38	**0.42**	0.32	0.38	0.32
S_{27}	0.50	0.46	0.54	0.44	**0.56**	0.50
S_{28}	0.36	**0.38**	0.30	0.34	0.29	0.31
S_{29}	0.63	0.66	0.64	0.69	0.70	**0.71**
S_{30}	0.58	0.61	0.58	**0.62**	0.61	**0.62**
S_{31}	0.50	**0.57**	0.44	0.55	0.46	0.50
S_{32}	**0.38**	0.36	0.34	0.30	0.31	0.29
S_{33}	0.42	0.42	0.39	0.48	**0.50**	**0.50**

Table 1 shows that most of the two and three stage circuits containing the scavenger unit have a recovery of more than 0.50. In terms of the cost of equipment (cell, pump, etc.), the scavenger unit needs more equipment than cleaner units in flotation circuit. As a result, achieving the required overall circuit recovery while using the minimum number of scavenger units can be economically desirable. For example, according to available recoveries, overall circuit recovery in the two stage S_3 and S_5 circuits is much more than the three stage R-S-SC and R-S-C circuits. Choosing these circuits will simplify and reduce the cost of the circuit design if the grade requirements are met. Generally, in circuits with a rougher-cleaner or rougher-scavenger structure that the middling stream step back one stage at a time, the optimal combination of the recovery in stages for maximizing the total circuit recovery, regardless of the number of stages in the circuit, follows a particular trend. In these types of circuits, the recovery from the rougher stage to the last cleaner or scavenger in the circuit is descending. As a result, the quantity and quality of the equipment used in this type of circuits also follow this trend. In a recent study, analysing 35 simple circuit designs showed that cleaning circuits and scavenging circuits have opposed behavior in level of recovery uncertainty [7]. The first type of circuits increases the level of recovery uncertainty in valuable particles, while they reduce the uncertainty in the recovery of gangue particles. The second types behave in an opposite fashion. These findings are in accordance with the result of this study that overall circuit recovery is sensitive to the circuit configuration and stage recovery. Although, the cleaning circuit increase the recovery uncertainty, but are desirable, since the scavenger unit needs more equipment and higher cost than cleaner units in flotation circuit. In Figure 4, countercurrent rougher-cleaner and rougher-scavenger circuit with five stages having one step back recycling streams are shown. The hypothetical recoveries are considered to be 0.30, 0.40, 0.50, 0.60, and 0.70 for different stages. In this case, there are 120 possible combinations for recoveries in stages.

The overall circuit recovery for all possible combinations was calculated according to the recovery of the stages. The highest total recovery for the five stage rougher-cleaner and rougher-scavenger circuits is presented in Table 2. As can be seen, increasing the number of stages lead to increase the overall circuit recovery in the rougher-scavenger systems and decrease in the rougher-cleaner systems. As an overall result, allocating recovery to different stages in these circuits follow a specific rule.

By making some changes to the arrangement of these circuits, for example, sending the scavenger concentrate to the final concentrate or the cleaner tail to the final tail or the return of the flows more than one step back, the process of allocating the recovery will change slightly. In the resulting circuits, however, the highest recovery in most cases is assigned to the rougher stage. Nevertheless, S6, S_{12}, and S_{13} circuits do not follow this rule. The transfer function of the S_6 and S_{12} circuits is such that the allocation of recovery in the stages is not important.

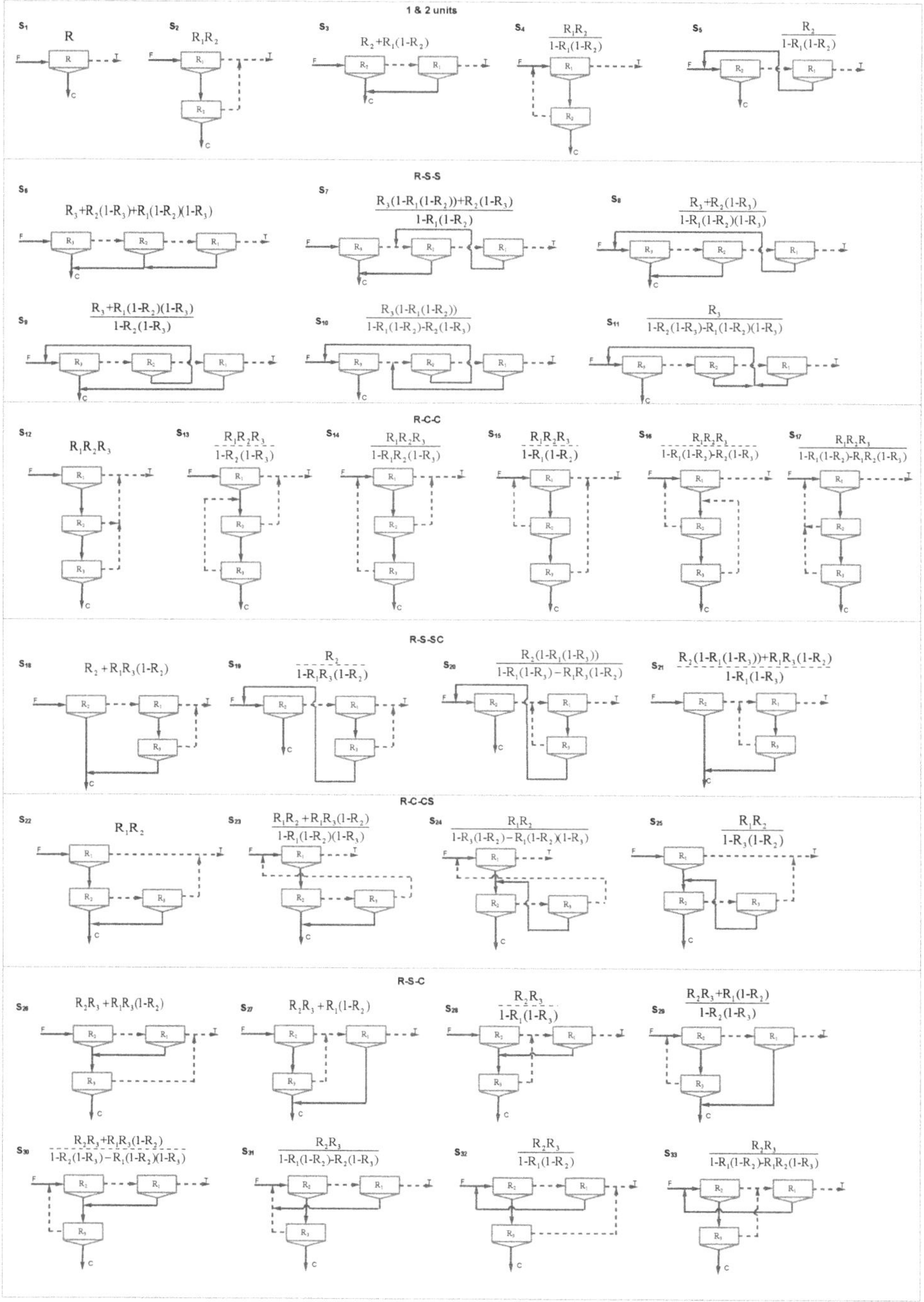

Figure 3. Simple circuit structures and their recovery function.

By creating cleaner-scavenger or scavenger-cleaner units, or using simultaneously the scavenger and cleaner units in a flotation circuit, the rule of recovery arrangement will no longer be the same as before. Although, in some of these configurations, the maximum total recovery of the circuit is achieved by the allocation of the highest recovery to the rougher stage, there are other cases that make it difficult to obtain a general rule. It was observed that the recycling of middling flows in the rougher-cleaner circuits would increase the recovery and in the rougher-scavenger circuits lead to reduce the recovery value. For example, S_4 circuit in which cleaner tail is returned to the rougher stage, in comparison with the S_2 circuit, the total circuit recovery increases by 30%, while in the S_5 circuit, which returns the scavenger froth to the rougher stage as compared to circuit 3, the overall circuit recovery decreases by 6%. The analysis of Table 1 shows that in most circuit configurations sending the scavenger stage concentrate to the final circuit concentrate increases overall circuit recovery. Using this solution in order to achieve higher recovery would be beneficial when the desired grade is met. The S_6 circuit, in which the concentrates produced from the scavenger stage is reported to the final circuit concentrate, the highest overall circuit recovery, 88%, is achieved. The opposite of this approach is to conduct the cleaner tail to the final tailing. In this case, although a high quality product will be obtained, the recovery loss will be tangible.

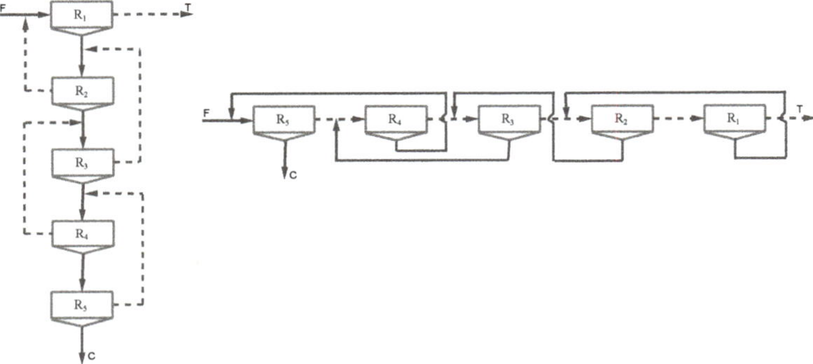

Figure 4. 5 stage countercurrent rougher-cleaner and rougher-scavenger circuits.

Table 2. Optimal recovery arrangement for five stage circuits.

Stage Recovery	R$_1$	R$_2$	R$_3$	R$_4$	R$_5$	Overall Recovery
Rougher-cleaner	0.70	0.60	0.50	0.40	0.30	0.29
Rougher-scavenger	0.30	0.40	0.50	0.60	0.70	0.93

As mentioned in the introduction, one of the criticisms of the circuit analysis method is the assumption of the identical recovery in the stages to evaluate the circuit configuration. Considering an identical recovery of 0.50 for the two and three stages circuits, the overall circuit recovery of all 33 circuits was calculated. The maximum total recovery of the circuits for the cases of identical and non-identical stage recovery is illustrated in Figure 5. Although there is a little difference between the overall recovery of the corresponding circuits, the general trend of overall recovery of the circuits is the same. As a result, the assumption of the identical recovery in the stages to evaluate and comparison of the circuit configuration alternatives and then generalize it to non-identical recovery in stages can be considered as an acceptable assumption.

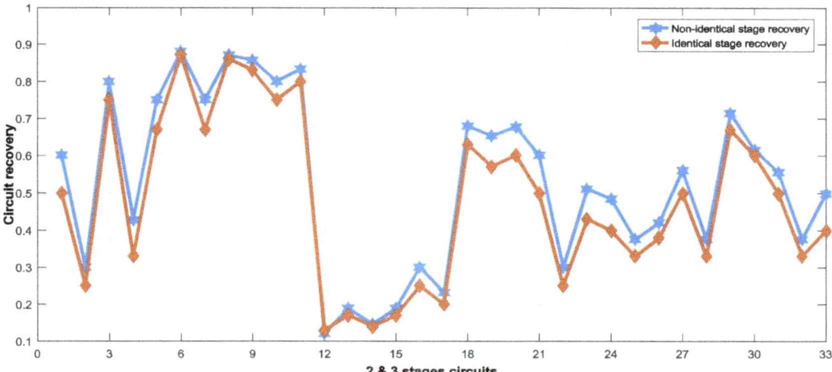

Figure 5. Comparison of the overall circuit recovery for two and three stages circuits in the case of identical and non-identical stage recovery.

4. Industrial Example

The circuit under investigation has six flotation stages, as shown in Figure 6. Valuable minerals in the plant feed include galena and cerussite. Sodium sulfide is used for the sulfidation of oxidized lead ore. Other chemical reagents are potassium amyl xanthate and sodium silicate, which are added to the circuit at various points in the plant. The dosage and points of addition are presented in Table 3. The flotation stages include rougher, scavenger, scavenger-cleaner, cleaner 1, cleaner 2, and cleaner 3, which have respectively 8, 10, 3, 3, 2, and 2 cells each with a volume of 1.1 cubic meters.

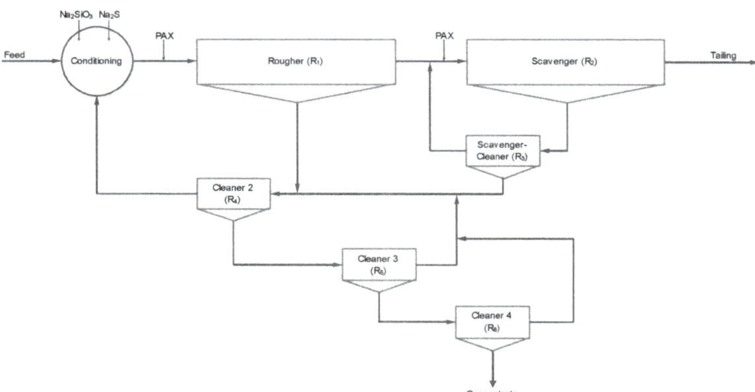

Figure 6. Lead flotation circuit.

Table 3. Reagents dosage and addition points.

Reagent	Addition Point	Dosage (gr/ton)
Sodium sulfide	Conditioning	760
	Rougher	300
	Scavenger	200
	Cleaner 1	40
potassium amyl xanthate	Pump	100
	Rougher	80
	Scavenger	40
sodium silicate	Ball mill	500

Grade and recovery of the flotation plant in 150 days of operation are shown in Figure 7. By sampling the input and output of each stage and calculating the recovery based on the two-product formula, the recovery values in the stages were 0.55, 0.70, 0.52, 0.61, 0.75, and 0.65, respectively. With regard to the evaluations that were carried out in the previous section, the first point that could be noted is the low recovery of the rougher stage in relation to the other stages. The transfer function of the circuit in Figure 6 is as follows:

$$\frac{C}{F} = \frac{R_1R_4R_5R_6(1-R_2(1-R_3))+(1-R_1)R_2R_3R_4R_5R_6}{1-[R_1(1-R_4)+R_2(1-R_3)+R_4(1-R_5)+(1-R_6)R_4R_5+(1-R_1)R_2R_3(1-R_4)]+R_1R_2(1-R_3)(1-R_4)+R_2R_4(1-R_3)(1-R_5)+R_2R_4R_5(1-R_3)(1-R_6)} \tag{12}$$

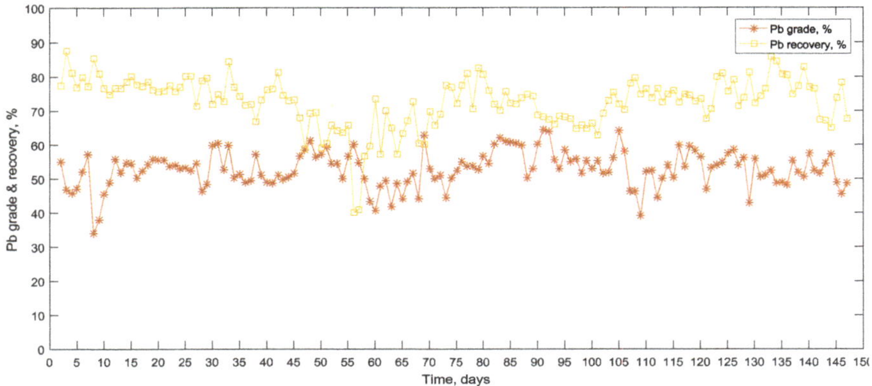

Figure 7. Daily grade and recovery data for lead flotation plant.

Regarding to recovery of each stage and the circuit transfer function, overall circuit recovery is 0.63. While considering Figure 6, for each stage recovery, there are 120 combinations of 720 possible combinations for the circuit with six stage. In Figure 8, all combinations and related recoveries are provided. A division based on the allocation of recovery to the first stage (rougher stage) has been created. From 120 to 240 and 480 to 600 the circuit recovery is higher than other combinations. These two parts belong to 0.70 and 0.75 recoveries for the rougher stage.

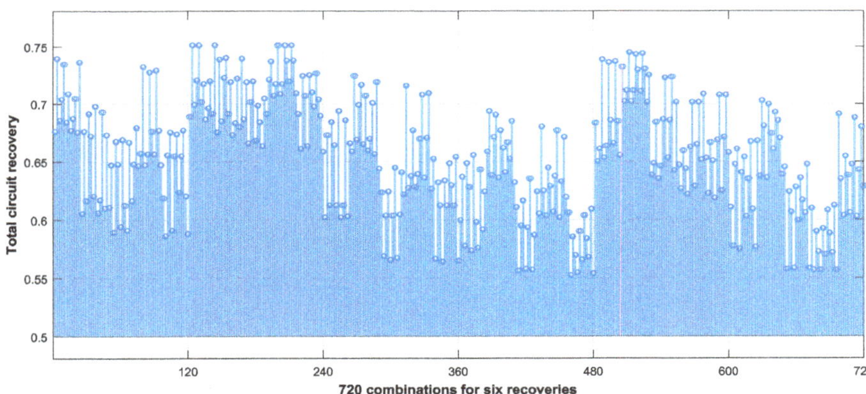

Figure 8. Overall circuit recovery for 720 possible combinations of stage recoveries.

Table 4 indicates the 12 recovery combinations that lead to the highest circuit recovery. The highest recovery was obtained when 0.75 is assigned to the first stage. In Section 3, it was shown that with a few number of stages it is possible to achieve high recoveries. This is especially important in the design phase of flotation circuits. Because the goal of flotation circuit design is to achieve optimum recovery at minimum acceptable grade and optimization with minimal capital and operating costs. After calculating the optimal recovery of the stages, design and operational parameters, such as the number of cells and residence time, can be determined. For example, given the optimal allocation of recovery in Figure 6, the required number of cells in the stages using the probability transfer function could be calculated. The probability of flotation of valuable and gangue minerals in each cell was analyzed and is considered as 0.20 and 0.01, respectively. The calculated required cell number are 6, 4, 3, 5, 5, and 4, respectively, in the rougher, scavenger, cleaner-scavenger, cleaner 1, cleaner 2, and cleaner 3. The kinetic transfer function can also be used to calculate the residence time in each cell/bank according to the kinetic constant measured from the industrial data.

Table 4. Maximum circuit recovery with regard to stage recoveries.

Rougher	Scavenger	Scavenger-Cleaner	Cleaner 1	Cleaner 2	Cleaner 3	Circuit Recovery
0.75	0.70	0.61	0.65	0.52	0.55	0.75
0.75	0.70	0.61	0.65	0.55	0.52	0.75
0.75	0.70	0.55	0.65	0.61	0.52	0.75
0.75	0.70	0.55	0.65	0.52	0.61	0.75
0.75	0.70	0.52	0.65	0.61	0.55	0.75
0.75	0.70	0.52	0.65	0.55	0.61	0.75
0.75	0.65	0.61	0.70	0.52	0.55	0.75
0.75	0.65	0.61	0.70	0.55	0.52	0.75
0.75	0.65	0.55	0.70	0.61	0.52	0.75
0.75	0.65	0.55	0.70	0.52	0.61	0.75
0.75	0.65	0.52	0.70	0.61	0.55	0.75
0.75	0.65	0.52	0.70	0.55	0.61	0.75

The overall circuit recovery can also be evaluated according to the circuit configuration or the removal and addition of a stage within the circuit. Because of the low grade of cleaner-scavenger tailing, close to the final tailing, this stream changed and conducted to the final tail (Figure 9). Sampling of the banks input and output after the change showed little variation in stage recoveries as 0.55, 0.73, 0.50, 0.65, 0.75, and 0.60, respectively. The final concentrate grade increased 4%.

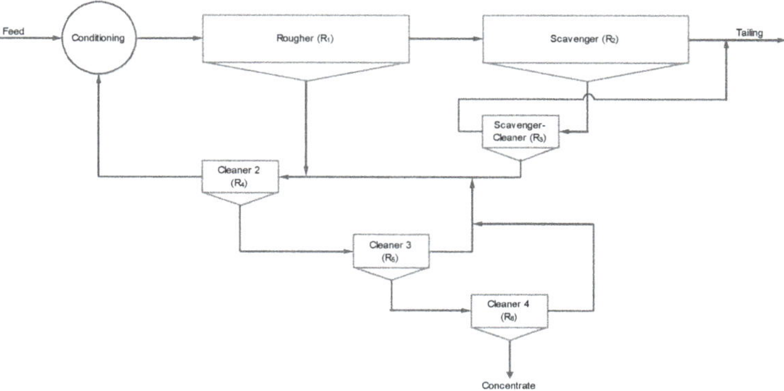

Figure 9. Alternative configuration, sending cleaner-scavenger tail to the final tail.

Nevertheless, the circuit recovery became 0.53 that drastically reduce when compared to the original circuit. If in this circuit the correct allocation of recovery in stages have used, they would be equal to 0.75, 0.65, 0.60, 0.73, 0.52, and 0.55. In this case, the total recovery of the circuit is 0.70. As another alternative, if the scavenger-cleaner stage is removed from the circuit (Figure 10) and the stage recoveries stay as the original circuit, the optimal allocation of recovery in the stages would be 0.75, 0.70, 0.65, 0.55, and 0.61. In this case, although the grade decreased by 2%, the total recovery of the circuit is increased to 0.83. As a result, most of the time, increasing the number of stages in a flotation circuit not only does not improve the circuit conditions, it can also lead to a loss of circuit efficiency. If it is possible to achieve optimal circuit recovery through three stages, while using a large number of stages in addition to imposing capital and operating costs will cause a loss of performance and reduce circuit recovery. Further stages can be taken to achieve the desired grade. As shown in the lead flotation circuit, by removing the scavenger-cleaner unit, it is possible to increase the total recovery of the circuit in the optimal allocation of recoveries in the stages from 0.62 to 0.83.

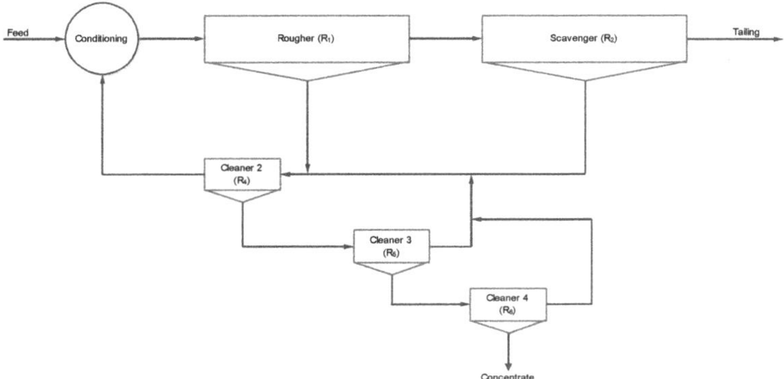

Figure 10. Elimination of cleaner-scavenger stage from the original circuit.

5. Conclusions

Circuit configuration plays an important role in the way of recovery arrangement for flotation stages. The present study utilized simple numerical examples to indicate various applications of stage recoveries to evaluate the flotation circuit. Observations show that:

(1) Stage recovery in some cases is strongly influenced by the structure of the circuit.
(2) In the countercurrent rougher-cleaner and rougher-scavenger circuits, the arrangement of recovery in stages always follows a particular rule.
(3) The assumption of the identical recovery in the stages is acceptable for the comparison and ranking of the circuit configurations to find the circuit with maximum recovery.
(4) Using many numbers of stages does not necessarily increase the circuit performance, and, in some cases, it may impose additional costs and cause the circuit inefficiencies.
(5) Based on the optimal recovery allocation for flotation stages, the design and operational parameters in the flotation circuit would be determined.

Author Contributions: V.R. and S.Z.S. designed the study; this paper was based on V.R.'s Ph.D. thesis; S.Z.S. directed the Ph.D. thesis, M.N. co-directed and H.A. consulted the Ph.D. thesis.

Funding: This research received no external funding.

Acknowledgments: The authors acknowledge support from the University of Tehran.

References

1. Mendez, D.A.; Gálvez, E.D.; Cisternas, L.A. State of the art in the conceptual design of flotation circuits. *Int. J. Miner. Process.* **2009**, *90*, 1–15. [CrossRef]
2. Sutherland, D.N. A study on the optimization of the arrangement of flotation circuits. *Int. J. Miner. Process.* **1981**, *7*, 319–346. [CrossRef]
3. Pirouzan, D.; Yahyaei, M.; Banisi, S. Pareto based optimization of flotation cells configuration using an oriented genetic algorithm. *Int. J. Miner. Process.* **2014**, *126*, 107–116. [CrossRef]
4. Guria, C.; Verma, M.; Mehrotra, S.P.; Gupta, S.K. Multi-objective optimal synthesis and design of froth flotation circuits for mineral processing, using the jumping gene adaptation of genetic algorithm. *Ind. Eng. Chem. Res.* **2005**, *44*, 2621–2633. [CrossRef]
5. Gharai, M.; Venugopal, R. Modeling of flotation process—An overview of different approaches. *Miner. Process. Extr. Metall. Rev.* **2016**, *37*, 120–133. [CrossRef]
6. Montenegro, M.R.; Gálvez, E.D.; Cisternas, L.A. The effects of stage recovery uncertainty in the performance of concentration circuits. *Int. J. Miner. Process.* **2015**, *143*, 12–17. [CrossRef]
7. Amini, S.H.; Noble, A. Application of linear circuit analysis in evaluation of mineral processing circuit design under uncertainty. *Miner. Eng.* **2017**, *102*, 18–29. [CrossRef]
8. Maldonado, M.; Araya, R.; Finch, J. Optimizing flotation bank performance by recovery profiling. *Miner. Eng.* **2011**, *24*, 939–943. [CrossRef]
9. Jamett, N.; Cisternas, L.A.; Vielma, J.P. Solution strategies to the stochastic design of mineral flotation plants. *Chem. Eng. Sci.* **2015**, *134*, 850–860. [CrossRef]
10. Montenegro, M.R.; Sepúlveda, F.D.; Gálvez, E.D.; Cisternas, L.A. Methodology for process analysis and design with multiple objectives under uncertainty: Application to flotation circuits. *Int. J. Miner. Process.* **2013**, *118*, 15–27. [CrossRef]
11. Noble, A.; Luttrell, G.H. The matrix reduction algorithm for solving separation circuits. *Miner. Eng.* **2014**, *64*, 97–108. [CrossRef]
12. Williams, M.C.; Meloy, T.P. On the definition and separation of fundamental process functions. *Int. J. Miner. Process.* **1989**, *26*, 65–72. [CrossRef]
13. McKeon, T.J.; Luttrell, G.H. Optimization of multistage circuits for gravity concentration of heavy mineral sands. *Miner. Metall. Process.* **2012**, *29*, 1–5.
14. Luttrell, G.H.; Kohmuench, J.N.; Stanley, F.L.; Trump, G.D. An evaluation of multi-stage spiral circuits. In Proceedings of the 16th International Coal Preparation Conference, Lexington, KY, USA, 27–29 April 1999; pp. 79–88.
15. Luttrell, G.H.; Kohmuench, J.N.; Mankosa, M.J. Optimization of magnetic separator circuit configurations. *Trans. Min. Metall. Explor. Inc.* **2004**, *316*, 153.
16. Gálvez, E.D. A shortcut procedure for the design of mineral separation circuits. *Miner. Eng.* **1998**, *11*, 113–123. [CrossRef]
17. Sepúlveda, F.D.; Cisternas, L.A.; Gálvez, E.D. The use of global sensitivity analysis for improving processes: Applications to mineral processing. *Comput. Chem. Eng.* **2014**, *66*, 221–232. [CrossRef]
18. Lucay, F.; Mellado, M.E.; Cisternas, L.A.; Gálvez, E.D. Sensitivity analysis of separation circuits. *Int. J. Miner. Process.* **2012**, *110*, 30–45. [CrossRef]
19. Calisaya, D.A.; López-Valdivieso, A.; Marcos, H.; Gálvez, E.E.; Cisternas, L.A. A strategy for the identification of optimal flotation circuits. *Miner. Eng.* **2016**, *96*, 157–167. [CrossRef]
20. Cisternas, L.A.; Jamett, N.; Gálvez, E.D. Approximate recovery values for each stage are sufficient to select the concentration circuit structures. *Miner. Eng.* **2015**, *83*, 175–184. [CrossRef]
21. Williams, M.; Meloy, T. Feasible designs for separation networks: A selection technique. *Int. J. Miner. Process.* **1991**, *32*, 161–174. [CrossRef]
22. Lynch, A.J.; Johnson, N.W.; Manlapig, E.V.; Thorne, C.G. *Mineral and Coal Flotation Circuits-Their Simulation and Control*; Developments in Mineral Processing No. 3; Elsevier Science Ltd.: New York, NY, USA, 1981.
23. Savassi, O.N.; Alexander, D.J.; Franzidis, J.P.; Manlapig, E.V. An empirical model for entrainment in industrial flotation plants. *Miner. Eng.* **1998**, *11*, 243–256. [CrossRef]
24. Kelsall, D.F. Application of probability in the assessment of flotation systems. *Trans. Inst. Mining Metall.* **1961**, *70*, 191–204.

25. Davis, W.J.N. The development of a mathematical model of the lead flotation circuit at the Zinc Corporation Ltd. *AIMM Trans.* **1964**, *212*, 61–89.
26. Meloy, T. Analysis and optimization of mineral processing and coal-cleaning circuits—Circuit analysis. *Int. J. Miner. Process.* **1983**, *10*, 61–80. [CrossRef]
27. Hu, W.; Hadler, K.; Neethling, S.J.; Cilliers, J.J. Determining flotation circuit layout using genetic algorithms with pulp and froth models. *Chem. Eng. Sci.* **2013**, *102*, 32–41. [CrossRef]

MDPI

St. Alban-Anlage 66

4052 Basel

Switzerland

Tel. +41 61 683 77 34

Fax +41 61 302 89 18

www.mdpi.com

Minerals Editorial Office

E-mail: minerals@mdpi.com

www.mdpi.com/journal/minerals

www.ingramcontent.com/pod-product-compliance
Lightning Source LLC
LaVergne TN
LVHW071515180726
843512LV00014B/1085